Springer Water

Series Editor

Andrey G. Kostianoy, Russian Academy of Sciences, P. P. Shirshov
Institute of Oceanology, Moscow, Russia

The book series Springer Water comprises a broad portfolio of multi- and interdisciplinary scientific books, aiming at researchers, students, and everyone interested in water-related science. The series includes peer-reviewed monographs, edited volumes, textbooks, and conference proceedings. Its volumes combine all kinds of water-related research areas, such as: the movement, distribution and quality of freshwater; water resources; the quality and pollution of water and its influence on health; the water industry including drinking water, wastewater, and desalination services and technologies; water history; as well as water management and the governmental, political, developmental, and ethical aspects of water.

More information about this series at http://www.springer.com/series/13419

Faiza Khebour Allouche · Abdelazim M. Negm
Editors

Environmental Remote Sensing and GIS in Tunisia

 Springer

Editors
Faiza Khebour Allouche
Laboratory of Phytopharmacy
and Weed Science, Higher Institute
of Agronomy-Chott Meriem
University of Sousse
Sousse, Tunisia

GREEN-TEAM Laboratory
(LR17AGR01), Higher Institute
of Agronomic Sciences-Chott Meriem
University of Sousse
Sousse, Tunisia

Abdelazim M. Negm
Faculty of Engineering
Zagazig University
Zagazig, Egypt

ISSN 2364-6934   ISSN 2364-8198 (electronic)
Springer Water
ISBN 978-3-030-63667-8   ISBN 978-3-030-63668-5 (eBook)
https://doi.org/10.1007/978-3-030-63668-5

This Springer imprint is published by the registered company Springer Nature Switzerland AG
The registered company address is: Gewerbestrasse 11, 6330 Cham, Switzerland

Preface

This volume came into conception to highlight the use of remote sensing (RS) and Geographical Information Systems (GIS) and their applications in Tunisia. This unique volume is authored by experts in the topic from Tunisia and other countries too to present the results and findings of their research work and the related state-of-the-art connected to the book title. The volume consists of five parts excluding the introduction and the conclusions parts. The book is comprised of 15 chapters written by more than 30 authors. **Part I** is an introduction to the environmental remote sensing and GIS in Tunisia where the editors present a general overview and highlight the technical elements of each chapter.

Part II of the volume is titled "**RS and GIS for Urban and Rural Applications**" and contains five Chaps. 2, 3, 4, 5 and 6. Chapter 2 is titled "**Approaching the Tunisian Human Environment by Using RS and the Dasymetric Method**" is devoted to demonstrating the development of the dasymetric method, based on the use of ancillary data, mainly on Remote Sensing and GIS to enhance the representation of population density in Tunisia at a small scale. Chapter 3 is titled "**The Role of GIS as a Planning Tool in a Tunisian Urban Landscape. Sfax City**" aims to find out the possibility of using Remote sensing and GIS techniques in contributing to urban sprawl surveys and to assess green spaces. Chapter 4 "**Spatio-Temporal Monitoring of the Meskat System Distribution in the Tunisian Sahel Region Using TM Landsat Images**." As an important anti-erosion technique used by Tunisian farmers, the Meskat system faced to dropout phenomenon is studied in this chapter using Landsat series performed by the Geometica software PCI to monitor its distribution in the Sahel of Tunisia. Chapter 5 "**Spatial Distribution of *Solanum Elaeagnifolium* in the Tunisian Arid Region Using GIS Tools**." In this chapter, authors attend to map and characterize the spatial distribution of an invasive weed called SOLEL using the GIS technique and thus to support the implementation of an appropriate control approach against it. The last chapter in Part II is titled "**PROMETHEE and Geospatial Analysis to Rank Suitable Sites for Grombalia Aquifer Recharge with Reclaimed Water**." A methodology was proposed in this chapter to help selecting the best suitable sites for aquifer recharge with reclaimed water using Multicriteria Analysis and the Geographic information system.

Part III is titled "**RS and GIS for Mapping and Modeling**" and contains two Chaps. 7 and 8. Chapter 7 is entitled "**Using RS and GIS to Mapping Land Cover of the Cap Bon (Tunisia)**." The main goal of this research is to develop a methodology that can identify, locate and map, at different scales and by using various modes of presentation, the land use as an overall information system. It contains different phases bringing together a useful synergy between GIS and remote sensing. Chapter 8 titled "**A GIS Based DRASTIC, Pesticide DRASTIC and SI Methods to Assess Groundwater Vulnerability to Pollution: Case Study of Oued Laya (Central Tunisia)**" emphasizes the use of three parametric approaches carried out by using the Geographic Information System as a tool for environmental studies and geospatial modeling of natural phenomena.

Part IV is titled "**RS and GIS for Natural Risks Applications**" is written in two Chaps. 9 and 10. In Chap. 9 with the title "**Mapping Environmental Risk Degradation Under Climate Stress and Anthropogenic Pressure: Case Study of Abdeladim Watershed, Tunisia**," the author aims to provide information on Environmental Risk Degradation using remotely sensed images such as aerial photography, thematic maps and Landsat satellite image in Kasserine district, Tunisia. In the chapter, three main tasks are identified and achieved including (a) erosion, (b) sensitivity to the degradation and (c) main factors of degradation. While, the authors of Chap. 10 under the title "**Application of Remote Sensing and GIS for Risk Assessment in Monastir, Tunisia**" propose a methodology for urban risk assessments using a multi-temporal remote sensing data from aerial photography, WorldView3 and Landsat applied to a coastal region, Monastir, Tunisia.

Part V has the theme "**Soil Degradation and Drought**" and is covered in two chapters. In Chap. 11 titled "**Monitoring of Land Use-Land Cover Changes and Assessment of Soil Degradation Using Landsat TM and OLI Data in Zarzis Arid Region**," the author tries to demonstrate the output for some works which have been done on monitoring and mapping spatial and temporal changes of an arid region using remote sensing and GIS techniques. Moreover, Chap. 12 under the title "**Drought Assessment in Tunisia by Time-Series Satellite Images: An Ecohydrologic Approach**" the author integrate LAI-MODIS time-series images processing into a water balance model for the simulation of water stress spatiotemporal dynamic.

Part VI contains four chapters, from 13 to 16 chapter. They are presented under the theme "**RS and GIS to Assess and Monitor Dry, Arid and Wetlands**." Chapter 13 is titled "**Monitoring of Dryland Vulnerability by Remote Sensing and Geoinformation Processing: Case of Wadi Bouhamed Watershed (Southern Tunisia)**." The chapter presents an overview of methods used for the assessment of dryland degradation vulnerability and an estimation of soil and water conservation techniques efficiency in a Tunisian watershed using remotely geo-information data. Chapter 14 is titled "**Assessing Tunisian Oasis Dynamics Using Earth Observation and Landscape Metrics: Case of Djerid and Nefzaoua Regions**." The authors attempt to highlight the benefits of the combination of the earth observation tools and landscape metrics for monitoring, evaluating and predicting the spatiotemporal changing patterns of oases are essential for managing oases for sustainability. Chapter 15 is titled "**Contribution of GIS in the Environmental Monitoring of a Tunisian**

Biosphere Reserve (Bou Hedma National Park)." Through the use of a multi-source processing and exploitation data (descriptive, map, field surveys, statistics, history, etc.), authors explain the implementation of a GIS database for characterizing and monitoring ancient and recent environmental sensitivity in BouHedma National Park. Additionally, Chap. 16 is presented under the title "**A Historical Look at the Spatiotemporal Dynamics of Tunisian Wetlands by Earth Observation.**" It aims to study the spatial and temporal evolution of the wetlands of the big Sfax using earth observation and GIS tools. The analysis of land use land-cover (LULC) temporal series of aerial photographs, satellite imagery between 1963 and 2018 showed that these wetlands underwent important changes explained by the installation of a polluting coastal industrial activity and the urban extension.

The last chapter in this book (**Part VII**) is the conclusions and recommendations. The chapter presents an update of the most recent findings, the most significant conclusions and recommendations of the chapters contained in the volume.

Special thanks are due to all authors who contributed to this volume, without their efforts and patience, it would not have been possible to produce this unique volume on Environmental RS and GIS in Tunisia. Also, thanks should be extended to include the Springer team who largely supported the authors and editors during the production of this volume. Special thanks are due to Andrey Kostoamoy and Alexis Vizcaino for their constructive comments. Additionally, Abdelazim M. Negm Thanks both ASRT of Egypt and CNR of Italty for their support offered through the bilateral collaboration framework.

Sousse, Tunisia   Faiza Khebour Allouche
Zagazig, Egypt   Abdelazim M. Negm
November 2019

Contents

Part I   Introduction

1   Introduction to "Environmental Remote Sensing and GIS
    in Tunisia" 3
    Abdelazim M. Negm and Faiza Khebour Allouche

Part II   RS and GIS for Urban and Rural Applications

2   Approaching the Tunisian Human Environment by Using RS
    and the Dasymetric Method 17
    Mohsen Dhieb, Taher Yengui, and Monaem Nasr

3   The Role of GIS as a Planning Tool in a Tunisian Urban
    Landscape, Sfax City 37
    Maha Bouhlel and Ali Bennasr

4   Spatio-Temporal Monitoring of the Meskat System
    Distribution in the Tunisian Sahel Region Using TM Landsat
    Images 59
    Asma Ben Salem, Asma El Amri, Soumaya M'nassri,
    Karem Chokman, and Rajouen Majdoub

5   Spatial Distribution of *Solanum elaeagnifolium* in the Tunisian
    Arid Region Using GIS Tools 77
    Najla Sayari, Faiza Khebour Allouche, Amal Laarif, and Mounir Mekki

6   PROMETHEE and Geospatial Analysis to Rank Suitable
    Sites for Grombalia Aquifer Recharge with Reclaimed Water 95
    Makram Anane, Rahma Souissi, Hanèn Faïdi, Rim Mehdaoui,
    and Khadija Gdoura

Part III RS and GIS for Mapping and Modeling

**7 Using RS and GIS to Mapping Land Cover of the Cap Bon
(Tunisia)** 117
Monaem Nasr, Hedi Zenati, and Mohsen Dhieb

**8 A GIS Based DRASTIC, Pesticide DRASTIC and SI Methods
to Assess Groundwater Vulnerability to Pollution: Case Study
of Oued Laya (Central Tunisia)** 143
Asma El Amrı, Makran Anane, Lotfi Drıdı, Manel Srasra,
and Rajouen Majdoub

Part IV RS and GIS for Natural Risks Applications

**9 Mapping Environmental Risk Degradation Under Climate
Stress and Anthropogenic Pressure: Case Study of Abdeladim
Watershed, Tunisia** 167
Olfa Riahi

**10 Application of Remote Sensing and GIS for Risk Assessment
in Monastir, Tunisia** 191
Felicitas Bellert, Konstanze Fila, Reinhard Thoms,
Michael Hagenlocher, Mostapha Harb, Davide Cotti,
Hayet Baccouche, Sonia Ayed, and Matthias Garschagen

Part V Soil Degradation and Drought

**11 Monitoring of Land Use-Land Cover Changes and Assessment
of Soil Degradation Using Landsat TM and OLI Data
in Zarzis Arid Region** 213
Katar Achraf, Hammouda Aichi, and Bouajila Essifi

**12 Drought Assessment in Tunisia by Time-Series Satellite
Images: An Ecohydrologic Approach** 233
Hedia Chakroun

**Part VI RS and GIS to Assess and Monitor Dry, Arid and
Wetlands**

**13 Monitoring of Dryland Vulnerability by Remote Sensing
and Geoinformation Processing: Case of Wadi Bouhamed
Watershed (Southern Tunisia)** 253
Najiba Chkir and Dalel Ouerchefani

**14 Assessing Tunisian Oasis Dynamics Using Earth Observation
and Landscape Metrics: Case of Djerid and Nefzaoua Regions** 285
Faiza Khebour Allouche, Ibticem Abidi,
Eric Delaître, Mohamed Saeid Desouky Abu-hashim,
Dalel Ouerchfeni Bousaida, Safa Hamad, and Ribh Riahi

**15 Contribution of GIS in the Environmental Monitoring
     of a Tunisian Biosphere Reserve (Bou Hedma National Park)** 303
     Olfa Riahi

**16 A Historical Look at the Spatiotemporal Dynamics of Tunisian
     Wetlands by Earth Observation** 329
     Balkis Chaabane and Faiza Khebour Allouche

Part VII   Conclusions

**17 Conclusions and Recommendations for "Environmental
     Remote Sensing and GIS in Tunisia"** 351
     Faiza Khebour Allouche, Abdallah Gad, and Abdelazim M. Negm

Part I
Introduction

Chapter 1
Introduction to "Environmental Remote Sensing and GIS in Tunisia"

Abdelazim M. Negm and Faiza Khebour Allouche

Abstract This chapter presents summaries of the chapters presented in this book. The chapters focus on the use and the role of remote sensing and Geographical systems (GIS) techniques in understanding, monitoring, and investigation of the environmentally related applications in Tunisia. Many case studies are presented. Wide range of topics are covered in the book including information extraction, environmental applications, remote sensing and GIS technologies for modeling, mapping and detecting land use/land cover changes, analyses of land degradation, dryland, drought, as well as to climate change, risks, groundwater vulnerability, and wetlands. All these topics are covered under five themes in 17 chapters including both the introduction and the conclusions chapters.

Keywords Planning · Monitoring · Spatial distribution · Decision support · Mapping · Vulnerability · Risks · Anthropogenic · Degradation · Drought · Geo-information · Remote sensing · Earth observation · GIS · Management · Environmental · Wetland · Arid

1.1 Background

Tunisia is located in North Africa between Morocco and Libya. In all North African countries, it is the only country that is known as "Green Tunisia." Wide of features are present in Tunisia. Remote sensing and GIS are used, in this book, to present most of the unique features of the Tunisia environment and discuss them including

A. M. Negm (✉)
Faculty of Engineering, Water and Water Structures Engineering Department,
Zagazig University, Zagazig 44519, Egypt
e-mail: amnegm@zu.edu.eg

F. Khebour Allouche
Laboratory of Phytopharmacy and Weed Science, Higher Institute of Agronomy-Chott Meriem,
University of Sousse, ISA CM BP 47, 4070 Sousse, Tunisia

GREEN-TEAM Laboratory (LR17AGR01), Higher Institute of Agronomic Sciences-Chott
Meriem, University of Sousse, Sousse, Tunisia

the related state-of-the-art review. Scientists from Tunisia, Europe, and Egypt put their great efforts for about two years to produce this unique book for all who are interested in Tunisia and its unique features. The book contains 17 different chapters in addition to the introduction (this chapter) and the conclusions/recommendations chapter which closes the book (Chapter 17).

In this chapter, a summary is presented to indicate briefly what will be provided in each chapter without mentioning much detail or going to conclusions and recommendations as they presented in the chapters and briefly at the conclusion chapter. For sure, the interested readers will consult the entire chapters for details.

It worth mentioning that this is the third book to be published in "Springer Water" which is follower to the book "Environmental Remote Sensing and GIS in Iraq" which was edited by Ayad Al-Quraishi and Abdelazim Negm (2020) [1]. and the book "Environmental Remote Sensing in Egypt" which was edited by Elbeih et al. (2020) [2].

1.2 Themes of the Book

Therefore, the book intends to address in more detail the following main themes:

- Remote Sensing and GIS for Urban and Rural Applications
- Remote Sensing and GIS for Mapping and Modeling
- Remote Sensing and GIS for Natural Risks Applications
- Remote Sensing and GIS for Soil Degradation and Drought Evaluation
- Remote Sensing and GIS to Assess and Monitor Dry, Arid and Wetlands.

1.3 Chapters' Summary

The next subsections present the main technical elements of each chapter under its related theme.

1.3.1 Remote Sensing and GIS for Urban and Rural Applications

This theme is covered in 5 chapters from 2–6. Chapter 2 is titled "Approaching the Tunisian Human Environment by using RS and the Dasymetric Method." The chapter presents the use of the dasymetric method in approaching the Tunisian human environment as it is more advantageous compared to the cartographic choropleth method, which is easy to use when portraying the density of the population. The method may lead to different and even unreal results depending on the degree of the inner heterogeneity of the studied spaces, their patterns, shapes, sizes, and mostly, land use categories. For spaces observing a high inner heterogeneity, it was proven that the choropleth method has many lacks and deficiencies related to the literature.

Today, an increasing number of cartographers prefer the dasymetric method, which is based on the use of the Remote Sensing technique, which yields more accurate data about the human environment. This method describes more closely the reality of the population distribution even though it reveals more challenging to implement practically. In most case studies, the dasymetric method was applied to large spaces, whether national or regional, scarcely to urban areas [3]. When using this method, the most crucial issue encountered is how to attribute new density values to the various land use classes as results of satellite images processing and whether this attribution applies equally to the whole territory or may vary over space and if so, how to handle it.

In this chapter, the density classes are weighed empirically with regards to the location of the land use classes over the Tunisian territory (North, Centre, South). An equation formula summarizes the overall calculation of the density of each new space unit corresponding to one land-use class and implies the possibility of drawing dasymetric maps. These maps may be achieved to portray the vast disparities characterizing Tunisian spaces' occupation: rural vs urban areas, oasis vs deserts, plains vs mountains, coast vs interior and so on depending on the map scale.

Chapter 3 discuss "The Role of GIS as a Planning Tool in a Tunisian Urban Landscape, Sfax City". It assesses the natural space consideration in the urban planning documents in a Tunisian urban landscape, Sfax City. This industrial city is subject to intense urban changes that reveal more, and more clearly its negative externalities especially as' it generate environmental issues such as the scarcity of green spaces. The recent planning choices to preserve the natural space of the city have not been able to solve issues [4]. Thereby, the first part of this chapter was devoted to the assessment of urban planning choices for green spaces by using the GIS tool, which is confirmed more and more as a guide for the best choice of urban development. The goal is to improve the evidence base of decision-making and facilitate urban planning. Initially, a GIS assessment of urban sprawl based on the diachronic approach was carried out. It showed the inadequacy of the strategy of the urban displacement plan of 1977, and the urban and social reality of Sfax City helped to trigger the peri-urbanization, whereas it wanted to preserve the agricultural areas "jnens." Then, the GIS tool was used to identify the green spaces issues and estimate the population concerned by these issues in the new context of sustainable development, which considered the creation of green spaces as one of the necessary choices for the reconciliation of Sfax with its natural environment. The second part of this chapter was rather interesting in the role of the local authorities and civil society in improving green spaces based on depth interviews conducted with the actors of the city. It should be noted that civil society plays a key role in preserving existing spaces and help people to appropriate the space by increasing acceptance and appreciation for green spaces.

Chapter 4 is about the "Spatio-Temporal Monitoring of The Full System Distribution in The Tunisian Sahel Region Using TM Landsat Images". It presents (i) a diachronic analysis on classified images and (ii) a spatio-temporal analysis of the rate change of Meskat system occupation. Therefore, it assesses the dynamic of the Spatio-temporal distribution of the Meskat in the Sahel of Tunisia (Region of Sousse) based on satellite images from Landsat Thematic Mapper in order to detect the land-use changes.

In Tunisia, various soil and water conservation strategies have been in use to control surface runoff and tackle soil erosion. The Meskat system that mainly allows the runoff mobilization for rainfed olive groves is the most commonly used anti-erosion technique in the Sahel region [5]. However, in the last decades, this technique has faced a dropout phenomenon. Hence, it is important to assess the spatio-temporal dynamics of the Meskat system, given that agricultural productivity is seriously threatened by the decline in Meskat system area and, eventually, the increase in soil moisture stress. In order to obtain a detailed analysis of Meskat system degradation, remote sensing seems to be the most effective technique to study the regression of this system. Hence, this research study is based on satellite images from the Landsat Thematic Mapper 5. Land use was performed by the PCI Geomatica software using supervised classification of images.

Chapter 5 focuses on "Spatial Distribution of *Solanum Elaeagnifolium* in The Tunisian Arid Region Using GIS Tools." The GIS tools are used to document the location and expansion of *Solanum elaeagnifolium*, a noxious invasive weed of the agro-ecosystems in Tunisia. Developing a mapping approach using GIS technology is considered the foundation for the development of a long-term strategic management plan to protect ecosystems biodiversity and prevent *Solanum elaeagnifolium* and other alien plant species invasion.

Alien plant invasions are getting more severe and widespread at an alarming rate around the world. *Solanum elaeagnifolium Cav.* is considered as one of the most widespread invasive weeds in Tunisia, invading a wide range of habitats and generate considerable negatives impacts [6]. Given its harmful threat, a management plan implementation is become crucial to control this weed. Thus, knowing the scale of *Solanum elaeagnifolium* infestation, and where it is located, can set management priorities and feasibility. For this purpose, our research consists of providing mapping surveys in a Tunisian arid region to estimate the actual extent of *Solanum elaeagnifolium* using geographic information system technology in the attempt to generate an accessible geodatabase and to develop a standardized mapping method to track and update weed population dynamics over time.

The last chapter in this section is Chapter 6 with the title "PROMETHEE And Geospatial Analysis To Rank Suitable Sites For Grombalia Aquifer Recharge With Reclaimed Water" It shows that combining multicriteria analysis and GIS helps to improve the management of treated wastewater reuse. It offers an objective way to select the best sites for phreatic aquifer recharge increasing the efficient use of reclaimed water and reducing its negative impact on the environment. Consequently, the authors established a methodology to help selecting the best suitable sites for aquifer recharge with reclaimed water. The tools used were the Multicriteria Analysis and the Geographic Information System. Grombalia shallow aquifer was chosen as a pilot site and the effluent of Bou Argoub wastewater treatment plant as reclaimed water.

The very first step was the conceptualisation of the multicriteria analysis in which each phase was defined. The authors begin by the identification of a set of constraints and their corresponding thresholds used to delineate the suitable sites. Then thirteen technical, environmental and economic criteria were selected based on literature and experts consulting. Weighting these criteria was carried out using the very know pairwise comparison method, AHP [7]. PROMETHEE II was chosen to aggregate the criteria performances and to get the index for ranking suitable site for aquifer recharge with reclaimed water.

To implement the multicriteria analysis GIS was used. Each constraint was first spatialized into a spatial dataset, then for each, a binary map using the disjunctive/conjunctive method was obtained according to the corresponding thresholds. The intersection of these maps gave the constraints maps which delineate the suitable sites for Grombalia aquifer recharge with reclaimed water. The next step was the spatialization of the criteria using different tools of GIS and spatial analysis. The values of each spatialized criterion were normalized using the fuzzy functions and then aggregated using PROMETHEE II method, considering the weight of each criterion.

1.3.2 Remote Sensing and GIS for Mapping and Modeling

This theme is coved by two Chapters, no. 7 and 8. Chapter 7 is about "Using RS and GIS to Mapping Land Cover of the Cap Bon (Tunisia)." The authors take advantage of the essential synergy between Remote Sensing and GIS and integrate images of different scales and dates into a single system. Besides, the establishment of a GIS of land use based on a systemic and systematic conceptual approach to hierarchical levels Hypergraph Based Data Structure (HBDS) [8], of Land-Use Land-Cover allows them a variable and very rich cartographic production both at the spatial level and at the thematic level. Also, the chapter presents the Land Use Information System (LUIS) by first creating the topological Geodatabase, hierachized and structured in such a way as to favor a multi-scale mapping of land use in the peninsula of Cap Bon: Purpose of the last paragraph of the chapter. However, the last two elements of the chapter (the creation of the Geodatabase and the multi-scale mapping of land use) represent the thematic and cartographic results of our work. Indeed the establishment of the Geodatabase is developed following a conceptual approach based on the theory of graphs and sets (HBDS). This conceptual work favors downstream thematic cartography very rich and very variable. Therefore the LUIS is not just a stack of layers, but rather hierarchical levels of land-use land-cover that can be displayed and manipulated if the user needs to perform a large number of actions. The GIS power supply is mainly provided by data from the supervised classifications and photo interpretations of remote sensing images, all according to the desired level.

Chapter 8 is titled "A GIS-based DRASTIC, Pesticide DRASTIC and SI Methods to Assess Groundwater Vulnerability to Pollution: Case Study of Oued Laya (Central Tunisia)." It evaluates the Oued Laya aquifer vulnerability to pollution through three parametric methods Standard DRASTIC [9], as intrinsic vulnerability, Pesticide DRASTIC, and Susceptibility Index as specific vulnerabilities, coupled with the Geographical Information System.

Groundwater pollution is a major issue because it is susceptible to contamination from land use and other anthropogenic impacts. Vulnerability and risk assessment has proved to be an essential step for preventing and controlling groundwater contamination. The Oued Laya phreatic aquifer located in the Centre East of Tunisia (Sahel region) is remarkably threatened by (i) urban pollution from an uncontrolled landfill and the wastewater treatment plant of Kalaa Sghira discharging poor treated water quality in Oued Laya, and (ii) agricultural pollution from the intensive fertilizer applications. This chapter aims to assess Oued Laya aquifer vulnerability to pollution and prepare a map that depicts the sensitive zones for contamination through three parametric approaches standard DRASTIC as intrinsic vulnerability, Pesticide DRASTIC, and Susceptibility Index (SI) as specific vulnerabilities. This assessment is carried out by using the Geographic Information System as a tool for environmental studies and geospatial modelling of natural phenomena. For each method, after deriving the geospatial layer corresponding to each environmental parameter, layers were aggregated to get the vulnerability index. A comparison between the adopted methods was carried out through groundwater nitrates content sampled in fifteen wells in the study area. The delineation of risk zones will guide the decision-makers to select appropriate landfill sites and manage the risk of groundwater contamination efficiently.

1.3.3 Remote Sensing and GIS for Natural Risks Applications

Also, this theme is covered in two chapters. Chapter 9 focuses on "Mapping Environmental Risk Degradation under Climate Stress and Anthropogenic Pressure: Case Study of Abdeladim watershed, Tunisia." The authors investigate the risk of degradation in an agricultural environment under climate stress and anthropogenic pressure, and therefore, the chapter aims to locate the most threatened agricultural land and identify the risk that threatens it. For example, Abdeladim watershed, one of most typical catchment area located in the south-west of Tunisian Dorsal [10] is distinguished by a set of bio-physical and anthropogenic characteristics that make it a fragile natural equilibrium. Indeed, climate factors (lower semi-arid), and anthropogenic (generally inadequate cultivation techniques and anti-erosion management) make it a vulnerable environment to degradation. The chapter involves (1) the study and monitoring of water erosion and, (2) the study of the erosion sensitivity of agricultural land in the watershed. Monitoring and cartography of the risk of water erosion have shown that the situation on the development of agricultural land in the catchment area of Abdeladim is mainly due to soil loss (by processes and mechanisms of water erosion) and by the degradation of the fertility of arable soils.

Also, Chapter 10 is an "Application of Remote Sensing and GIS for Risk Assessment in Monastir, Tunisia." That developed a standardized procedure to provide reliable data and information on disaster risk trends as well as urban growth in the area of Monastir, Tunisia based to similar work [11]. The chapter presents a methodology for urban risk assessment, emphasizing the role of remote sensing and GIS and the lessons from the application. In a study case for the coastal City of Monastir, Tunisia, a multi-temporal optical remote sensing and spatial analysis have been used to support the assessment of current and future exposure, vulnerability, and risk associated with flash floods and coastal erosion. The chapter focuses on the options arising from the use of remote sensing and GIS for urban risk assessments and discusses the transferability of the method to other urban settings. Overall workflow for risk assessment is shown including the collection and processing of input data, the modelling of urban growth and flash flood events, and present risk assessment results in their spatial context. The methodology part explains aspects of data acquisition and management, hazard, and exposure analysis that leads to vulnerability analysis and risk assessment. Furthermore, the results of the flash flood and coastal erosion analyses are presented as the major inputs for the risk assessment within the pilot municipality.

1.3.4   Soil Degradation and Drought

The soil degradation and drought theme are covered by two chapters 11 and 12. Chapter 11 is titled "Monitoring of Land Use-Land Cover Changes and Assessment of Soil Degradation Using Landsat TM and OLI Data in Zarzis Arid Region." It demonstrates the efficiency of remote sensing data in assessing land use and land cover changes, with a good understanding of these changes for the 2007–2014 period based-on Land Change Modeler (LCM) method. Modelling and predicting land-use land cover provide valuable information [12] concerning natural resources degradation in southern Tunisia regions Land Use/Land Cover Change (LUCC) is recognized as a crucial driver of environmental change on all spatio-temporal scales. This paper focuses on the monitoring of land use dynamics underpinning climate change and on the spatio-temporal assessment of the vigor of olive groves in southern Tunisia, based on Land Change Modeler (LCM) method. Tunisian arid ecosystems are undergoing accelerated change due to natural and anthropogenic disturbances. Remote sensing dates have been used as a tool for monitoring desertification, land degradation, and landscape management activities. Zarzis arid region, located in southeastern Tunisia, is well-known for its olive groves with great socio-economic value. However, severe climatic factors and anthropogenic activities made these ecosystems vulnerable. In this context, estimating land-use land-cover changes in the region is crucial. In our study, two Landsat satellite images were used to perform the analysis, acquired in

2007, and 2014. Several processing steps of Landsat images were performed to deliver the resulting maps: Atmospheric Correction, Principal Components Analysis, the Iterative Self-Organizing Unsupervised Classifier (ISOCLUST) module, Land Change Modeler (LCM) under IDRISI Selva environment. The results have revealed changes in land use cover categories over the study period change analysis, and transition maps indicate an improvement in the vigor of olive groves in the region.

Chapter 12 is devoted to "Drought Assessment in Tunisia by Time-Series Satellite Images: An Ecohydrologic Approach." It is known that in the Middle East and North Africa (MENA) regions drought episodes highly control water availability and, consequently, the functioning of both forested and cultivated ecosystems. The ecohydrological approach [13] is used in this chapter. It represents a relatively new trend in the holistic assessment of these limited water resources ecosystems. It explains the equilibrium between the components of the soil-vegetation-climate complex.

On the other hand, during the last two decades, the models used in the assessment of drought causes and manifestations combine more and more indicators from multisensors satellite images. Therefore, the ecohydrology concepts and their methodological basis are first explained, and then, a general review of biophysical and energetic variables derived from remote sensing data at the regional scale is exposed. We proposed various models using time-series satellite images for drought indices determination at the regional scale. A case study of the humidity canopy through the analysis of various spectral vegetation at the ecoregion level is presented and discussed. Also, the ecohydrology equilibrium assessment by remote sensing is investigated through the integration of times-series LAI-MODIS into a water balance modelling and the analysis of the water stress index effect on the ecohydrological equilibrium. Finally, the data quality issues related to the time-series image quality and their effects on drought assessment are discussed.

1.3.5   Remote Sensing and GIS to Assess and Monitor Dry, Arid and Wetlands

The last theme in the book is covered in Chapters 13–16. Chapter 13 is titled "Monitoring of Dryland Vulnerability by Remote Sensing and Geoinformation Processing: Case of Wadi Bouhamed Watershed (Southern Tunisia)." It aims to apply an easy and reliable methodology based on both remote sensing field knowledge to investigate the land degradation processes observed in the watershed of Wadi Bouhamed, as a part of dryland areas of Southern Tunisia. This investigation allows identifying the several controlling factors concerning natural conditions and human pressure.

Land degradation is a key ecological process that affects all the Tunisian country, in particular, its southern part [14] even if less than 15% of the Tunisian population lives in these arid regions. The threat of desertification ranks among the most critical environmental problems and has significant impacts on human well-being.

Unfortunately, soil degradation in dryland is a severe and long-lasting irreversible disturbance that prevents the ecosystem recovery unless comprehensive artificial restoration measures. Southern Tunisia, as a part of the North African steppe, has economic activities that are based on agro-pastoralism, underwent significant degradation since the eighties. The combination of severe droughts and an exponential increase in livestock had a catastrophic impact on pastoral resources. In this arid area, land degradation risk is exacerbated by inappropriate human practices.

Chapter 14 is devoted to "Assessing Oasis Dynamics Using Earth Observation Tools for Predicting Future Management in Arid Regions." Since monitoring, evaluating, and predicting the spatiotemporal changing patterns of oases, is essential for managing oases for sustainability [15]. This chapter focuses on the use of a method combining landscape metrics and human dimension to understand the dynamics of the oasis in the south of Tunisia. Future sustainable management of oasis is related to the good selected archives of historical dynamics, assessment of actual stats by using earth observation data, and GIS tools. In this study, we combined remote sensing and landscape metrics to monitor the dynamics of Tunisian oasis landscapes, in particular, to characterize land use in the Nefzaoua and Djerid regions. The methodology is based on the analysis of NDVI maps and then through the metrics index analysis using Patch Analyst for oasis class. The application covers a period of thirty-six years from 1979 to 2015 from the processing of Landsat images series. Results reveal an unequally distributed oasis progression in space, with a slightly different rate over the period considered. The increased oasis area was mainly due to uncultivated land cover. However, the analysis of spatial metrics suggests a transition from an accelerated fragmentation phase of the oasis landscape, to an expansion phase by the continuous spreading of existing oasis surfaces. A novelty of this study is that the results observed for landscape metrics can be correlated with a human dimension. The use of Geographical Information System based metric information system and spatial analysis with Inverse Distance Weighted interpolation enabled the future mapping of oasis extension in South Tunisia.

Following Chapters 14, 15 comes to discuss the "Contribution of GIS In the Environmental Monitoring of a Tunisian Biosphere Reserve (Bou Hedma National Park)." It aims to present the analysis of environmental dynamics in a biosphere reserve since the Pleistocene using a GIS and to propose a method of studying the sensitivity of the environment to degradation.

For example, Bou Hedma National Park is home to the last acacia raddiana forest north of the Great Sahara, which has given it significant ecological value and allowed it to be classified as a biosphere reserve by UNESCO [16]. The environment of Bou Hedma National Park is a legacy of a long and complex evolution. Monitoring this evolution has allowed us to analyse and understand the old (since the Pleistocene) and current environmental dynamics and to estimate the sensitivity to the potential degradation of the environment.

Our study has shown that the cumulative effect of the arid climate, limited natural resources, and the sometimes excessive exploitation of these resources has made it a degrading environment. Indeed, the analysis of the biophysical and anthropogenic characteristics (in a GIS) of the Bou Hedma National Park shows that it is threatened by two types of potential degradation. The first is the potential degradation of soil quality. The second is potential degradation through water erosion. This sensitivity calls for a number of measures to be taken to safeguard and enhance the natural environment. It is essentially a question of changing the way in which land is used and.

The method based, on the one hand, on the monitoring and analysis of the biophysical and anthropogenic characteristics of the medium in GIS (since the Pleistocene) is appropriate to be used as a protocol for studying environmental dynamics in other arid climate national parks.

The last chapter in this theme will present "A Historical Look at The Spatiotemporal Dynamics of Tunisian Wetlands by Earth Observation." In this chapter, the authors aim at studying the spatial and temporal evolution of the wetlands of the big Sfax using the tools of the GIS and photo-interpretation approach. The multi-date analysis of their landscape dynamics allows understanding the evolutionary sense, the development strategies, and the conservation of the natural environment adopted for the city.

Wetlands are transitory areas between terrestrial and aquatic systems characterized by their high ecological, economic and social values [17]. In this research, we propose to follow the landscape dynamics of the big Sfax wetlands through the visual interpretation of aerial photographs and satellite imagery. In Sfax, located in the center of Tunisia, wetlands are classified according to the Ramsar typology into three categories: marine wetlands, continental wetlands, and artificial wetlands. The multi-date analysis of their landscape dynamics allows understanding the evolutionary sense, the development strategies, and the conservation of the natural environment adopted for the city. Varied data are used during this study. Indeed, we used scanned topographic maps, panchromatic aerial photographs dating 1963 and 1982. For the year 2003, satellite imagery taken from the satellite 'IKONOS 2' is used. As a result, in our study resulting from a lack of recent data, the use of high-resolution archive imagery available in the free software Google Earth to map the wetlands of big Sfax in 2018 was used.

The Interpretation of land use maps for the years 1963, 1982, 2003, and 2018 highlights changes in the spatial extension of big Sfax wetlands during the period of 55-years. These variations are discontinuous over time. There is a trend of regression and disappearance of certain types of wetlands is especially on the north coast of the study area around the 1970s. Big Sfax wetlands have undergone a regressive evolution and an anthropization tendency.

The book ends with the conclusions and recommendations chapter numbered 17. In the conclusion chapter, an update of the literature is made to cover some of the interesting topics which are related to the themes of the book [18–28].

Acknowledgements The writers who wrote this chapter would like to acknowledge the authors of the chapters for their efforts during the different phases of the book including their inputs into this chapter.

References

1. Al-Quraishi A, Negm AM (2020) Environmental remote sensing and GIS in Iraq. Springer International Publishing, Cham, 529 p
2. Elbieh S, Negm AM, Kostianoy A (2020) Environmental remote sensing in Egypt. Springer International Publishing, Cham
3. Wright JK (1936) A method of mapping densities of population. Geogr Rev 26:103–110
4. Charfi F (2016) Sfax 2030 strategy: from strategic vision to action plan. Final Report, ADSS, 180 p (French)
5. Majdoub R, Ben Salem A, Khlifi S, M'Sadak Y (2011) Traditional anti-erosion management (Meskat): exploitation of runoff water and improvement of soil characteristics. Proceedings of the Euro Mediterranean scientific engineering congress. Algeciras, Spain, pp 159–165
6. Sayari N, Brundu G, Mekki M (2016) Mapping and monitoring an invasive alien plant in Tunisia: Silverleaf nightshade (*Solanum elaeagnifolium*) a noxious weed of agricultural areas. Tunis J Plant Prot 11:219–227
7. Saaty TL (1980) The analytic hierarchy process. Ed: McGraw-Hill, New York, NY
8. Bouillé F (1977) An universal model for database simultaneously shareable, portable and distributed. PhD thesis State of Science (speciality: mathematics, mention: IT). Pierre and Marie Curie-Paris University VI, p 44
9. Aller L, Lehr JH, Petty R, Bennett T (1987) DRASTIC: A standardized system to evaluate groundwater pollution using hydrogeologic settings. J Geol Soc India 29(1):23–37
10. Khebour F, Labiadh M, Richard J-F, Temple-Boyer E (2002) Water and landscape, a typology of the small watersheds of the Tunisian Dorsale. IRD Tunis, p 34
11. Eckert S, Jelinek R, Zeug G, Krausmann E (2012) Remote sensing-based assessment of tsunami vulnerability and risk in Alexandria, Egypt. Appl Geogr 32(2):714–723. https://doi.org/10.1016/j.apgeog.2011.08.003
12. Mashame G, Akinyem F (2016) Towards a remote sensing-based assessment of land susceptibility to degradation: examining seasonal variation in land use-land cover for modeling land degradation in a semi-arid context. ISPRS Annals of the Photogrammetry, Remote Sensing and Spatial Information Sciences, Volume III–8, 2016. XXIII ISPRS Congress, 12–19 July 2016, Prague, Czech Republic
13. Chakroun H, Benabdallah S, Lili Chabaane Z (2015) Spatial decision support systems integrating ecohydrology in water limited resources regions. Adv Environ Res 39:215p
14. UNCCD (2013) A stronger UNCCD for a land-degradation neutral world. www.unccd.int
15. Yuchu X, Gong J, Sun P, Gou X (2014) Oasis dynamics change and its influence on landscape pattern on Jinta oasis in arid China from 1963a to 2010a: Integration of multi-source satellite images. Int J Appl Earth Obs Geoinf 33:181–191
16. http://whc.unesco.org/en/tentativelists/5384/
17. Daoud-Bouattour A, Muller SD, Ferchichi-Ben Jamaa H, Ben Saad-Limam S, Rhazi L, Soulié-Marsche I, Rouissi M, Touati B, Ben Haj Jilani I, Gammar AM, Ghrabi-Gammar Z (2011) Conservation of Mediterranean wetlands: interest of historical approach. Biol Rep 334:742–756 (French)
18. Daoud A (2013) Feedback on the floods in the agglomeration of Sfax (southern Tunisia) from 1982 to 2009: from prevention to risk territorialization. Rev Géogr l'Est 53(1–2). http://rge.rev ues.org/4630 (French)
19. Majdoub R, Khlifi S, Ben Salem A, M'Sadak Y (2013) Impacts of the Meskat water harvesting system on soil horizon thickness, organic matter, and canopy volume of olive tree in Tunisia. Desalinization Water Treat 52:2157–2164
20. Karem A (2014) A dot focal of Ramsar in Tunisia. Presentation on wetlands in Tunisia. Directorate of Forest Conservation. Ministry of Agriculture, Water Resources and Fisheries (French)
21. Mehdi L, Weber C, Pietro F, Selmi W (2014) Evolution of the place of plants in the city, from green space to the greenway. VertigO. http://journals.openedition.org/vertigo/12670; https://doi.org/10.4000/vertigo.12670 (French)

22. Petiteville I, Ward S, Dyke G, Steventon M, Harry J (2015) Satellite earth observation intérieure support of disaster risk reduction. Special 2015 WCDRR. Edition: CEOS and ESA. 84 p. http://www.eohandbook.com/eohb2015/files/CEOS_EOHB_2015_WCDRR.pdf
23. Garschagen M, Romero-Lankao P (2015) Exploring the relation-ships between urbanization trends and climate change vulnerability. Clim Change 133(1):37–52. https://doi.org/10.1007/s10584-013-0812-6
24. Attri P, Chaudhry S, Sharma S (2015) Remote sensing and GIS based approaches for LULC change detection-a review. Int J Curr Eng Technol 5(5):3126–3137.
25. Camacho Olmedo MT, Paegelow M, Mas JF, Escobar F (2017) Geomatic approaches for modeling land change scenarios. an introduction. Lecture notes in geoinformation and cartography, 1–8. https://doi.org/10.1007/978-3-319-60801-3_1
26. Hirche A, Salamani M, Boughani A, Belala F, Essafi B, Gashut EH, Hourizi R, Grandi M, Ain Hamouda T (2017) Land degradation and restoration: The North African experiences. Geophysical Research Abstracts 19, EGU2017-11898
27. USAID (2018) Climate risk profile Tunisia. Available from: https://www.climatelinks.org/sites/default/files/asset/document/Tunisia_CRP.pdf
28. Adham A, Wesseling JG, Riksen M, Ouessar M, Ritsema CJ (2019) Assessing the impact of climate change on rainwater harvesting in the Oum Zessar watershed in Southeastern Tunisia. https://doi.org/10.1016/j.agwat.2019.05.006

Part II
RS and GIS for Urban and Rural Applications

Chapter 2
Approaching the Tunisian Human Environment by Using RS and the Dasymetric Method

Mohsen Dhieb, Taher Yengui, and Monaem Nasr

Abstract Among cartographic population density methods, the choropleth method is obviously one of the most used. Two reasons explain this statement: the cartographer's ease of implementation and the reader's ease of understanding. Nevertheless, in many cases, this cartographic method may lead to misleading and erroneous results. This happens especially when the case studies reveal high various inner densities, because of the use of calculated density means, the numbers, shapes and sizes of the counting numbers. For spaces observing a high inner heterogeneity, it was proven that the choropleth method has many lacks and deficiencies related in the literature. A few cartographers prefer the dasymetric method based on satellite images since this latter describes more closely the reality of the population distribution even though this method is more difficult to implement. In most cases studies, the dasymetric method was applied to large spaces, whether national or regional, scarcely to urban spaces. The purpose of the chapter book is to give a real idea on the human environment of Tunisia. This will be achieved through the elaboration of a 1:1000000 scale dasymetric map of the population of Tunisia. This map should be the first since similar work on the issue was never achieved on Tunisia. It would portray the great disparities characterizing spaces' occupation: rural vs urban areas, oasis vs deserts, plains vs mountains, coast vs interior and so on.

Keywords Dasymetric method · Human environment · Population mapping methods · Land cover/land use · Tunisia

M. Dhieb (✉) · T. Yengui · M. Nasr
GEOMAGE Team-Laboratory "SYFACTE", Faculty of Arts and Humanities of Sfax, University of Sfax, 1168, 3000 Sfax, Tunisia

M. Nasr
e-mail: monaem.nasr@flshs.usf.tn

M. Dhieb
Department of Geography and Geographic Information Systems,
Faculty of Arts and Humanities, King Abdulaziz University, Jeddah, Saudi Arabia

© Springer Nature Switzerland AG 2021
F. Khebour Allouche et al. (eds.), *Environmental Remote Sensing and GIS in Tunisia*, Springer Water, https://doi.org/10.1007/978-3-030-63668-5_2

2.1 Introduction

Traditionally, geographic variables are continuous or discontinuous in geographical space, depending on their location, nature, characteristics and relationships to the environment. In related literature, mostly physical geographic variables are presented as continuous whereas human ones are presented as discontinuous. Nevertheless, it is hard to set rigorously such scheme today since many phenomena whether physical or mostly human do not obey rigorously to such dichotomy. Many geographic variables are more or less continuous or discontinuous, depending particularly on the area and scale study, and the degree of generalization adopted. This is the case of the density of population in Tunisia, which may be viewed more or less continuous over space. Its representation with classical methods such as the choropleth method, emphasizing the principle of cartographic generalization, and based on fixed enumeration units such as the administrative subdivisions (regions, governorates, delegations or sectors), tends to simplify a much more complex and tedious reality. The point is that our understanding and assimilation of geographic variables and therefore, decisions we make are highly geographically dependent on the way we perceive their distribution over space. Moreover, this perception relies greatly on the cartographic representation we realize. This statement applies to population distribution as well as to natural phenomena. All things being equal, the question of scale plays a crucial role. On a given scale, the reader should have a clear idea of the greater or lesser continuity of the geographic variable. That is why the author recommends the use of a method adapted to the environment to portray the population distribution. The dasymetric method, based on the use of ancillary data, mainly on Remote Sensing and GIS today fits this need. Indeed, high-resolution satellite images reveal various classes of density within the same enumeration units and therefore overcome the deficiencies in portraying the map description of the human environment with classical methods such as the choropleth method.

The issue is to adjust such technique to the map scale and to affect weighing ratios to the various land use classes determined by satellite image treatment. That is what it is attempted in this presentation with a calculation formula borrowed and adapted from ArcGIS software to model the calculation of new densities, from cells in a raster model format to a vector format. The author assumes that such formula should enhance the representation of population density in Tunisia at a small scale. It should also help various categories of users as planners and geographers to implement spatial decisions when and where the accurate location of the population is needed.

2.2 Portraying the Density of Population: An Overview of the Methods in Use

When it comes to represent the density of population on maps, the most used cartographic method today's is the choropleth method. As its name indicates, *choropleth* signifies representation on a map the quantities (*plethos*) of a phenomenon in space units cut out generally beforehand (*khoré*). This very common type of statistical mapping is also the most criticized by cartographers [1]. The critique addresses the way population distribution is viewed over the method which supports as most of the other cartographic methods do the principle of discretization or simplification at the expense of continuity [2]. This is the case of most of the classical methods of thematic cartography such as the choropleth and dot maps, and cartograms. These methods consider that population density is a discontinuous phenomenon. So, if we consider an urban area, the general case is that it bends around the edges or borders of the urban organism with more or less rhythm. However, sometimes it is abrupt when it comes, for example, to barriers or physical constraints: the existence of a relief, a hostile environment, a marshy area, a desert, or a polluted area.

Among all cartographic methods representing population densities, the choropleth method is of the most frequent, probably because of its ease of use and because the geographic data is mostly reported by enumeration units. At least, there are two major criticisms addressed to this type of cartographic representation: on the one hand, the spatial units on which this method is based do not observe in general a geographical homogeneity; it is even common to deal with units of various shapes, areas, human and geographic contents, whatever the level (Fig. 2.1).

On the other hand, in the cartographic literature, it was proven that there is not only one method to divide a series of values, to choose the optimal breaks, nor an ideal number of classes which determine the cartographic output, in terms of length, distribution, equilibrium and human perception limits (Fig. 2.2). The issue is that the decisions we make may vary according to several factors related to the context of use and mainly space and time and the results might be quite different. It is generally admitted that there are only discretization methods that respond, in one given context of use, to a certain number of concerns and interests and which prove to be better than others in this context. They cannot, therefore, be available for all situations.

All things being equal, the choropleth map does not reflect necessarily reality. To be convinced of this, we should compare a choropleth representation of population density with a high-resolution satellite image of the given area (Fig. 2.3).

The color tint used to portray the density average in the choropleth map reveals various land use classes, which means logically that these classes do not bear homogeneous density. The extreme cases are highly populated areas such as urban entities are portrayed similarly as totally inhabited areas such as the desert. This is shown by the example of Tozeur governorate through the presentation of a classical density of a choropleth population map based on delegations and its transformation into a dasymetric map considering the various land use occupations [3] (Fig. 2.4).

Mapping inner variations inside the enumeration units is often neglected. Geographic reality is much more variable and nuanced than shown by the choropleth map with its emptiness, its gradations, and sometimes its jerks and abruptness. Besides, the administrative or other boundaries of these units may simply be solely "views of the mind" based on political, electoral, historical or other considerations. These limits exist only in the minds of their creators. In Tunisia, we can go smoothly from one governorate to another without noticing any change in the landscape and environment unless a road sign reminds us that we have just crossed a boundary

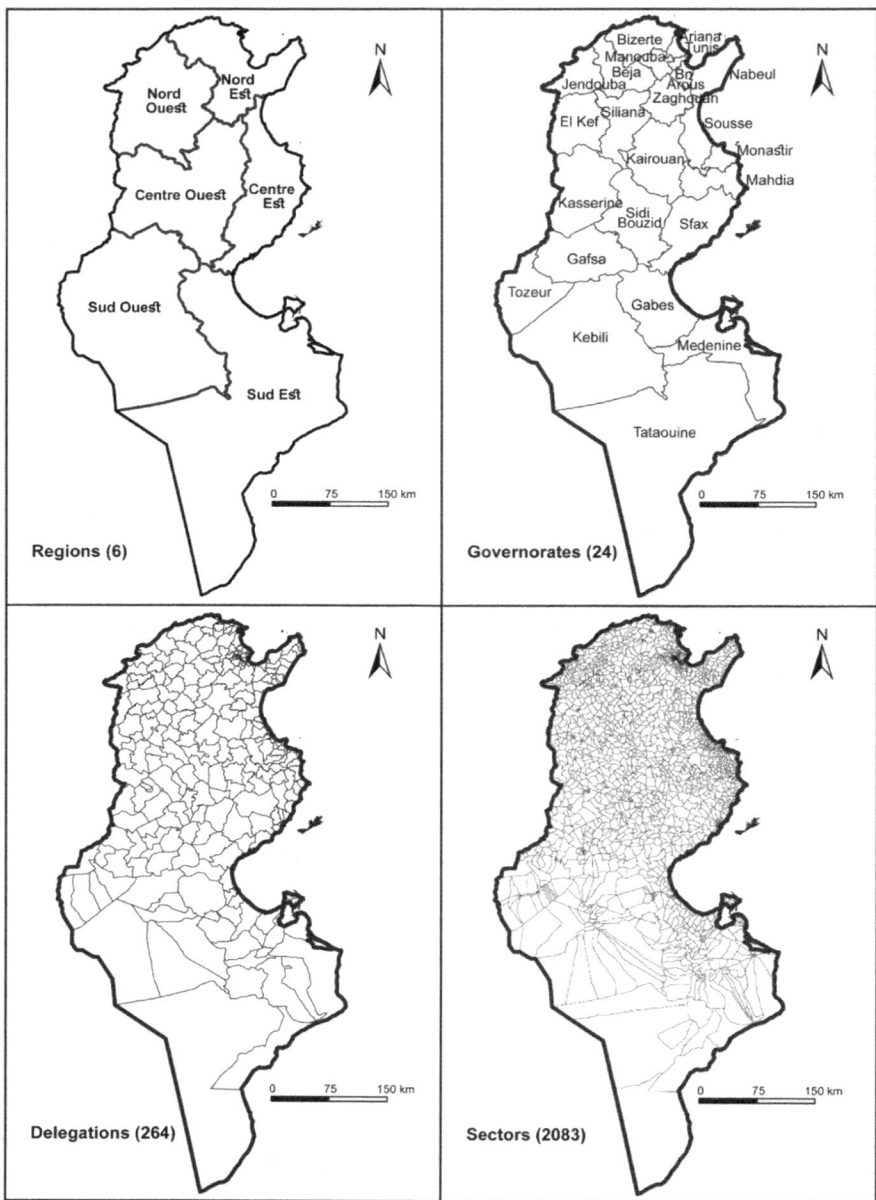

Fig. 2.1 The territorial Tunisian subdivisions: the total number of enumeration units is written between brackets (*Source* Authors work)

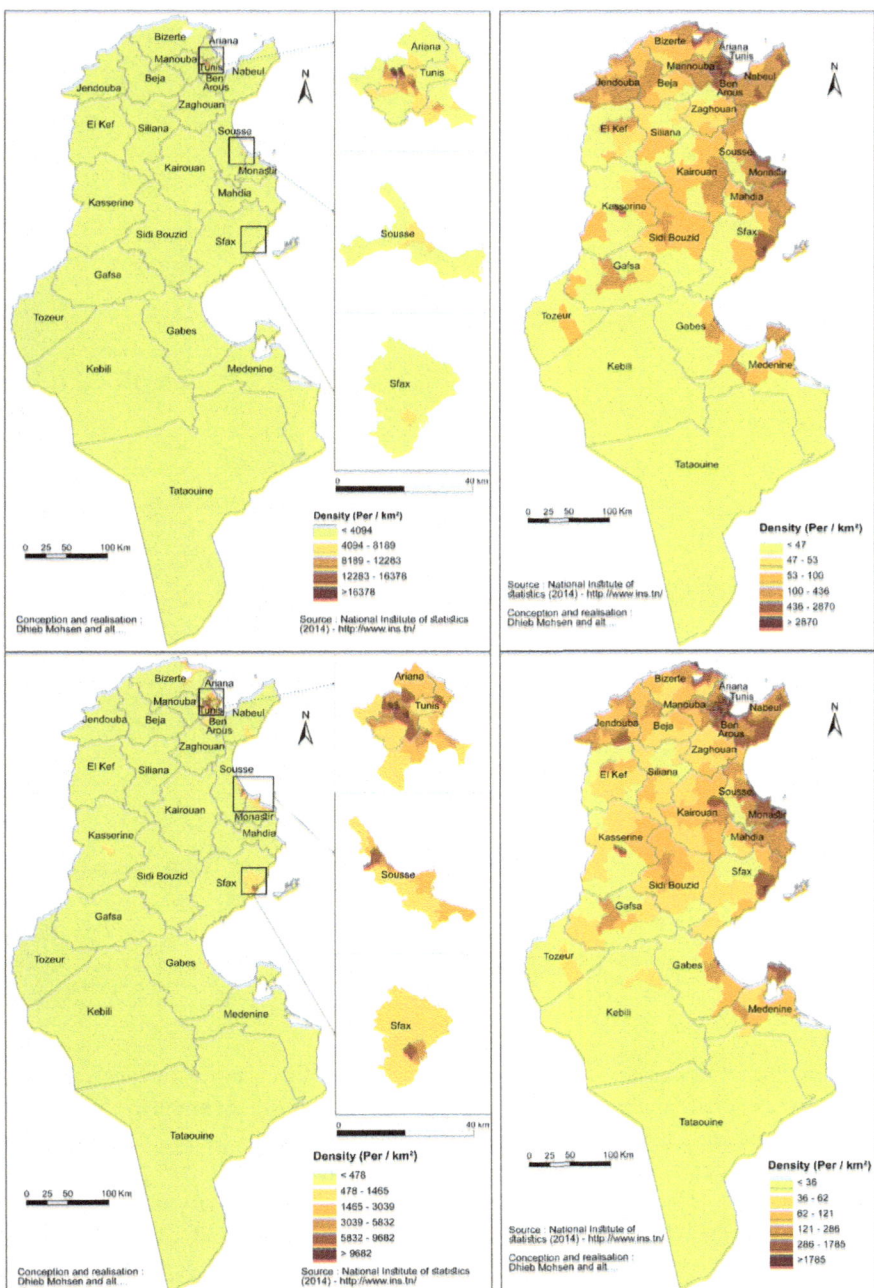

Fig. 2.2 Density of population portrayed by the choropleth method using various classification methods (*Source* Authors work)

Fig. 2.3 A satellite image
(MRSID) showing the
various land use classes in
the Tunisian territory and the
density classes in urban areas
(*Source* Authors work)

between two governorates, delegations or sectors in this case, for instance. Therefore, the choropleth method poses not only a statistical problem but also a graphical
one: the limits are viewed as artificial space discontinuities!

This final statement on the deficiency of the choropleth method to portray conti-
nuities and discontinuities over space, also applies to other cartographic classical
methods in use such as the dot method or cartograms. However, the isopleth method
was intended theoretically to avoid such deficiency since it portrays discontinuous
phenomena [4] and represents the space with a succession of graded classes of
intensity. Theoretically, the isoline method is supposed to overcome the deficiency
of the choropleth method by representing geographical phenomena continuously,
smoothed, but always in stages. It is supposed to be abstracted from space units in

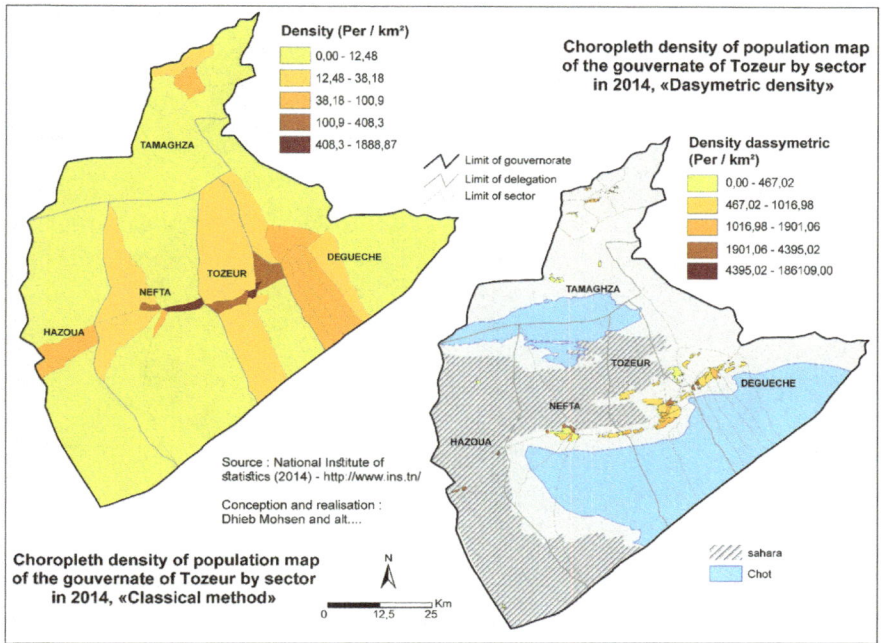

Fig. 2.4 A choropleth map showing the calculated average density of population and a dasymetric map showing accurate data considering the land cover and using the weighing calculation formula of new densities within new space units (*Source* Authors work)

the case of the isometer, whereas in the case of the isopleth, in the beginning, it is based precisely on measurements or averages calculated at the level of these units (Fig. 2.5). Thus, a continuous gradation does not represent the jumps and ruptures in the environment, whether natural or anthropogenic, caused by the geographical variables. Nevertheless, this method fails to suggest a certain discontinuity in density classes which exist really. Probably, the most convenient way to solve this difficulty is to use the dasymetric method, issued by the choropleth method.

2.3   The Dasymetric Method

Basically, the dasymetric method uses areal symbols to classify volumetric data [5, 6]. It substitutes the original enumeration units by new ones based on geographic content and extracted from the use of ancillary data and more adapted to the real population densities. Calculations are generally done to determine the new density values for the new space units realizing a certain geographic homogeneity and therefore sensibly neighboring densities measurements. Among all researches on the issue, Tobler set

Fig. 2.5 An isoline density of population map (*Source* Authors work)

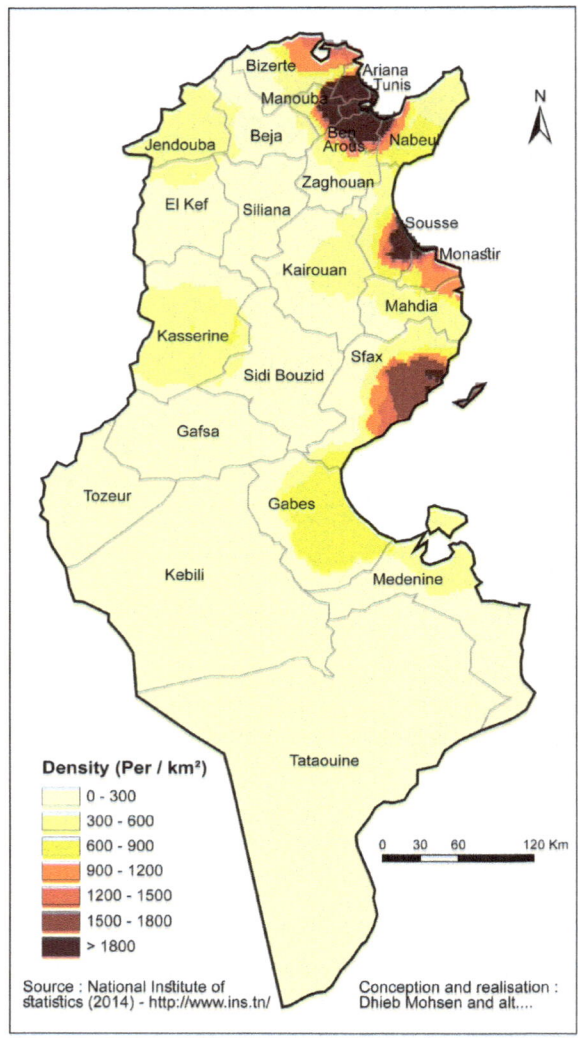

the rule of pycnophylactic property which states that the sum of the estimated population in the new spatial methods created by the dasymetric must equate the originally encoded zone of each population in the original choropleth map, whatever the ancillary data used to identify and set homogeneous density areas [7]. In the past, various data sources were used such as topographic and thematic maps, aerial photographs. Today's, Remote Sensing is the favored technique used to determine the land use categories and GIS reveal very useful in layer manipulation and calculation.

According to Mennis and Hultgren [8] we can retrieve the first idea traces of the method in George Poulett Scrope work or/and Henry Drury Harness [9]. However, this method was created at the beginning of the twentieth century by Veniamin Petrovich Semenov-Tyan-Shansky who invented the term *dasymetric* which means literally "measuring density" [10] and popularized later by the American cartographer J. K. Wright on Cape Code [11]. Therefore, the dasymetric maps are generally defined as maps which show the population density within new created areas. These new areas do not fit the traditional administrative boundaries, but they tell map users about accurate variation of densities: concentration, rarefaction; smooth or rough changes; gradients and breaks…etc.

Since the dasymetric method seems to be tedious and cumbersome, only a few attempts to implement it through various spaces, whether national, regional, rural or urban, with a certain degree of success, were realized [12–15].

The method was not widespread as previewed, probably because of the difficulties to implement it since it requires accurate ancillary data and its treatment. Today, with the possibilities yielded by Remote Sensing in identifying data land use, more and more geographers are using this technique because of its ability to extract populated and unpopulated space areas. Variations within the application of the method are set and discussed. To simplify the presentation scheme and present the variants of the method, the most common subdivision distinguishes between the bipolar and the multivariate types, first. Second, subdivisions are done between the various ways to set the new densities. Third, if this method was applied mostly to calculate population densities, it may theoretically be applied to other statistical data such as elevation, pressure…etc. Fourth, the method is also applied to population estimates on unknown data over small areas characterized by lack of data on population to support variety of administrative functions as well as planning.

2.4 Related Cartographic Literature

Many authors wrote on the dasymetric method. Some of them evaluated it through case studies applications. Only, a few selected works are set in the present review. Most of the researches prove that this method yields by far to better results than choropleth and graded circles maps for instance even when used in urban spaces, such as the research conducted on Jeddah City [16].

An interesting application of the dasymetric method was realized after Bielecka [15]. It combines the bipolar and the multivariate variants of the technique. Applied to a part of Poland, and departing from the Corine Land Cover database, the author excluded first the unpopulated areas from the calculations of the densities. Second, she attributed fixed coefficients to determine population ratios for the remaining land cover classes, based on homogeneous estimates on the 'comunes'. However, Goodchild and Lam [17] argued that what is presented and referred as 'binary' as a remote sensing application describing a particular dasymetric mapping technique method is confusing.

Mennis defined thoroughly the technique widely used in the classical literature, representing the continuous but irregular population variations over space in the natural world [14]. This technique is opposed to the choropleth method where boundaries are set in known enumeration units based on administrative or jurisdictional divisions of the territory. Thus, these original boundaries may not be suitable for the real distribution of the phenomena. They do not correspond to portray accurately changes in density as suggested by the choropleth method. In another research of the same author [18], the technique was used to "estimate population distribution to facilitate a variety of type analyses, including criminal justice, environmental risk, public health accessibility to social services and historical analysis". This is to show that this technique is not used only to portray density of population, but also to extract estimates.

Mohammed Alahmadi and al. presented another case study in Riyadh, Saudi Arabia [19]. In this study, population distribution estimates were obtained by downscaling detailed residential land-use classes. The proposed method was applied to fine spatial resolution remotely sensed imagery [19]. Three statistical regression models were combined with two dasymetric areal interpolation models leading to six-classes [19]. The authors prove that dasymetric mapping approach is more accurate than binary dasymetric mapping [19].

Most of these research studies are based on data furnished by satellite images treatment and processing whether classified or unclassified. Remote Sensing is considered today as the most prominent technique used to define new boundaries and the new space units among other collection data techniques to help to portray population settlement. The various land use classes issued from the supervised or unsupervised classification of the satellite images have strong links with population densities which may be approached empirically or even calculated.

However, GISs reveal powerful tools to manage and process data calculations. This was proven after Eicher and Brewer study [13]. The authors state first that dasymetric maps which display statistical data in meaningful spatial zones are preferable to choropleth maps that show data by enumeration zones because dasymetric zones represent more accurately the underlying data distributions [13]. Second, they tested five dasymetric mapping methods, including methods derived from work on areal interpolation and resulting dasymetric maps of the six socio-economic variables were produced for 159 counties [13]. In this research, both polygonal (vector) and grid (raster) dasymetric methods were tested; map accuracy was evaluated using both statistical analyses and visual presentations of error [13]. Finally, a repeated-measures analysis of variance showed that the traditional limiting variable method had a significantly lower error than the other four methods [13].

Other applications were concerned only by the raster model. The point is to set a value density for each pixel. For instance, an approach to linking remotely sensed data and areal census data was conducted after K. Chen [20]. The goal is to identify the correlations between zone-based census data and pixel-based remotely sensed data. In this study [20] three levels (i.e. macro, medium and micro) of underlying land classification are proposed: (1) non-residential/residential area; (2) land use/cover or density within the residential area; and (3) individual block or dwelling. The dasymetric approach demonstrates the transformation of areal census data from arbitrary zonation to true geography using an accurate delineation of nonresidential and residential areas from remotely sensed imagery [20].

2.5 Advantages of the Dasymetric Method Applied to the Tunisian Territory

Tunisia is the smallest country of Maghrebian states. Its population was estimated 10982754 inhabitants in 2014, but by now, it should go beyond 11 million. Nevertheless, a few spaces of this country are populated, particularly the urban coastal areas with 67.78% of the entire population residing in cities, assimilated to the communes [21]. The Great Tunis concentrates approximately 24% of the whole population for an area of 2000 sq.km. The other 76% of the Tunisian population are sparsely distributed in the great part of the Tunisian territory [22]. The urban rate of urbanization is very high since it reaches 68% in 2014 (INS 2014 [23]). The country's population is tremendously unequally populated, and the average density of inhabitants does not have much meaning since the greatest part of Tunisians lives close to the eastern and northern coastal strip. The proximity of cities and transportation networks also play a crucial role in fixing populations.

To complete the description of population distribution, a few areas are populated in the oasis, in cultivated irrigated perimeters or rich rural areas. Indeed, the country is composed of chains of mountains, coastal and inner plains and narrow irrigated areas, semi-arid lands, and arid and hostile desert (Sahara). Winter is rainy and cold whereas, summer has high daily temperatures and limited rainfall. Variations on vegetation from North to South are remarkable. Departing from this general description, our hypothesis is that dasymetric method may yield good results to portray the Tunisian density of population.

The dasymetric method (dasymetric = density measurement) relies on setting new steepest boundaries between density classes. Therefore, data classification based on satellite images is crucial. A balance should be ensured between the number and types of variation factors and the degree of accuracy. This must be done through an optimization process. The example of the governorate of Tozeur tested and cited above shows how we can get two different maps when using the classical choropleth method and a dasymetric one. This is done by substituting the unreal one class density map for each sector by a new binary class map highly contrasted based on a spatial dichotomy between the populated areas (oasis) and the surrounding desert and chotts. However, this is the simplest case study. In general, there are more than 2 land use classes in one enumeration unit.

2.6   Methodology and Methods Implementation

The number of land use classes we should consider when using the dasymetric method to represent densities of population in Tunisia, depends greatly on the scale study, on the map goal and users and therefore the generalization degree adopted. All these factors can influence this decision. However, all things being equal, the greater is the scale, the greater is the number of land use classes considered and the more detailed is the representation. If we apply this cartographic rule to realize a small-scale map for the whole Tunisian territory such as 1 to 2,000,000, we need only the basic land use classes that deliver the greater density contrast and recognizable from the satellite imagery.

The procedure of establishing densities for the New Spatial Units (NSU) for the dasymetric map is a 6 steps-process as shown in the following chart (Fig. 2.6).

1. Calculating densities for the class Urban areas by dividing the population of the communes by their areas, for each delegation.
2. Calculating densities for the classes Forests and Oasis by dividing their population calculated on the basis of a choropleth map and modulated by their relative density. The relative density should consider the orientation gradients set for these two classes.

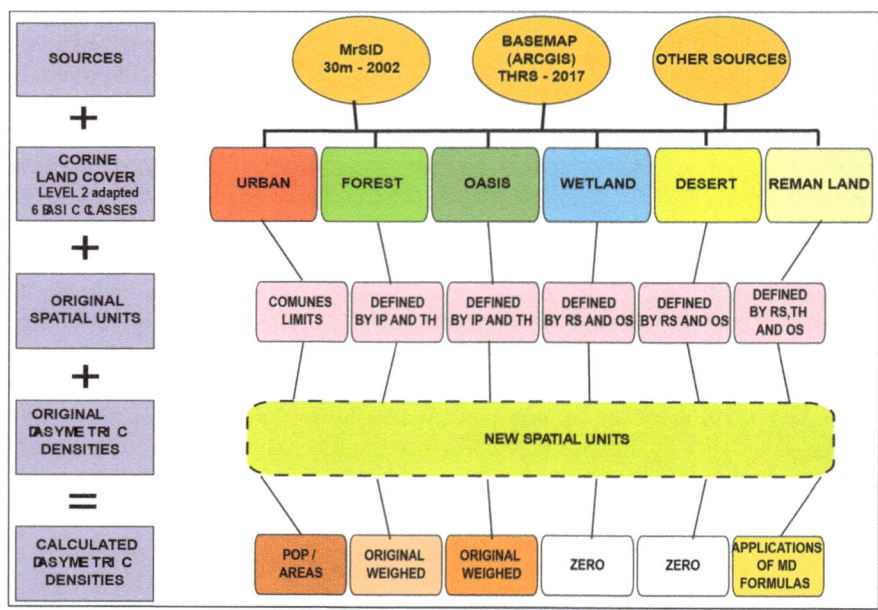

Fig. 2.6 A flow chart showing the methodology used to calculate the new densities related to the new spatial units to implement the dasymetric method (*Source* Authors work)

3. Summing the population of Urban, Forests and Oasis within each delegation and calculating the remaining population by subtracting from the total population of the delegation the population of the classes: Urban, Forest and Oasis.
4. Calculating the Remaining area for the Remaining NSU by subtracting the total delegation area from the sum of the areas of the classes Urban, Forests, Oasis, Wetlands and Desert for each delegation.
5. Calculating the density of the NSUs by dividing the remaining population in each delegation by its area.
6. Establishing a multivariate dasymetric map by establishing breaks inside class densities for the six land use classes ranging from 0 (for Wetlands and Desert) to the highest value.

Departing from the MRSID (Multi-resolution Seamless Image Database for Landsat) 30 m and a Basemap THRS for ArcGIS 2017, we chose 6 land use classes inspired from Corine Land Cover taxonomy but adapted to the Tunisian case. Corine Land Cover was initiated in 1985 in European countries as an inventory of land uses classes derived from high-resolution satellite images. It was applied first to European countries but generalized later in other countries all over the world.

Depending on the scale, 5 hierarchical levels of land use classification were set. Since Tunisia is located near the southern part of Europe, and since the scale study is small for the national level to implement the dasymetric method, we may consider the choice of the level 2 as logical. The following 6 land use classes were defined, slightly different from Corine Land Cover level 2: Urban, Forest, Oasis, Wetlands, and Desert; the sixth class is formed by the remaining spaces which do not belong to any of the precedent classes, mainly formed by rural or inhabited areas (Fig. 2.7).

Of course, if we change the map scale, we may consider other detailed land use classes and finest Corine Land Cover levels.

In addition to these data, we set that urban population is defined through Census since we have urban population by enumeration units, i.e. delegations in the case study [22]. The point is to determine weighing values for each remaining land use class. We considered that the classes Desert and Wetlands are unpopulated, so they were excluded. Then, we considered that Oasis may contain 10–20% of the average density of population of the enumeration unit; that forest may contain up to 50% of the average density. These values were set here empirically as indicative values. Attempts to calculate them more accurately through a detailed image treatment may be conducted to determine averages on the whole Tunisian territory, but this operation goes beyond the goal of this research. Also, decreasing gradients of relative density of population were set depending on the location of (NSU) in the Tunisian territory. The orientation of the gradients is added to assign varying values over space. These values were set empirically from Tunisian geographic literature review and field studies [24]. Extreme values are retrieved at the beginning and ending points of the Tunisian territory (delegations in the present case). Averages are in the midways (Table 2.1).

This table summarizes how Relative Density RD is calculated over the various New Space Units NSU determined after land use classification and identification of

Fig. 2.7 Tunisian land cover occupation (adapted to 1: 2000000 map after Basemap ArcGis THRS 2017)

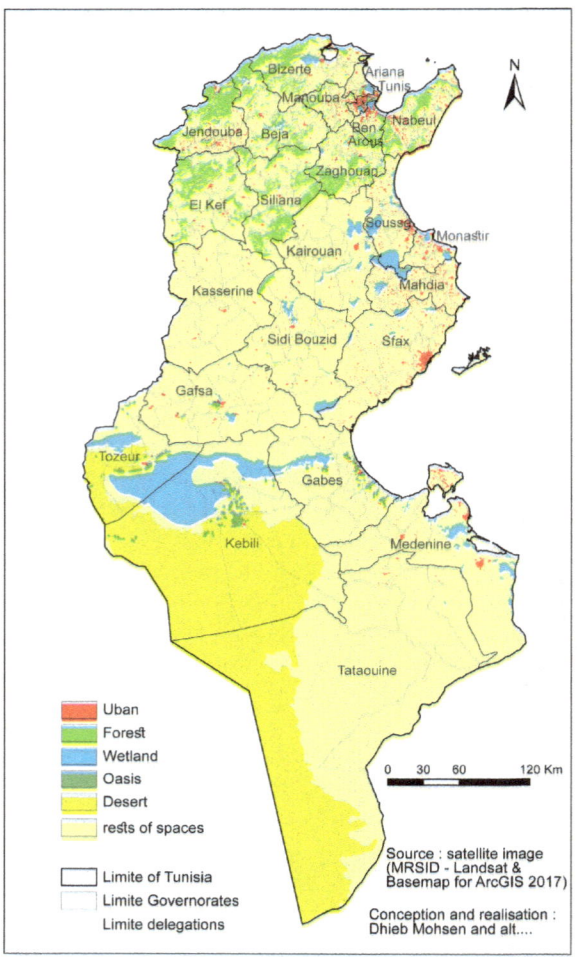

the 6 classes of Urban, Forest, Oasis, Sahara, Waterland and R Rest of Space over Original Space Units OSU.

2.7   Results and Discussion

We will take two virtual examples to verify the procedure; the first d1 is located in the north and the second d2 in the south of Tunisia.

Example 1 A delegation d1 located in north Tunisia has an average density of 100 h/km^2 (i.e. 100000 inhabitants for 1000 km^2).

Table 2.1 Relative densities of Tunisian land classes territorial new spaces units

Land-Cover Code	Descriptions	Relative Density (R_A) = % of mean density over OSU	Orientation Gradient
U	Urban (Cities, what is considered as communal population)	50–70%	*N-S and E-W*
F	Forest and "parcours"	0–50	*N-S*
O	Oasis	10–20	*S-N*
RD	Desert (Sahara)	0	–
W	Water/wetland	0	–
R	Rest of land	15–25	*S-N and W-E*

Where: **RD** Relative Density (in % of mean density); **A** Area; **U** Urban land use class; **F** Forest; **O** Oasis; **S** Sahara; **W** Water; **R** Rest of Space; New Space Units NSU; Original Space Units OSU

d1 is composed of 2 urban portions Up1 and Up2, counting respectively 30000 and 20000 inhabitants and 70 and 30 km^2); 1 Forest of 40 km^2 and a Waterland of 50 km^2.

Therefore:

Total Urbans are $30000 + 20000 = 50000$ inhabitants;

Total Urban Area is $70 + 30 = 100$ km^2.

Therefore

Urban density is $50000/100 = 500$ h/km^2.

Forest density is $(100000/1000) \times 0.5 = 50$ h/km^2;

(0.5 is a gradient affected to Forest because this delegation is in the north of Tunisia as shown in Table 2.1);

Forest population is $50 \times 40 = 2000$ inhabitants;

Remaining uncalculated area is $1000 - (100 + 50 + 40) = 810$ km^2;

Remaining uncalculated population is $100000 - (50000 + 2000) = 48000$;

Therefore

Rest of space density RDd1 is $48000/810 = 59.3$ h/km^2.

Example 2 A delegation d2 in south Tunisia has an average density of 50 km^2 (i.e. 50000 inhabitants for 2000 km^2).

d2 is composed of 1 Urban portion, counting 20000 inhabitants for 20 km^2); 1 Oasis of 40 km^2 and a Desert portion of 500 km^2.

Therefore:

Urban density is $= 20000/20 = 100$ h/km^2.

Oasis density is: $(20000/2000) \times 0.2 = 2$ h/km^2;

(0.2 is affected as gradient to Oasis in this delegation located in the south of Tunisia);

Its population is $0.2 \times 2000 = 400$ inhabitants.

Therefore:

Remaining uncalculated area is: $2000 - (20 + 40 + 500) = 1440$ km^2;

Remaining uncalculated population is: $50000 - (20000 + 400) = 29600$

Therefore: The Rest of space density is: $29600/1400 = 20.5$ h/km^2.

This calculation procedure may apply to sectors as well as delegations or other administrative levels. When applied to detailed units, it gives better results. Using this procedure, we set this procedure of calculation for the governorate of Tozeur, its delegations and its sectors and we obtained the resulting dasymetric map which is more realistic and gives an idea of the real distribution shown in Fig. 2.3. Compared to the satellite image, it fits the land use classes.

Planners, geographers and common users may take great advantages when using the method at a convenient scale. The ascendant information quality and quantity of spatial information outputs from the satellite image processing may enhance accuracy of the density of calculations. Through this test, the dasymetric method proved to be an interesting alternative to the problematic choropleth map. Our knowledge towards the density is more accurate and our endorsement raised questions about the enumeration units we often use in cartographic reasoning.

2.8 Conclusion

Studying the Human Environment of Tunisian Territory needs a convenient choice of optimal cartographic means and methods of portraying population distribution. Yet, as shown in the dedicated literature, this common cartographic method in use is the choropleth method because of its ease to apply often leads to big biases towards reality.

For all these geographical reasons, portraying the distribution of population must be as accurate as possible and give common readers or professionals accurate idea of the reality. The dasymetric method is seen as one alternative solution. In the past, cartographers based their interpretation on sources like topographic or thematic maps to extract new spatial entities instead of the original ones to base their density calculation. However, today, RS yields today a higher accuracy on the human environment when dealing to classify land use data. RS images may provide an alternative solution as data sources, even though results depend on data classification. As far as we know, this method was not applied for Tunisia, at least for the whole country, probably because it seems difficult to implement.

Using dasymetric method in mapping densities also reveals very useful when compared to the classical methods used in similar cases such as the choropleth method, the isopleth method, the dot map method, the cartograms and other cartographic methods describing population distribution. In various fields of science, and when appropriately approached, the dasymetric method is far superior to these methods to describe accurately the Human environment.

The authors think that dasymetric mapping will be used in the future more and more and applied to not only in academic milieu but applied studies dealing with topics such as environment, transportation, planning, agriculture, sustainable development or other topics requiring knowledge about accurate population settlement. Spatial population distribution constitutes a crucial issue in all spatial studies and difficulties of its implementation will be Multi-resolution Seamless Image Database reduced.

2.9   Recommendations

Many planning decisions in Tunisia which may impact the regional development of the regions, as well as attribution of special subventions accorded to low life level populations, are made departing from calculations of the average population density of corresponding administrative units.

The authors recommend changing this approach. It is worth to establish dasymetric population maps for Tunisia at convenient scales based on the land use distribution. These scales depend not only on the real needs of users but also on the data accuracy, especially the spatial resolution of images and population numbers in terms of the spatial units concerned and the data actuality.

The authors recommend also refinement of the calculation formula used in this research for the whole territory to adapt it to different case studies depending on their location and their characters.

The new calculation formulas may be implemented and dasymetric maps produced automatically on regular basis and applied for all territorial decisions whatever the domain: equipment, care, transportation, planning, environment…etc.

The knowledge of the population location using the dasymetric method may help all planners, geographers and also political authorities and private sector when preparing planning documents or linking population distribution with other themes by using GIS.

Moreover, all present and future projects implying spatial dimension should consider not only the real current population distribution and its contrasts, but also its future trends. Geographical and planning studies may help this issue.

References

1. Olson J (1975) Autocorrelation and visual map complexity. Ann Assoc Am Cartographers 75(2):189–204
2. Andrienko G, Andrienko N, Savinov A (2001) Choropleth maps: classification revisited. In Proceedings 20th international cartographic conference—ICA'2001, August 6–10, 2001, Beijing, China. 1209–1219. https://www.researchgate.net/publication/228959242_Cho ropleth_maps_Classification_revisited/citation/download
3. Dhieb M (1991a) Deux exemples de conception de planches d'atlas: La répartition de la population en Tunisie. ICA Proceedings "Mapping The Nations", Bournemouth (In French)
4. Robinson A-H. and al., (1985) Elements of cartography, 5th edn. United States of America, 544 p
5. Dent BD (1999) Cartography: thematic map design, 5th edn. WCB/McGraw-Hill, Boston, MA
6. Slocum T et al (2003) Thematic cartography and geographic visualization, 2nd edn. Upper Saddle River, NJ: Prentice Hall 5th edn. WCB/McGraw-Hill, Boston, MA
7. Tobler W (1979) Smooth pycnophylactic interpolation for geographical regions. J Am Stat Assoc. 74: 519–530. https://www.jstor.org/stable/2286968?seq=1#page_scan_tab_contents
8. Mennis J, Hultgren T (2006) Intelligent dasymetric mapping and its application to areal interpolation. Cartogr Geogr Inf Sci 33(3):179–194. https://www.researchgate.net/publication/228 653393_Intelligent_Dasymetric_Mapping_and_Its_Application_to_Areal_Interpolation
9. Robinson AH, (1955) The 1837 maps of Henry Drury Harness. Geogr J 121(4):440–450
10. Petrov A (2012) Setting the record straight: on the Russian origins of dasymetric mapping. Cartogr J 49(2):256–264
11. Wright JK (1936) A method of mapping densities of population. Geogr Rev 26:103–110
12. Langford M (2006) Obtaining population estimates in non-census reporting zones: an evaluation of the 3-class dasymetric method. Comput Environ Urban Syst 30:161–180. https://www.researchgate.net/publication/220666265_Obtaining_population_estimates_ in_non-census_reporting_zones_An_evaluation_of_the_3-class_dasymetric_method
13. Eicher CL, Brewer CA (2013) Dasymetric mapping and areal interpolation: implementation and evaluation. Cartogr Geogr Inf Sci 28(2):138. https://www.tandfonline.com/doi/abs/10.1559/ 152304001782173727
14. Mennis J (2003) Generating surface models of population using dasymetric mapping. Prof. Geogr 55(1):31–42. https://www.tandfonline.com/doi/abs/10.1111/0033-0124.10042
15. Bielecka E (2005) A dasymetric population density map of Poland In: Proceedings of the international conference July 9–15 Spain CD: A Coruna
16. Hamza MH, Thubaiti A, Dhieb M, and al (2016) Dasymetric mapping as a tool to assess the spatial distribution of population in Jeddah City (Kingdom of Saudi Arabia). Urban Stud 4(3):329–342. https://www.scirp.org/Journal/PaperInformation.aspx?PaperID=70931
17. Goodchild MF, Lam N (1980) Areal interpolation: a variant of the traditional spatial problem. Geo-Processing 1:297–312. http://www.geog.ucsb.edu/~good/papers/46.pdf
18. Mennis J (2009) Dasymetric mapping for estimating population in small areas. Geogr Compass 3:727–745. https://www.researchgate.net/publication/229649191_Dasymetric_A pping_for_Estimating_Population_in_Small_Areas/citation/download
19. Alahmadi MS, Atkinson P, Martin JI (2012) Estimation of the spatial distribution of urban population using remotely sensed satellite data in Riyadh Saudi Arabia. https://www.geos.ed. ac.uk/~gisteac/proceedingsonline/GISRUK2012/Papers/presentation-31.pdf
20. Chen K (2002) An approach to linking remotely sensed data and areal census data. Int J Remote Sensing 23(1):37–48. http://cartesianos.com/geodoc/icc2005/pdf/oral/TEMA5/Session%209/ ELZBIETA%20BIELECKA.pdf
21. Dhieb M (1991b) La représentation cartographique des densités de population urbaine: Aspects méthodologiques. Rev Tunis Sci Soc, 110 Tunis, CERES (In Arabic)
22. Dhieb M (1995) La répartition spatiale de la population tunisienne d'après les premiers résultats du recensement de 1994. Rev Tunis Géogr, 28 FSHS de Tunis (In French)

23. The National Institute of statistics (2014) Recensement Général de la Population et de l'Habitat 2014, vol 9. http://www.ins.tn/fr/search-stat?text=recensement%202014&f0.=field_theme%3A16
24. Belhédi Amor (1992) L'organisation de l'espace en Tunisie, Faculté des sciences humaines et sociales de Tunis, 1992

Chapter 3
The Role of GIS as a Planning Tool in a Tunisian Urban Landscape, Sfax City

Maha Bouhlel and Ali Bennasr

Abstract Cities in Tunisia are affected by a rapid urban expansion, which has had a significant impact on green space structure. The recent trend of urban policies to preserve green spaces in the context of sustainable urban development has not been able to solve issues. This chapter discusses the problem of green areas reduction and degradation through the case of Sfax city. As an industrial city which suffers from pollution and scarcity of natural space, Sfax needs all its green potential to ensure a better living environment for its citizens. Some spaces still exist but they suffer from deteriorated and degraded conditions and also from the lack of security. Through the use of GIS, the chapter intends to comprehend the spatial effect of urban mutations on green spaces and to assess the natural space in urban planning documents in a city considered as a laboratory for sustainable development. Specific attention is given to the role of local actors, especially civil society, in preserving green spaces and improving citizens' living environment.

Keywords GIS · Nature · Green spaces · Urban sprawl · Urban planning

3.1 Introduction

Globally, green space issues has increased over recent decades under the pressure of rapid urban expansion. Throughout the world cities are experiencing rapid change as a result of a strong process of urbanization. This strong growth in size and density is mainly related to the detriment of natural areas, which limits access to nature and can increase exposure to certain environmental hazards, such as air and noise pollution, in addition to The tendency of increasing of natural desasters [1, 2]. However, it should be noted that recent studies, such as Haaland and Konijnendijk [3], confirm

M. Bouhlel (✉)
Faculty of Letters and Human Sciences, Syfacte Laboratory, University of Sfax,
10 Rue de Caire, 1001 Tunis, Tunisia
e-mail: b2l.maha@yahoo.fr

A. Bennasr
Faculty of Letters and Human Sciences, Syfacte Laboratory, University of Sfax,
BP 304 Sakiet-Ezzit, 3021 Sfax, Tunisia

© Springer Nature Switzerland AG 2021   37
F. Khebour Allouche et al. (eds.), *Environmental Remote Sensing and GIS in Tunisia*, Springer Water, https://doi.org/10.1007/978-3-030-63668-5_3

that urban densification process, presented as an alternative for sustainable urban development, is no less problematic and can pose threat to urban green spaces. In the face of this degradation of green spaces under the pressure of strong urban changes (expansion/densification), recent studies have shown that green space must be a key consideration in urban planning if the health of a city and its people are both considered important. Urban greening is considred as most important in order to keep the connection with nature and for people health [1, 4]. «*Urban green space, such as parks, playgrounds, and residential greenery, can promote mental and physical health and reduce morbidity and mortality in urban residents by providing psychological relaxation and stress alleviation, stimulating social cohesion, supporting physical activity, and reducing exposure to air pollutants, noise and excessive heat*» [5]. Thus, preserving the urban biodiversity and reinforcing the role of nature in the city is now the main concern of sustainable urban planning policies, particularly in cities that suffer from the scarcity of green spaces, such as Sfax city that is characterized by a weak and degraded natural vegetation. It is most often a discontinuous steppe that allows only poor protection for the different rocky outcrops. The degraded nature of vegetation cover is due to the rarity and the irregularity of rains, the great heat, the strong evaporation, as well as the weakness of grounds especially on the hills there. This was due to the man who has exploited this natural environment since the first century of historical times, as shown by many archaeological vestiges that demonstrate the continuity of human presence. The scarcity of green spaces in the city was accompanied by a massive consumption of its agricultural space following the fragmentation of family farmland «Jnens», especially with the breakup of patriarchal family, «*It's a successive process; At each generation there is a phase of dividing the jnens between the heirs: population growth and the break-up of traditional family into autonomous households helped to accelerate the fragmentation process*» [6]. This article discusses the role of green spaces in urban planning and policy in Sfax city especially that land allocation priorities to urban green cover are usually neglected or readily negotiated in the countries that are in transition [7] as is the case in Tunisia. The use of GIS will allow identifying the green spaces issues in Sfax city, thereby improving the evidence base of decision-making.

3.2 GIS and Planning of Green Spaces

Urban planning is «*the technical, social and political process concerned with the design, development and maintenance of land use in an urban environment*» [8]. It is the method of accuracy and organization that allows public authorities to guide and control the urban development through the development and implementation of planning documents. Spatial and geographic information plays a fundamental role in the planning of urban space. Thereby, the use of Géographic Information System (GIS) is indispensable. Urban planning is one of the main applications of GIS which is increasingly becoming an important component of planning support systems [9]. It allow to gather and organize, to manage and combine, to develop and present information geographically located, contributing in particular to the management of space. A GIS is a decision support that can facilitate urban planning and becomes a guide for the best choice of irreversible urban devolpment [10, 11]. It becomes a key tool for identifying urban development issues for the different components of the city such as green spaces. During the twentieth century, urban planning allowed permanently to integrate the planted spaces in the policies and the urban practices in particular in the developed countries [12]. Thereby, green space have become a vocabulary of urban planning. It designates lands not yet built, planted, wooded or agricultural. However, the characterization of green spaces is still inaccurate, and their evaluation methods remain little known by the designers of urban planning. This can explain that the concept of green spaces is recent. It is involved in the urban environment because of the non-stop of building spaces in the cities. Vegetation was thus reduced practically to its simple visual aspect, and in the best cases to its role as a climatic regulator.

Nevertheless, the scale of the environmental problems and the spreading of urban areas make it an obligation to take into consideration nature in the design and management of urban environment today. The creation of green spaces in the city only for hygienist and aesthetic reasons seems outdated, and nature takes intrinsic value in the context of sustainable development [13]. «*The presence of nature in the city has become a key factor in assessing the quality of urban life*» [14]. This presence plays many roles, including the preservation of biodiversity, the limitation of urban sprawl, and its role as an essential element in public health (air quality, leisure and recreation areas) and as a source of social bond thanks to the green spaces or family gardens. In addition, the essential role of nature is to mitigate and adapt our cities to climate change in accordance with Laile and al. [15], «*Nature has a property of thermal regulation, but also to maintain permeable soils that allow infiltrating the rainwater and preventing the risk of flooding while restoring the water tables*». Accordingly, these urban green spaces contribute to the quality of the living environment and the attractiveness of cities.

3.3 GIS Assessement of Natural Space Planning in Sfax City

The consideration of nature in the planning of Sfax city has gone through two major phases: the first phase during the 1970s/1980s was mainly concerned with the preservation of the agricultural area of Sfax city. Then the second phase that has started since 1990 and has been interested in the preservation of natural environment and the creation of green spaces, not only to meet aesthetic and social requirements but also to allow the city to take its place as a world economic metropolis. The use of GIS allows assessing urban planning choices in both phases.

3.3.1   GIS Assessement of Urban Sprawl and Consumption of Agricultural Spaces in Sfax City

The urban planning documents of Sfax city, particularly those of the post-independence period (2nd generation of urban plans), have been trying since the beginning of the 1960s to preserve the semi-rural character of the towns, especially with the established zoning by the urban displacement plan 1977. The zone NA (high density) and NB (low density) benefited, almost more than three times, from authorizations to build compared to those fixed by the plan of development. Moreover, the actors of the city never admitted the preservation of the semi-rural character of the belt of agricultural spaces especially with the deficiency in means of control which was responsible for the peripheral urbanization outside the communal tax base where a good part of the housing is illegal [16]. The inadequacy of the strategy of the urban displacement plan of 1977 and the urban and social reality of Sfax city helped to trigger the peri-urbanization whereas it wanted to preserve the jnens. As a result, Sfax city has not been able to regulate the urban system as a living system and its natural environment. This difficulty is shown by the unbridled artificialisation of soils in the Sfax city, guided by the spread of urban fabric at the expense of agricultural and vegetable areas.

3.3.1.1   Dataset and Methods

GIS allows apprehending the urban sprawl of Sfax city at the expense of its agricultural area. the research will adopt a diachronic approach. To observe the phenomenon of urban sprawl in a diachronic way, we propose to set up an original methodology based on specialized data. The ideas is to establish an evolution over a given period based on the coverage of the same territory on differents dates.

Dataset:

The sources available for spatially understanding the phenomenon of urban sprawl are numerous and varied (satellite images and aerial photographs). The perception of land use is more precise in aerial photographs. However, it is easier to manipulate a satellite image than several hundred photographic tiles. Thereby, after examining the databases available to date, the selected data source is LANDSAT. The satellite images obtained from this program offer several advantages: first, they cover the entire study area and are available on different dates. Secondly, these images are ortho-rectified. Ortho-rectification is a geometric correction of the images which aims to present them as if they had been acquired from the vertical. In practice, this involves rendering the image acquired by the satellite stackable on a map. Third and last, LANDSAT images are free of rights; they are distributed free of charge and without a license to use. To acquire these images, you simply have to go to the Lan dsat.org site.

The images chosen are those that contain few artifacts, such as clouds or shadows, which complicate image processing. Moreover, the favored shots are those recorded in spring or summer, when the photosynthetic activity is high.

The first image available dates from 1972, it is available on Landsat 1 (23 July 1972–6 January 1978). Unfortunately this image is unusable because of the strong presence of clouds on Sfax city. As a result, we chose to apprehend the spread from 1987, which is the second image available on Landsat TM.

Finally, 4 satellite images of 30 m resolution were selected: a Landsat TM (4) image (1987), a Landsat ETM+ (7) image (2001), a Landsat ETM+ (7) image (2006), and a Google Earth image (2014) (Fig. 3.1). The image was processed in order to

Image: LANDSAT TM (1987) true color band 3, 2, 1   **Image: LANDSAT ETM + (2001) true color band 3, 2, 1.**

Image LANDSAT ETM + (2006) true color band 3, 2, 1.   **Image: Google Earth 2014**

Fig. 3.1 Satellite images selected to measure urban sprawl in Sfax City (*Source* Landsat.org; www. google.com/intl/fr/earth/)

improve its parameters in order to obtain an excellent homogeneity of the colors of luminance and contrast, and then it was recalibrated by taking the Landsat image as a reference.

The size of the images was then reduced to the study area (coordinates in decimal degrees: 10,897 East/34,957 North/10,568 West/34,618 South).

The properties of each image are:

- Projection: Carthage UTM, Zone 32 N
- Ellipsoid: Clarke_1880_IGN.

Methods:

The satellite images collected will make it possible to measure the urban sprawl of Sfax city. to do so, a well-defined methodology was adopted. work was started by downloading spectral bands from 1 to 3 for Landsat satellite images (1987, 2001, 2006) from Landsat.org.

On "Envi 5", composite images in true colors of the Grand Sfax were produced using the spectral bands 3, 2, 1. Composite images in true colors are obtained by combining the red Landsat channels (band 3: $0.63 - 0.69$ μm), green (band 2: $0.52 - 0.60$ μm) and blue (band 1: $0.45 - 0.52$ μm). This combination gives an image resembling a color photo (Fig. 3.2).

Then an image of the area dating from 2014 was downloaded from Google Earth. the captured image was georeferenced and its resolution was matched to the resolution of the Landsat images. an automatic unsupervised classification was performed for the images on Envi 5.

4 unsupervised classifications was obtained from the Landsat images as well as the Google Earth image. The next step is to prepare the data in a GIS to correct the errors. In order to map four (04) states of the urban sprawl of Grand Sfax in 1987, 2001, 2006 and 2014. Only the built-up areas was retained (Fig. 3.3).

The classifications obtained have been exported to Shapefile, and for each date the zones that concern urbanization have been extracted. Than, manual verification and correction of the shapefiles was performed on Arc-Map. software to refine the results. At this level the work consists of manually checking and cleaning the shapefiles obtained from the unsupervised classification. It is a question of removing the superfluous polygons, as well as representing graphically the non-remotely-sensed built surfaces. In addition, it also involves modifying the polygons that underestimate or overestimate the built surfaces. In the following example Fig. 3.4 the polygons located on plots corresponding to bare soil in the satellite image have been eliminated. While buildings not detected by remote sensing have been vectorized. Finally, the data obtained are presented on the Urban Sprawl map.

3.3.1.2 Urban Sprawl Detection in Sfax City

The diachronic assessment of the evolution of the built-up areas on 4 dates 1987, 2001, 2006 and 2014 (Fig. 3.5) showed that the built-up areas went from a little less than 5000 ha in 1987 to 13,152 ha in 2001, to reach 15,792 ha in 2006 and 19,739 ha

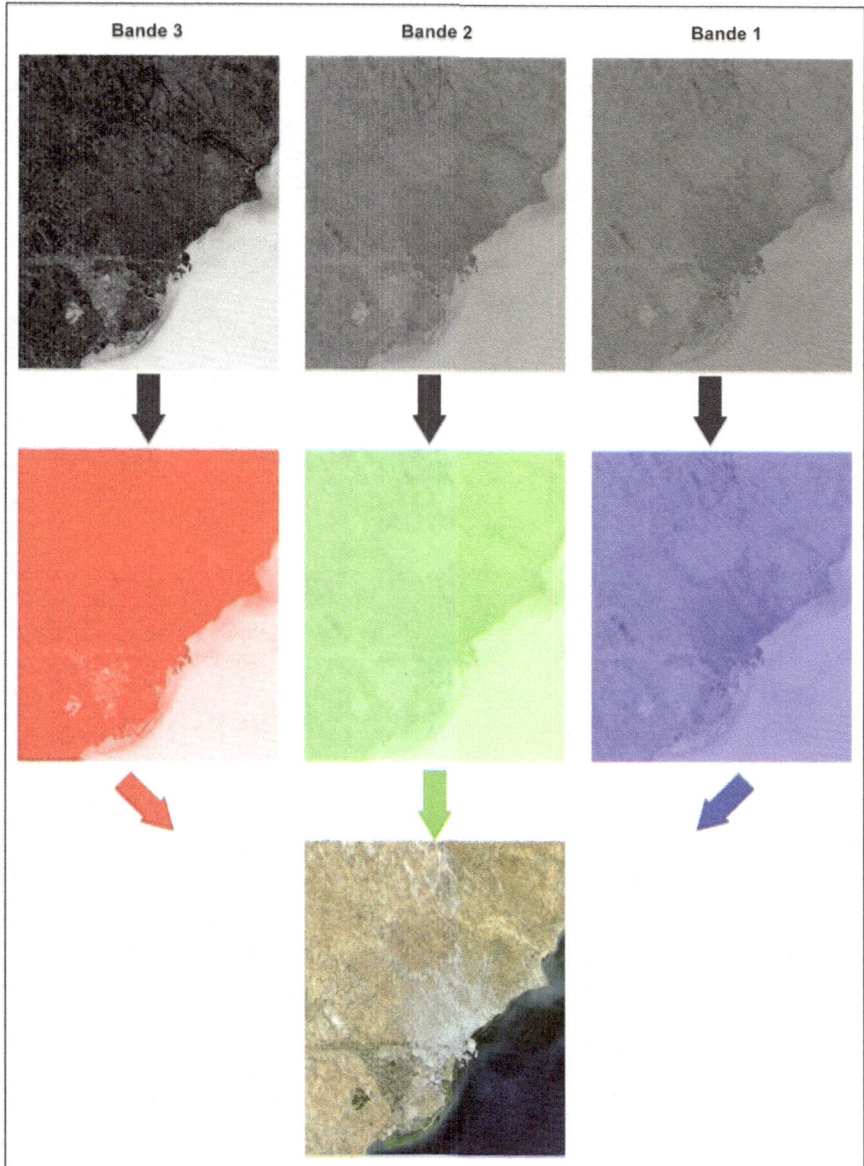

Fig. 3.2 Principle of colored composition Sfax city 2006

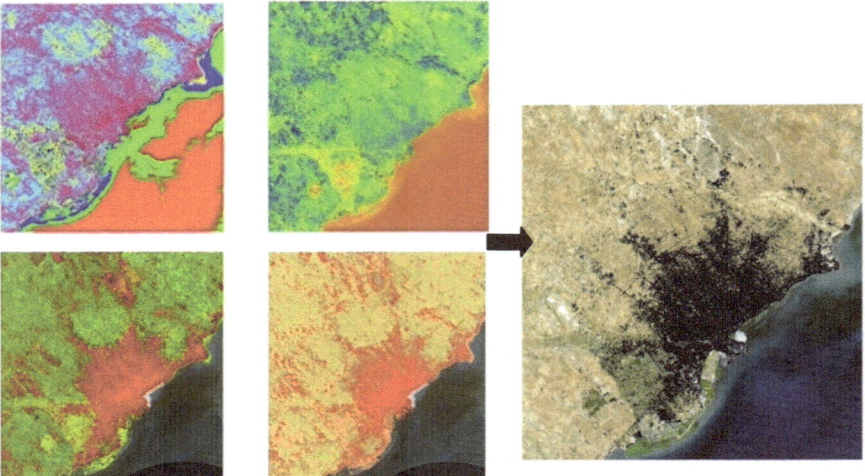

Fig. 3.3 Principle of isolation of built-up areas Sfax city 2006

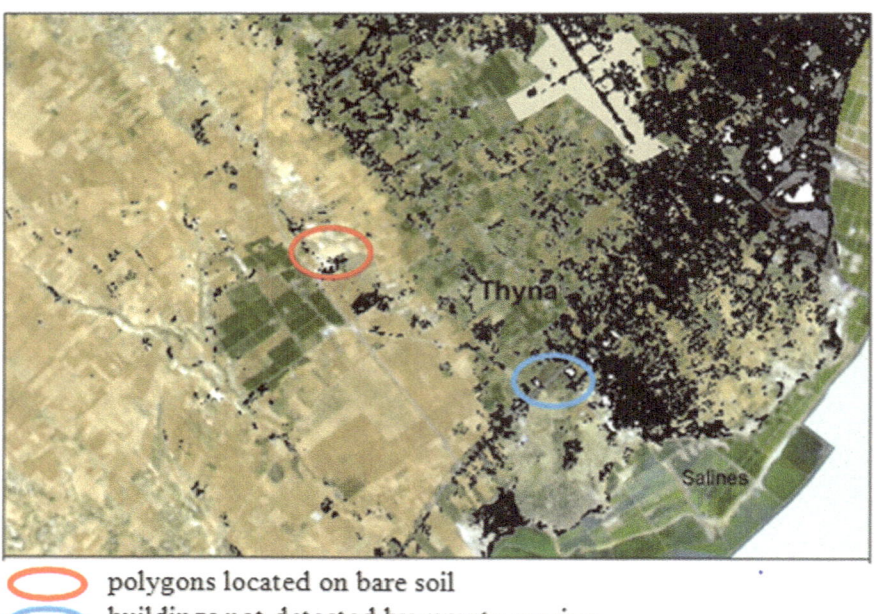

⬭ polygons located on bare soil
⬭ buildings not detected by remote sensing

Fig. 3.4 Examples of manual correction of shapefiles

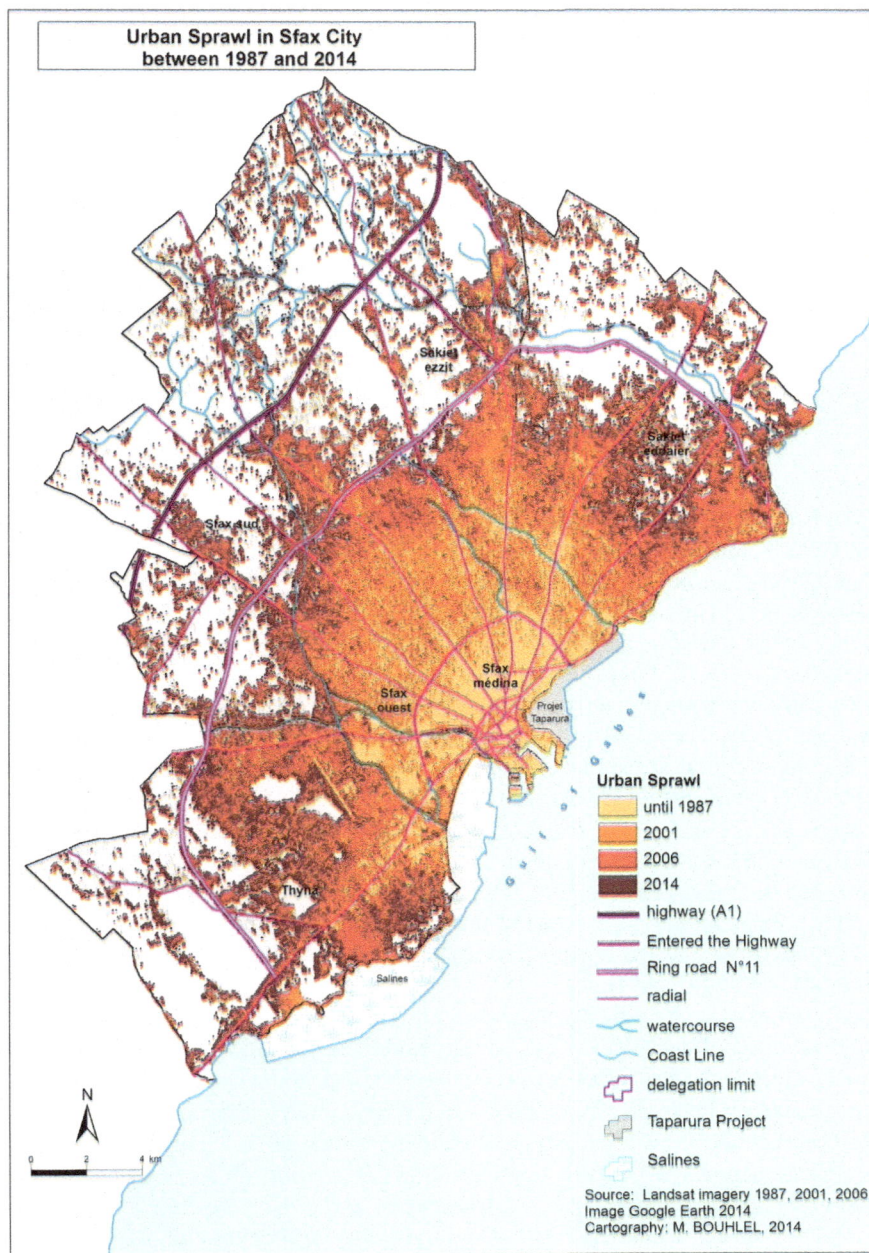

Fig. 3.5 Urban sprawl in Sfax city

Table 3.1 Surfaces built in Sfax city in 1987, 2001, 2006 and 2014

Year of reference	1987	2001	2006	2014
Area in ha of built	4099	13,152	15,792	19,739
Rate of the area occupied by the building of the total area of Grand Sfax (%)	7.3	23.5	28.2	35.2
Total area in ha of Grand Sfax	56,009	56,009	56,009	56,009

Table 3.2 Evolution of built-up areas in Sfax city from 1987 to 2014

Period	Gross evolution (%)	Annual change (%)
1987–2001	220.9	8.68
2001–2006	20.1	3.73
2006–2014	25.0	3.79
1987–2014	79.2	6

in 2014, i.e. respectively 7.3, 23.5, 28.2, 35.2% of the total territory of Grand Sfax (56,009 ha) (Table 3.1).

In 27 years, the built area of Sfax city has increased by 79.2%, an average growth of 6% per year. The most significant change was recorded between 1987 and 2001, during which the built area increased by 220.9%, an increase of 8.6% per year (Table 3.2).

This strong growth is mainly related to the detriment of agricultural areas. Between 1987 and 2014, more than 15,600 ha, which the majority of them are made up of Sfaxian agricultural space, have been waterproofed. The rhythm of this growth is impressive: on average, between 1987 and 2014, 579 ha were consumed each year, a rate of 1.5 ha per day. The rate is more frightening when it is reduced to a shorter time scale: 661 m^2 per hour or 11 m^2 per minute. At this cadence, 5790 ha will be consumed by 2024, 11,580 ha by 2034 and 17,370 ha by 2044.

This urbanization, poorly planned and managed, generated some challenges. It's energy intensive, contributing dangerously to climate change. Recent studies, notably those of Daoud [17] and Dahech [18], Daoud and Dahech [19, 20], have shown the repercussions of such artificialization on the environment of Sfax city. The unbridled artificialization of urban space in Sfax city and the gradual increase in the ratio of built-up areas/natural surfaces have been the main factors in the rise of temperatures. This warming is becoming more intense in the city center and the peri-central zone, the densest areas of the city, which has led to the formation of the urban heat island. This phenomenon can be explained by the fact that «*Built-up surfaces modify the energy balance near the ground. They store more heat during the day (after the artificialization of the surface, the amount of sensible heat increases at the expense of latent heat) and return it at night in the form of telluric radiation*» [18]. This rise in temperature is not noticeable in the zone of jnens where the suburban habitat and family gardens dominate. Indeed, as indicated by Clergeau, [21], «*wooded, agricultural and gardened areas mitigate heat island and air pollution (microclimatization). Also, they help to reduce CO_2 emissions (carbon sinks)*».

Similarly, the artificialization of soils in Sfax city has aggravated the physical vulnerability of the city to the risk of flooding, as shown by the floods of September 2009 (105 mm in 35 min) or September 2013 and December 2013 A rainy episode that once went unnoticed and without damage, causes now large and very violent runoff, due to the increased flow transfer rates, on areas occupied by housing, infrastructure and even collective equipment [17]. This problem is all the more noticeable in the central zone where the waterproofing of soils is more intense.

3.3.2 GIS Assessment of Green Space Consideration in the Sustainable Development Context

The consideration of green spaces in the urban planning of Sfax city has lifted an aspiration to a better living environment in a city that aims to be a Mediterranean metropolis.

At the end of the twentieth century, the green spaces are becoming included in the urban planning of Sfax city. Towards 1998 and following the alarming report of the serious environmental situation of Sfax city (pollution, coastal degradation, lack of green spaces), the improvement of environment quality becomes one of the major options around which articulates the layout of the city. The master plan of the city [22] considered the creation of green spaces as one of the necessary choices for the reconciliation of Sfax with its natural environment. As a result, a green plan was envisaged for the period 1995–2010, including essentially the green spaces planned by the plan of big urban projects in Sfax city covering a total area of 127.6 hectares, the most important of them are the green spaces envisaged in the Taparura project (60 hectares) and the thyna park (57 hectares). In addition to promoting new green spaces on beyond the ring road 11 to provide areas of landscape interest with recreational and cultural functionality, The master plan has estimated the need for green space in Sfax at 740 hectares to reach the rate of 10 m^2/inhabitant of the public green area per person recommended by World Health Organization. According to this planning document, this objective requires the planning of 127 hectare envisaged by the planning of the big surbans projects of Sfax city as well as the creation of new green spaces on 540 hectares beyond the ring road kilometer 11.

The development strategies of Sfax city 2016 [23] and 2030 [24] will be essentially limited to the valuation of what exists. The actions planned to be undertaken and implemented concern mainly the Thyna urban park.

In the development strategy of the Sfax city 2016, the realization of green areas, urban parks and green spaces is among the priority projects for the improvement of the living environment of citizens in order to increase the attractiveness of city,

Table 3.3 Actions planned in planning documents

Development strategy of Sfax city 2016	Sfax strategy 2030
• Completion of the Thyna Urban Park Development Project • Project of valorization of the ornithological wealth: ornithological center, circuits of visit…	• Development and equipment of Thyna and El Khalijurbanparks • Establishment of the natural and cultural pole of Thyna: observatory of birds, ornithological center… • Provision of green plan municipalities

considered as one of the strategic objectives to make Sfax a competitive and attractive Mediterranean metropolis.

The Sfax strategy 2030 focuses more on the assets of city with a faunistic and floral richness and a rich landscape linked to the beauty of its olive grove. However, Sfax is penalized by several constraints, including the delays in the projects planned and not yet realized, such as the creation of natural and cultural center of Thyna that has been programmed since the 1990s. This delay has meant that the actions that have been planned for 2016 are almost the same actions planned to be committed and realized by 2030 (Table 3.3).

The idea of staffing municipalities with green plans reflects the concern for sustainable development. The goal is to offer residents a qualified living environment through the preservation of its environment and the enhancement of the natural landscape of the municipality.

GIS tool allows to assess the results of the various actions planned to improve the place of nature in the city. To carry out this evaluation we choose buffer zone and mesh methods. Generally, the public gardens in Sfax city are distinguished by their advantageous central position except the Thyna Park and the El-Khalij Garden which is respectively 10 km and 5 km from the city center (Fig. 3.6).

However, the rapid urbanization of the city that has escaped all attempts to thwart it, and the failure to plan the urban evolution of this city (densification, verticalization …) generated multiples forms of inequalities, exclusion and deprivation.

The buffer zone technique is one of the most important transformations available to the GIS. In case of Geographical Information Systems, the units of buffering are points, lines, and polygons. Buffer operation refers the creation of a zone of a specified width around a point, a line or a polygon area. It is also referred to as a zone of specified distance around coverage features. Thereby, buffer allows to visually identifying the spaces distant from any nearby green space in Sfax city. From (Fig. 3.7) we can identify areas distant from gardens and public parks more than 3 km away. However, this technique does not allow estimating the population rate concerned by this inequality of access to green spaces in terms of distance. So, to approximately quantify the Sfaxian population concerned by the lack of green spaces, we used the mesh method (Fig. 3.8). It has shown that more than 50% of the total population suffers from the absence of any nearby public green space. This inequality of access to nature is accentuated by inequalities of mobility that come mainly from the penalization of inhabitants by distance. Tariff policy makes people

Fig. 3.6 Localisation of gardens and parks in Sfax city

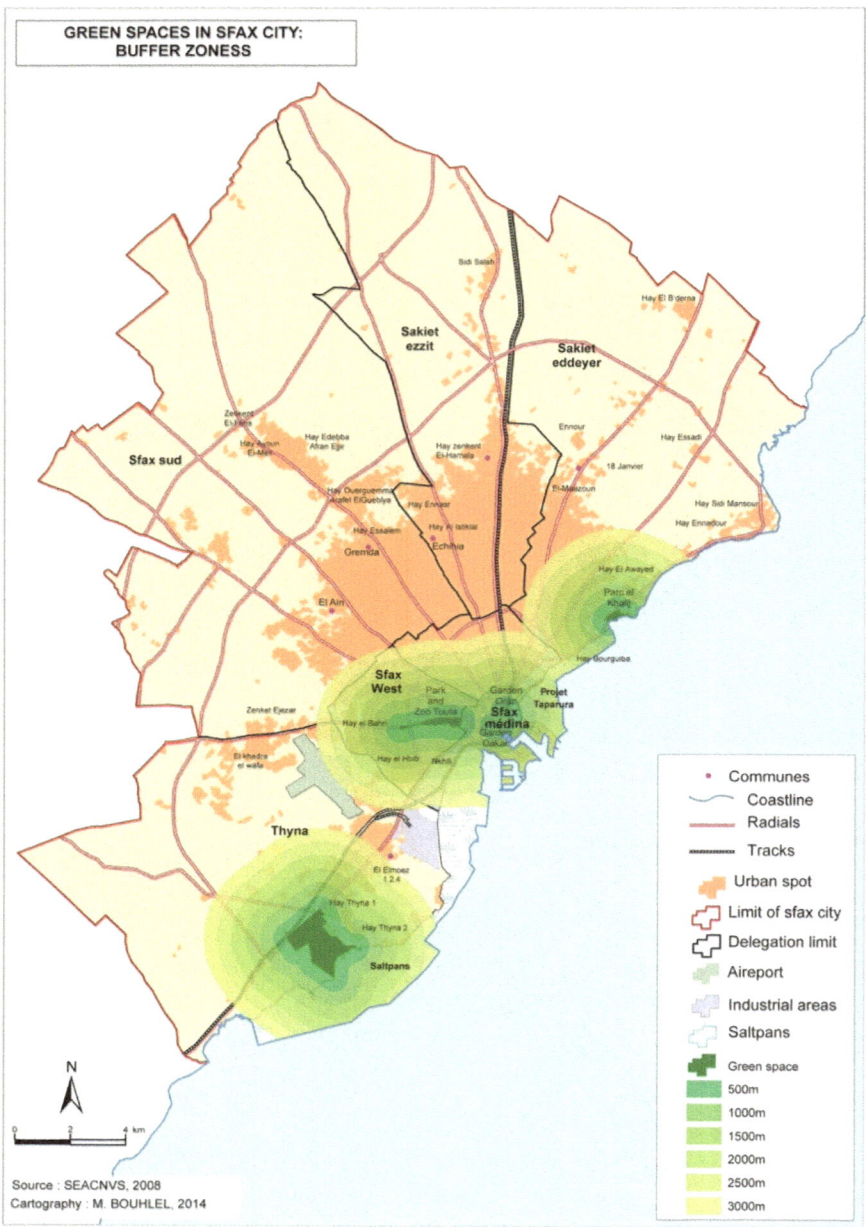

Fig. 3.7 Green spaces in Sfax city: Buffer Zones

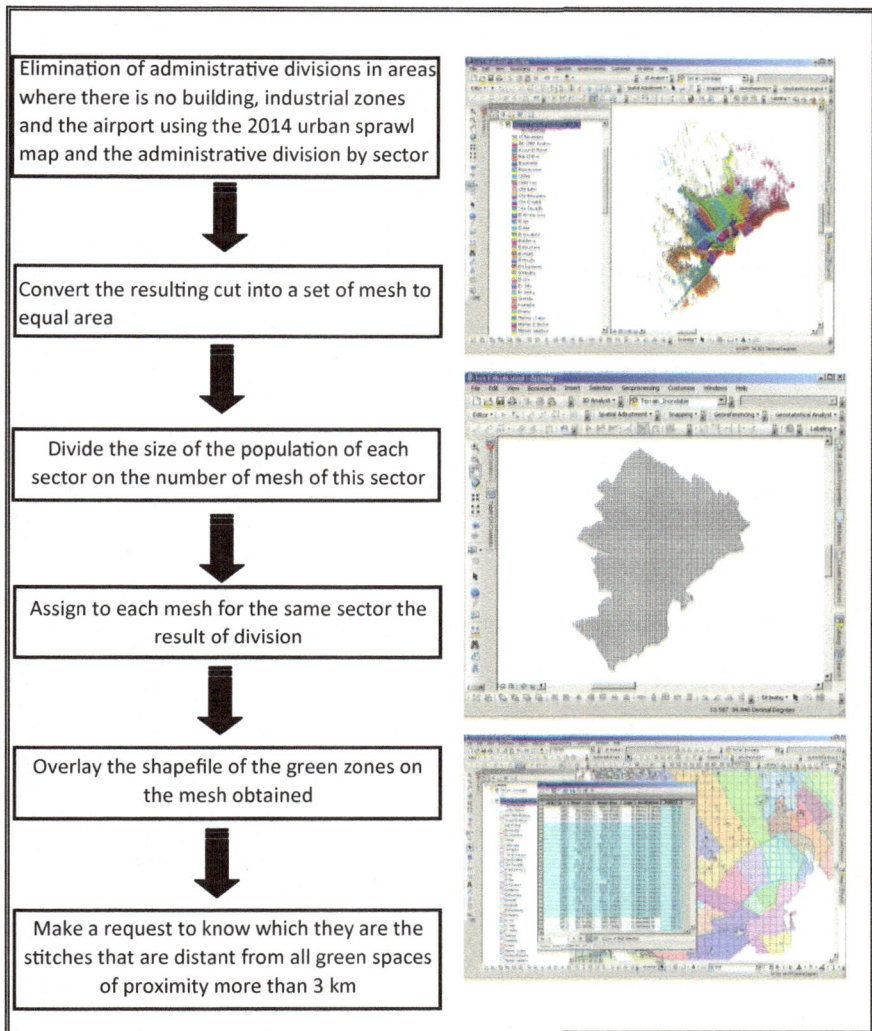

Fig. 3.8 Main steps of the mesh method and estimation of the population concerned by the absence of local green spaces

who live further away from the center pay more for their travel. The penalization of the inhabitants of the periphery by the distance is accompanied by long waiting times which varies between one hour during the school year and one hour 30 min during the holidays in the fourth section between 11 and 17 km against 20 min and 30 min in sections between 0 and 11 km. The inhabitants concerned by the low frequency of passage live in suburban popular neighborhoods like El Aouabed, El Khazzanet, Essghar, Sidi Salah, AinTorkia, Arafet ... This idea confirms the role of urban sprawl as generator of environmental inequalities but also social inequalities since these same neighborhoods are generally characterized by a high degree of precariousness and where the majority of the population do not have access to the private car.

The additional potential generated by the creation of Thyna Urban Park and El Khalij Garden did not solve the problem. The Thyna Park was envisioned as part of the 1998 master plan. Since it was considred by the various documents as new lung for the city. As for the garden of El-Khalij, it emerged in the 1990s from the will of the municipality of Sfax on a space gained on the sea following the first embankment work on the northern coast supposed to host the big urban project of Taparura. However, these two spaces, which are supposed to improve the place of nature in the city and offer new entertainment areas for its inhabitants, suffer from several problems caused by their deteriorated environmental state and a strong sense of insecurity exacerbated by the 2011 revolution and the events that followed. The exploration of the two areas shows a desolate landscape full of waste with a major under-supply of street furniture (bench, litterbins, lighting, etc.), as well as the absence of entertainment activities. In Thyna, the only recreation area in the park has closed due to the lack of visitors (Fig. 3.9).

Thyna urban park   El-khalij garden

Areas that are overwhelmed by litter

The leisure area « Aqua City » that has been closed because of the lack of visitors

Fig. 3.9 Deteriorated environmental state of Thyna Urban Park and El-Khalij garden

3.4   Green Spaces Issues in the Actors' Discourses

To better understand the green spaces management issues in Sfax, in depth interviews were conducted with the actors of the city. Two types of actors were selected: a representative of local authorities and another of civil society.

3.4.1   Complicated Mission of Local Authorities

To better understand the place occupied by green spaces and nature in the urban planning of the city of Sfax in depth interview was conducted with Deputy Director of Planning and Development of the municipality of Sfax.

How much do you take into consideration the place of nature in the planning of the city?

> The consideration of nature and green spaces in the urban planning of Sfax city is essentially limited to the preservation of the existing and indicated spaces since the 1983 urban development plan. The last Urban Development Plan 2002 only emphasizes the preservation of these spaces. We cannot talk about the creation of new green spaces because of the absence of state land reserves. (Interview Deputy Director of planning and development of the municipality of Sfax)

The weakness of the state's land resources in Sfax city and the consideration that this city is self-sufficient [25], led to the early disengagement of the state in the urban development of this city. Indeed, neither the public authorities nor the local authorities have done enough for the living environment, the reception and leisure facilities. This explains the lack of creation of new green spaces.

> However, it should be noted that Tunisian law is very rigid in terms of green spaces, we can't change their vocation or even displace them. Moving a green space requires a presidential order. It can be said that the green spaces in the urban development plan are unmanageable areas. (Interview Deputy Director of planning and development of the municipality of Sfax)

This rigidity in the management of green spaces comes from article 20 of code of spatial and urban planning: «… *A green space having acquired this vocation by the effect of a management plan, can lose it only by decree taken on the proposal of Minister in charge of Town Planning, after the opinion of the Minister of the Environment and Spatial Planning*». This article comes within the framework of the code of spatial and urban planning which ensures sustainable development and the right of citizens to a healthy environment.

> However, the revision of existing Urban Development Plan will give back the land designed as green space to its owners including those who have building permits. (Interview Deputy Director of planning and development of the municipality of Sfax)

The revision of existing Urban Development Plan may limit more and more land planned to be future green spaces in the Grand Sfax, which confirms the lack of coordination between the various departments of public service and the lack of inter-ministerial work.

Are there still solutions to increase the potential of green spaces in Grand Sfax?

> Owners of subdivisions of more than one hectare are forced to dedicate an area for green space. However, these subdivisions are not numerous and existing, yet they meet the standards leaving plots devoted to nature, these parcels generally remain empty land without development. At this level, the big urban project TAPARURA is considered as an opportunity to overcome the errors, especially that the citizens of Sfax have begun to claim their needs for green spaces whose scarcity is considered a handicap for their well-being. (Interview Deputy Director of planning and development of the municipality of Sfax)

The role of large urban projects in the return of nature to the city can be important. The example of Taparura is very significant. The urban travel plan of this project devotes 11.6% of the total surface area for green spaces that will cover a total of 260 hectares which will increase the rate of green space per person, an estimated growth of 3 m^2. However, the major projects already implemented, such as The Sfax-El-Jadida Project, show that landscaping of green spaces does not seem to be a priority for developers who are essentially dominated by a spirit of gain and maximum profitability. The partial layout of these spaces confirms this idea. Even for the BIPs where the undeveloped empty land testifies to the place of nature in the urban planning of the city of Sfax but also in all the Tunisian cities. As a result, we can wonder about the completion of green spaces recommended by the planning of Taparura project, which is already finding it difficult to start because of investment. Similarly, the consideration of nature in the planning of these projects often comes at the margin, invoked mainly for procedures dictated by the law (state of play, impact studies, environmental assessment, etc.). Its recommendations are formulated independently of the objectives set by developers or their sponsors.

3.4.2 An Active and Influential Civil Society

In the post-revolution period we have witnessed an incredible boom in community life while seeing the evolution of civil society. This process was confirmed and anchored by the article n ° 139 of the new Tunisian constitution promulgated on January 27, 2014: «*the local authorities adopt the mechanisms of the participative democracy and the principles of the open governance in order to guarantee the widest participation of the citizens and civil society in the preparation of development projects and regional planning and monitoring their implementation in accordance with the law*» . The problems with the green spaces in Sfax city show the incapacity of the public authorities to adopt a perennial solution. In this situation, civil society plays a leading role in preserving green spaces and improving citizens' living environment.

The interview with the president of the association of Sfax El Mezyana was able to show the growing role of civil society in the transformation of the living environment of the inhabitants.

What is the main objective of the association?

Our main goal is to make Sfax a pleasant city to live in, and it is in this context that we have started to organize events of many kinds (cleanliness campaigns, monthly collective breakfasts in neglected green spaces…), moreover, the idea of Sfax El Mezyana was at the beginning only a campaign of cleanliness organized in October 2010 by the young people, the event took the name of Sfax El Mezyana and it is in November 2011 that the idea became an association. (interview president of Sfax El Mezyana, 2017)

What is the role of green spaces in your actions concerning the living environment of citizens?

Green spaces occupy a primordial place in the various actions carried out by the association. Improving the living environment of citizens is essential. The example of the urban park of Touta is a real success for us …Touta was a poorly attended area, not frequented due to its bad reputation, we organized monthly breakfasts that managed to attract the Sfaxiens back to this important green potential, and after six months Touta has turned from an unpopulated green space to a real space of sociability where the municipality can not manage the growing number of these visitors. Then we brought animals like the Lion couple. The municipality was obliged to intervene so it launched a whole project of retyping and redevelopment of the park of 300,000 dinars (benches, bins, animals, implantation, …, this experience has led us to conquer all the green spaces abandoned despite their locations central (Wahran Garden, Monji Bali Garden, etc.) through tree planting, pollution control, etc. (Interview President of Sfax El Mezyana, 2017)

Currently, the Touta Zoo is, according to the data collected, a real success for the civil society represented by the association Sfax El Mezyana who initiated monthly collective breakfasts (Fig. 3.10) to improve the bad image that the park suffered from. Today we cannot manage the growing number of visitors that can reach up to 5000 people per day.

What are your strategies for encouraging citizens to preserve green spaces?

Every citizen must be an agent of a positive change in their environment, for example through the project "Madrasty El Mezyana" (my beautiful school) we try to encourage students to be positive agents in the improvement of their schools, likewise for the project "Houmty El Mezyana" (my beautiful neighborhood), where we seek to encourage young people from neighborhoods to be positive agents and no longer passive citizens who are still waiting for

Fig. 3.10 Monthly collective break fast organized in the park by the association of Sfax El Mezyana (*Source* https://www.facebook.com/Sfax.El.Mezyena/)

state intervention, on the contrary a citizen must be a creative agent to improve his living environment, as it requires the principle of sustainable development "think globally, act locally", the local starts from home, the motivation of neighborhood youth will be ensured by setting up principal members belonging to Sfax El Mezyana in each neighborhood to attract young people and to enhance the participative spirit in the new generation. (Interview President of Sfax El Mezyana, 2017)

The strategies of the civil society in Sfax that were strongly influenced by the good practices of sustainable development had a principal objective which was encouraging the participative spirit. However, it should be noted that awareness, which is important and indispensable, can not be the only way to preserve nature and improve the living environment of citizens. The restitution laws remain an effective means to annihilate bad practices transmitted from father to son.

3.5   Conclusion

The reduction and deterioration of green spaces accompanied by disastrous environmental consequences have drawn attention to the importance of nature in the city. The city is no longer perceived as a space hostile to nature. Sfax is concerned by this new trend which aims to consider green spaces as one of the key elements required to achieve sustainable urban development. They represent public values and have an important role to play for the wellbeing and health for the inhabitants, especially in a city where pollution is one of these main characteristics. However, this research based on GIS assessment has shown that neither urban policies nor urban documents have done enough for the protection, creation and development of these spaces, even the most recent documents, based on the principles of sustainable urban development. The actions implemented such as the creation of the urban park of Thyna was failed. However, it should be noted that civil society plays a key role in preserving existing spaces, and help people to appropriate the space by increasing acceptance and appreciation for green spaces.

3.6   Recommandations

Urban green spaces should be seen as an essential part of urban infrastructure not an «add-on», a «nice to have», or a luxury [26]. Green space must be a key consideration in urban planning. Therefore, an Urban Green Space Strategy must be implemented for the development, planning, and management of green spaces in Sfax city. It is a strategic document for achieving a number of specific objectives, with a long-termperspective. A suitable approach or method should be developed to attaining goals and resolving specifics issues. Moreover, urban planning policies, as well as different urban planning documents, must aim at the preservation of nature and the creation of newly planted spaces, particularly in the case of a city where the chemical industries have been responsible for the heavy chronic pollution until now. In Sfax, as elsewhere, we must show a growing interest in green spaces while conceiving them as vital spaces, close and accessible to the inhabitants of the city. The GIS tool will be used to put in place adequate urban planning and management, which takes into consideration the natural space management issues in Sfax city.

The current trends that have been observed in Sfax city suggest an increasing degradation in the general quality of existing urban green spaces. The majority of green spaces have very poor design innovation, desolate equipment, bad accessibility, low ecological value, etc. Thus, more attention should be paid to increasing the quality to encourage more people to cherish the space.

A sufficient political and financial support from national or local authorities, other funding agencies or private investors, must be available for a green space management able to reserve that process.

References

1. Smaniotto C, Erjavec I-S, Mathey J (2008) Green spaces—key resources for urban sustainability. The green keys approach for developing green spaces. UrbaniIzziv 19(2):199–211
2. Najihah A, Corstanje R, Harris J, Brewer T (2017) Impact of rapid urban expansion on green space structure. Ecol Ind 81:274–284
3. Haaland C, Konijnendijk Van den Bosh C (2015) Challenges and strategies for urban green-space planning in cities undergoing densification: a review. Urban For Urban Green 14(4):760–771
4. Khoshtaria TK, Chachava NT (2017) The planning of urban green areas and its protective importance in resort cities (case of Georgian resorts). Ann Agrar Sci 15:217–223
5. World Health Organization (2016) Urban green spaces and health: a review of evidence. Regional Office For Europe
6. Megdiche T (2010) The evolution of the social division of space in Sfax, Urban sprawl: an uncontrollable process? University Presses of Rennes, pp 207–219 (French)
7. Anguluri R, Narayanan P (2017) Role of green space in urban planning: outlook towards smart cities. Urban For Urban Green 25:58–65
8. Maarseveen M-V, Martinez J, Flacke J (2019) GIS in sustainable urban planning and management: a global perspective, CRC Press, By Taylor and Francis Group, Boca Raton, 364 p
9. Yeh AG-O (1999) Urban planning and GIS, geographical information system, 2nd edn, vol 2, Manage issues appl 62:877–888
10. Lee J, Tian L, Erickson LJ, Kulikowski TD (1998) Analyser les politiques de gestion de la croissance avec des systèmes d'information géographique. Environ Planif B: Planif concept 25(6):865–879
11. Dhaoui I (2014) Urban sprawl: The GIS and remote sensing data assessments, MPRA Paper No. 87650, posted 29 June 2018 18:40 UTC. [Online]. https://mpra.ub.uni-muenchen.de/87650/
12. Mehdi L, Weber C, Pietro F, Selmi W (2014) Evolution of the place of plants in the city, from green space to the green fabric, *VertigO*—Electron J Environ Sci [Online], 12(2) I September 2012, posted on February 10, 2014, URL:http://journals.openedition.org/vertigo/12670; DOI : https://doi.org/10.4000/vertigo.12670. Accessed Feb 11, 2018 (French)
13. Clergeau PH (2008) Preserving nature in the city. Ann Min—Responsib environ, 4/2008 52:55–59 (French)
14. Donadieu P (2013) Make way for nature in the city. The need for new professions. Métropolitiques (French). http://www.metropolitiques.eu/Faire-place-a-la-nature-enville.html

15. Laille P, Provendier D, Colson F (2013) The benefits of plants in the city—synthesis of scientific work and analysis method. Plante & Cité, engineering of nature in the city (French)
16. Dlala H (1996) Le Grand Sfax: recent morpho-functional dynamics and improvements. Ann Géogr [Online]. 1996, t.105, 590:369–394 (French). http://www.persee.fr/web/revues/home/prescript/article/geo_00034010_1996_num_105590_20745
17. Daoud A (2013) Feedback on the floods in the agglomeration of Sfax (Southern Tunisia) from 1982 to 2009: from prevention to the territorialization of risk. Rev Géogr l'Est [Online], 53(1–2) I 2013, [Online] September 16, 2013, http://rge.revues.org/4630. Accessed Jun 22, 2014. (French)
18. Dahech S (2012) Evolution of the spatial distribution of air and surface temperatures in the agglomeration of Sfax between 1987 and 2010. Impact on energy consumption in summer, Climatology, special issue' climates and climate change in cities', pp 11–33 (French)
19. Daoud A, Dahech S (2009) Climate change and urban governance: The case of the agglomeration of Sfax (Southern Tunisia), Fifth Urban Research Symposium (French)
20. Daoud A, Dahech S (2012) Resilience of the agglomeration of Sfax (Southern Tunisia) in the face of climate change: assessment test, climatology, special issue 'climates and climate change in cities', pp 109–126 (French)
21. Clergeau PH (2007) An ecology of the urban landscape. Ed. Apogée, Rennes. nature in the city, 36 p (French)
22. Ministry of the Environment and Local Development (1998) Master plan for the development of Sfax city. Final report, AUDEC, 223 p (French)
23. Ministry of the Environment and Regional Planning (2010) Grand Sfax development strategy: phase II, study of popular urban areas, final report, Architecture and innovation, 581 p (French)
24. Charfi F (2016) Sfax 2030 Strategy: from strategic vision to action plan. Final report, ADSS, 180 p (French)
25. Bennasr A, Megdiche T, Verdeil E (2013) Sfax, laboratory of sustainable urban development in Tunisia? Environ Urbain/Urban Environ 7:a83–a98 (French) http://id.erudit.org/iderudit/1027728ar
26. Orr Sh, Paskins J, Chaytor S (2014) Ucl policy briefing—October 2014. Valuing urban green space: challenges and opportunities

Chapter 4
Spatio-Temporal Monitoring of the Meskat System Distribution in the Tunisian Sahel Region Using TM Landsat Images

Asma Ben Salem, Asma El Amri, Soumaya M'nassri, Karem Chokman, and Rajouen Majdoub

Abstract The Meskat system, which is an ancestral anti-erosion practice, has faced abandonment in the last decades. The present study is intended to monitor the spatial and temporal evolution of the Meskat system in the region of Sousse (Tunisian Sahel). Satellite images from the Landsat Thematic Mapper (TM) 5 sensor were used for four years of 1987, 2003, 2007 and 2011. The detection of land use changes was applied with the Geomatica Focus software, based on a supervised classification. Analysis of land use validation and classification highlighted nine land use classes, with an average global success and Kappa indexes of, 86 and 84%, respectively. Results of the space-time evolution of the Meskat system showed that it is mainly divided among three watersheds: Sabkha Halek El Menjel, Wadi Laya El Hammam and Wadi Hamdoun. The diachronic analysis of these three watersheds classification between 1987 and 2011 revealed an expansion of urban areas at the expense of the Meskat system. Indeed, urban areas recorded an increase of 1, 8, and 4%, while areas occupied by Meskat system were decreased by 6, 8, and 13% for the Sabkha Halek El Menjel, Laya El Hammam Wadi and Hamdoun Wadi watersheds, respectively.

Keywords Remote sensing · Landsat TM 5 · Supervised classification · Land use · Meskat

A. Ben Salem
Sylvo-Pastoral Institute of Tabarka, University of Jendouba, ISP Tabarka 345, 8110 Jendouba, Tunisia

A. El Amri · S. M'nassri (✉) · R. Majdoub
Higher Institute of Agronomy-Chott Meriem, University of Sousse, ISA CM BP 47, 4042 Sousse, Tunisie

K. Chokman
National Institute of Scientific Research, Quebec G1K 9A9, Canada

© Springer Nature Switzerland AG 2021
F. Khebour Allouche et al. (eds.), *Environmental Remote Sensing and GIS in Tunisia*, Springer Water, https://doi.org/10.1007/978-3-030-63668-5_4

4.1   Introduction

Soil and land resources are facing with severe physical, chemical and biological degradation due to various natural and anthropogenic factors such as drought [1], desertification [2], erosion [3, 4] and urbanization [5]. In recent years, several studies focused on land quality and land degradation effect on agricultural productivity. However, conventional methods (surveys, thematic maps, Meta data, etc.) used to study the degradation of soil resources are either insufficient or ineffective [6]. In order to better guide the research, the use of more advanced techniques, such as remote sensing, is more effective to obtain a precise analysis of the whys and wherefores of these degradation phenomena [7].

Remote sensing has been widely used to study soil erosion and water and soil conservation in Tunisia. Bouchnak et al. [8] used remote sensing data to study the evolution of gulley formation in Central Tunisia. They identified the most sensitive rock units to gully formation, namely geological formations made of gypsiferous marl, and landscape location, mainly the glacis. Identification of erosion control practices with the use of the Geographic Information Systems (GIS) was conducted by Anane et al. [9], they mainly focused on sites that had been landscaped by small hill dams. Aiming to increase farmer's awareness to the importance of the Jessours system (mounds of compacted soil along Talwegs that make up mountainous streams); Abdelli et al. [10] identified, using remote sensing, 87 and 94% of the Jessours of Wadi Hallouf and Wadi Jir's watersheds indicating these types of data to locate such hydraulic structures. GIS and remote sensing have also been used to diagnose abnormalities affecting water and soil conservation practices (WSP). Indeed, Baccari et al. [11] used these tools to analyze the risks of cracks' occurrence on contour ridge benches in Tunisian semiarid regions. In the same context, Fourati et al. [12] proceeded to the visual interpretation of high resolution satellite images to detect operation anomalies in the WSP in the Sfax region.

In the Tunisian Sahel region of Sousse, the Meskat system, a traditional erosion control practice for water harvesting and soil conservation, is the most common technique of soil-water practice [13]. It is mainly used for olive production, considered the main farmers income [14]. In the last decades, changes in land use have been observed. These changes are related to farmland decrease and regression of the Meskat system. The diagnosis of the operational state and monitoring of the distribution of this WSP are almost nonexistent. The main studies are those of Dallemand et al. [15], Houimli [16], and Bouchiha and Besbes [17]. The first study focused on mapping land use in Central Tunisia by using a combination of microwave (SEASAT and SIR-A) and optical data (SPOT, Landsat MSS and TM). Analysis of this data set illustrates their potential uses for specific observations, for the management of rural spaces, including urban and agricultural zones, WSC, moisture in a depression and topography. The second study, mainly, addresses the situation of coastal agriculture by referring to map databases (topographic maps of 1962 and 1994). This study reveals the emergence of an urban extension on coastal farmland such as Akouda, Hammam Sousse and Chott Mariem. Bouchiha and Besbes [17] studied urban expansion in different regions of the Tunisian Sahel by referring to Landsat MSS, TM and ETM+ satellite images. The comparison between a supervised and hybrid classification highlights the evolution of the urban area's size over the years. Their study reveals that hybrid classification provides sufficiently consistent results. It indicates that the total area occupied by urban spaces increased by 310% in the last 33 years. It should be noted that these have not focused on the diachronic study of the Meskat system.

This chapter aims to study the spatio-temporal distribution of the Meskat system. To achieve this goal, a diachronic analysis was conducted on classified images, as well as a spatio-temporal analysis of the rate of change of occupation by the Meskat system.

4.2   Meskat System

The Meskat system is a technique used to collect runoff water that has been used since Roman times. The operational principle of this impluvium system is to collect upstream rainwater (Fig. 4.1). The water is directed through distributors to irrigate

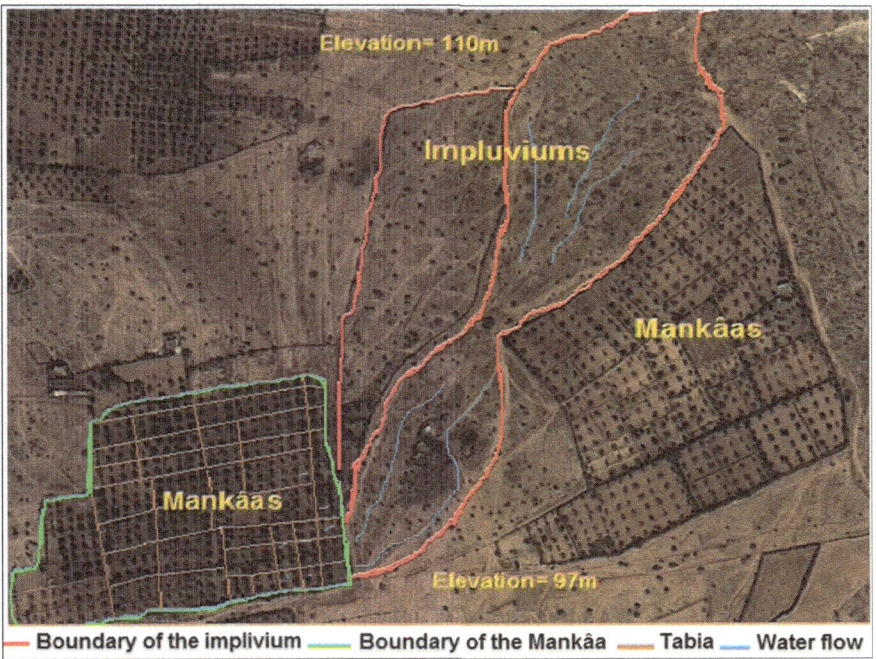

Fig. 4.1 Spatial configuration of the Meskat system in the hilly landscape. *Source* @Google earth modified, 2013

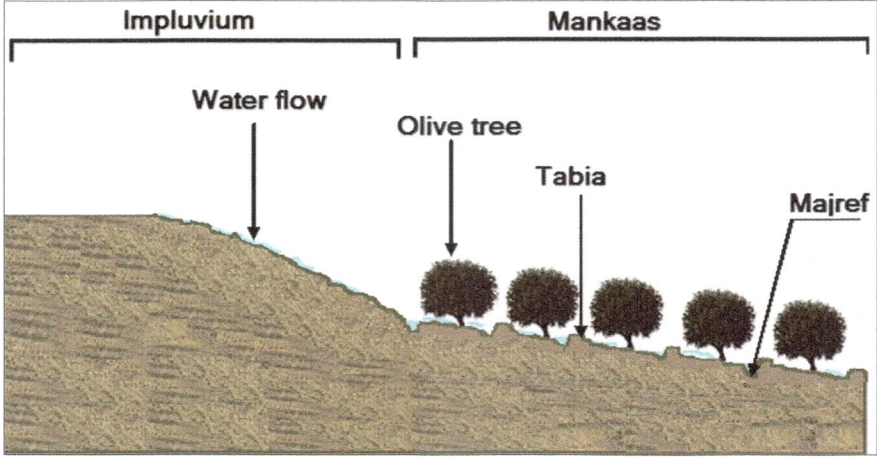

Fig. 4.2 Cross section of the Meskat system

the olive trees planted in downstream plots, called Mankaas. At this level, the runoff is intercepted by mounds known as "Tabias" and directed from one plot to another through small openings (majref) [18].

The operating performance of the Meskat system essentially depends on the ratio between the dimension of the impluvium and that of the mankaa. According to Snane et al. [19], the optimal ratio is 2/3 for the first and 1/3 for the second. The impluvium is characterized by the dominance of the limestone layer, the development of spontaneous plants (Artemisia, Rosemary, Alfa, etc.) and the existence of old runoff water harvesting structures such as "Fesguias" or "Majels". The mankaas are mainly occupied by olive trees (Fig. 4.2). The impluvium sizing essentially depends on rain and runoff occurrence. Its slope varies between 3 and 10%, while the general slope of the mankaas is of 1–2%. The alternation between impluvium and mankaa surfaces forms a mosaic [20].

4.3 Recognition of the Meskat System Processing in Tunisian Sahel Region: Thematic Focus and Methodology

4.3.1 Study Area

The Sousse region, located on Eastern Tunisia's coastline, is composed of seven watersheds (Fig. 4.3). These watersheds make up the tributary of the three main local wadis, namely from North to South, Wadi Blibene, Wadi El Hallouf and Wadi Hamdoun. The Sousse region has a coastline of about 90 km in the Golf of Hammamet

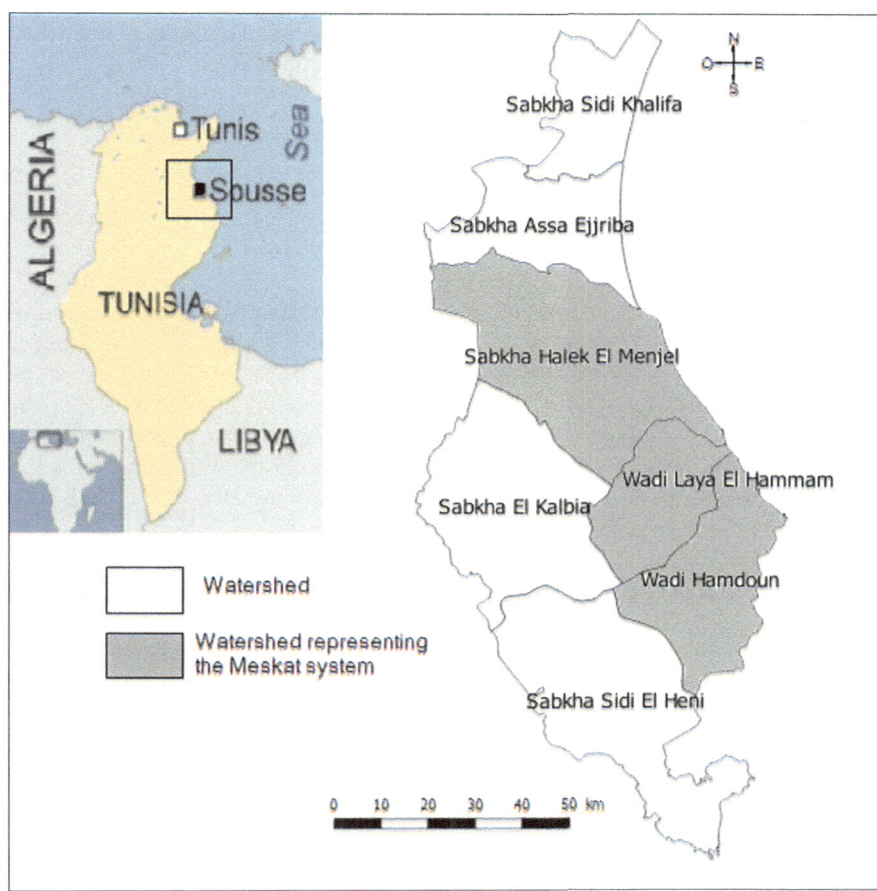

Fig. 4.3 Location of the study area and its delineation into watersheds. *Source* Ben Salem, 2017

(Mediterranean). The farmland area covers 205,000 ha. This area constitutes a transition zone between coastal Sahel and Central Tunisia, and is predisposed to both continental and maritime effects.

The landform of the Sousse region is dominated by low altitudes, hilly areas in the Southwest and mountain areas in the North-West of the Governorate (Hmax = 350 m). A corrugated landscape marks the coastline's landform, with more or less rugged slopes that are often cobbly upstream. The bed rock is exposed in the highest part of the area (>20 m above sea level) and with an average slope of 4–10%. The study area belongs to the arid and semiarid bioclimatic stages. The average annual rainfall is 350 mm with an average annual temperature of 20 °C. Winds are generally moderate with a speed less than 15 km/h. This study essentially covered the watersheds of the Sousse region, where the Meskat system is the most widely used as water harvesting practice (Watershed of Sabkha Halek El Menjel, Watershed of Wadi Laya El Hammam and Watershed of Wadi Hamdoun).

4.3.2 Exploitation of Satellite Images

4.3.2.1 Data Collection

Thematic geographic data were used for watershed boundaries and erosion control practices in order to delimit the study area. In addition, Landsat Thematic Mapper (TM) 5 satellite images were downloaded from the United States Geological Survey (USGS) site/Global Visualization Viewer (Glovis). According to Irons et al. [21], the choice of this type of image is justified by its potential to locate agricultural areas with minimum uniformity (30 × 30 m resolution) and by the availability of seven spectral bands [22].

4.3.2.2 Pre-processing of Satellite Images

This operation creates a mask and eliminates unnecessary areas. To ensure high quality data, the images were georeferenced and an atmospheric correction was made using the algorithm 'Atcor-2' to minimize atmospheric effects [23]. This correction takes into account sensor degradation, radiometric calibration, observation angle and sun angles [24].

4.3.2.3 Classification of Land Use

The Normalized Difference Vegetation Index (NDVI) was calculated by applying the subtraction to addition ratio of near-infrared and red bands [25]. Similarly, the Soil-Adjusted Vegetation Index (SAVI) was obtained by calculating the same ratio with plant cover data [26]. The Brightness Index (BI) is an indicator of the soil's brightness reflectance and is related to surface soil moisture and salt presence. It was calculated by using the same bands. The Building Index (Ibt) was assessed by applying the subtraction to addition ratio of the blue band and the BI index. In addition, a principal component analysis (PCA) was applied to every processed image in order to distinguish the informative bands [27]. A calibration sampling was also performed. For this purpose, random polygons representing every (real) visualized class on Google Earth were sorted. Several trials were made until the images showed sharp classes. These classes were set according to the image analysis resulting from the calculated indices. In this regard, it should be noted that the latter are correlated with climate change and comply with previous studies [24, 28].

4.3.2.4 Validation of the Classification

Validation of the resulting classification was obtained by selecting new polygons using Google Earth, which has a very high spatial resolution. Indeed, a limited number of pixels were chosen for each type of occupation in well-known and visited areas. Some scientists prefer to use the following equation based on multinomial distribution (Eq. 1) to determine a priori the required number of pixels (N) for the validation of the classification of each component [27].

$$N = \frac{B}{4b^2} \tag{1}$$

where $B = \frac{\alpha}{k}$ determined from χ^2, with $1°$ of freedom; **k**: the number of classes; **b**: the percentage of accuracy (usually 5%).

A confusion matrix was obtained between the selected classes (calibration samples) and the real classes (validation samples). This type of matrix allows calculating the Kappa index (K) and the Global Success index (GS). The Kappa index (Eq. 2) expresses the proportional reduction of the error obtained by a classification, compared with that obtained by a completely random classification. The Global Success index (Eq. 3) expresses the success rate of the classification (overlay of the same classes).

$$K = \left. N^2 - \sum_{i=1}^{r}(x_{i+}x_{+i}) \middle/ N \sum_{i=1}^{r} x_{ii} - \sum_{i=1}^{r}(x_{i+}x_{+i}) \right. \tag{2}$$

where **xii**: number of observations in row i and column i; **xi+** and **x + i**: marginal totals of line i and column i, respectively; r: number of rows in the confusion matrix; **N**: total number of observations.

$$GS = \left. \sum_{i}^{r} x_{ii} \middle/ N \right. \tag{3}$$

4.3.3 Methodology

Thematic geographic data of Sousse region scaled 1:25000 were applied. The selected images dates are 5 August 1987; 17 August 2003; 27 July 2007; and 5 August 2011. They were selected from the driest season to ensure their sharpness [17]. The study area is located in two different scenes (p191r035 and p190r036), which explains the need for a mosaicking operation prior to analysis. The study area extraction was performed according to the UTM, WGS 1984 zone 32 N projection system by using the PCI Geomatica software (2012). The classification was of the supervised type, given our knowledge of the study area. First, four classifier indices were calculated, thereby producing images with color levels that vary from one class to another (Table 4.1).

Table 4.1 Calculation results of the NDVI, SAVI, BI and Ibt indices

Class	NDVI	SAVI	BI	Ibt (*100)
Urban Spaces (US)	(−0.009)–0.20	0.13–0.20	120–262	1–9
Dense Agricultural Areas (DAA)	0.30–0.60	0.05–0.30	86–154	(−8)–(−5)
Moderately Dense Agricultural Area (MDAA)	0.18–0.30	0.03–0.25	80–105	(−0.9)–(−1)
Bare Soil (BS)	(−0.01)–0.02	0–0.02	110–140	(−0.07)–1
Bare Limestone Soils (BLS)	0.008–0.03	0.01–0.06	145–204	(−7)–(−4)
Dry Sabkhas (DS)	(−0.09)–(−0.06)	−0.13–(−0.09)	105–256	1–5
Wadis (W)	(−0.35)–0.08	0.2–0.6	90–132	(−1)–(−0.3)
Turbid Waters (TW)	(−0.1)–(−0.06)	(−0.1)–(−0.06)	90–132	6–10
Humid Sabkhas (HS)	0.12–0.14	0.15–0.2	85–90	5

The Meskat system was identified as a result of the distinction between moderately dense agricultural areas (MDAA) and bare limestone soils (BLS) from the land use maps. This classification was validated by using a recognized area. Indeed, polygon-shaped samples were selected around each plot. At this level, the polygons represent a set of pixels for the impluvium and the mankaas. Given the fact that each land use category has a specific spectral signature, FOCUS software was used to search within satellite images for the typical pixel signature of the impluvium and the mankaas. A confusion matrix was then developed between the control samples, from which the classification had been identified, and the validation samples for the impluvium (I) and mankaas (MK) classes.

The spatio-temporal evolution analysis of the Meskat system is based on the detection of land use changes. This analysis was carried out by comparing four validated classified images of 1987, 2003, 2007 and 2011. Areas containing Meskat system and urban classes were compared. The confusion matrix was then used to compute the number of pixels in each class for each of the four years. A correction was made to consider errors of omission and commission. The variation from one year to another was thus calculated.

4.4 Results and Discussion

4.4.1 Land Use Mapping

4.4.1.1 Typology of the Generated Classes

Nine classes of land cover were defined for the supervised classification of the satellite images (Table 4.2). These classes can be grouped into three groups (urban,

Table 4.2 Land use classification of in Sousse region

Group	Class	Description
Urban areas	Urban area (UA)	Buildings, factories, roads (impervious surfaces)
Agricultural areas	Dense areas (DAA)	Vegetable and forage crops
	Moderately dense areas (MDAA)	Fruit tree, extensive cultures
	Bare soil (BS)	Fallows
	Bare Limestone Soils (BLS)	Non-cropped hills (Impluvium)
Wetlands	Wadis (W)	River system
	Turbid Waters (TW)	Urban wastewater and oil mills, etc.
	Humid Sabkhas (HS) Dry Sabkha (DS)	Sabkha El Kelbia, Assa Jerida Dry saline Sabkha: Sabkha Echraita

agricultural and wet areas). These classes were identified after having determined the different classifiers. In fact, analysis of the images representing each class index showed that the NDVI varied from (-0.009)–0.6; the SAVI from (-0.13)–0.6; the IB from 80 to 262 and the Ibt from (-800)–1000. These indices varied in wide intervals and this can be explained by occupation variations, which lead to reflectance changes. These results are similar to those found by Bouchiha and Besbes [17]. These authors showed that, in the Tunisian Sahel, the NDVI of urban and agricultural classes is respectively, less than 0.6 and greater than 0.55.

4.4.1.2 Validation of the Classification

In order to validate the classification, a minimum of 765 pixels is required. Therefore, the optimal number of pixels relative to each class is 85. Analysis of the confusion matrix showed a good similarity between features of the same class. This is also demonstrated by the calculated global success and Kappa indices (Table 4.3). The results revealed the lowest percentage of errors (CE) and (OE) occurred at the level of the class (DS), it is 0%. This makes it possible to conclude that this class is well distinguished. The success of class distinction (DS) is explained by the sharpness of their spectral reflectance. In addition, the SH, UA and TW classes have the highest validation (CVI) and purity (CPI) indices. The superposition between the classes appeared more strongly in the DAA, MDAA, LBS, BS and W classes. Similarly for the W class, despite the existence of a branched hydrographic network, some Wadis were hidden in the map occupation. This class has the highest OE value, 44% in 2011.

The average indices of GS and K are 86 and 84%, respectively. The performance is inferior for the year 2007, due to atmospheric and climatic conditions during the

Table 4.3 Validation indices of classified images at Sousse region

Date	Index	Class								
		HS	DS	UA	DAA	MDAA	BLS	BS	W	TW
1987	CPI (%)	99	100	90	74	81	89	96	82	100
	CVI (%)	92	100	94	71	82	92	87	88	98
	OE (%)	1	0	10	26	19	11	4	18	0
	CE (%)	8	0	6	29	18	8	13	12	2
	GS (%)	88								
	K (%)	86								
2003	CPI (%)	100	100	99	83	88	87	96	76	100
	CVI (%)	99	100	98	86	74	93	97	96	100
	OE (%)	0	0	1	17	12	13	4	24	0
	CE (%)	1	0	2	15	26	7	3	4	0
	GS (%)	90								
	K (%)	88								
2007	CPI (%)	100	100	91	84	68	56	93	74	100
	CVI (%)	100	100	98	72	50	80	90	95	100
	OE (%)	0	0	9	16	32	44	7	26	0
	CE (%)	0	0	2	28	50	20	10	5	0
	GS (%)	76								
	K (%)	73								
2011	CPI (%)	98	100	93	86	93	85	83	56	100
	CVI (%)	99	100	97	75	89	91	97	92	100
	OE (%)	2	0	7	14	7	15	17	44	0
	CE (%)	1	0	3	25	11	9	3	8	0
	GS (%)	90								
	K (%)	88								

Class **HS**: Humid Sabkha; **DS**: Dry Sabkha; **US**: Urban spaces; **DAA**: Dense agricultural area; **MDAA**: Moderately dense agricultural area; **BLS**: Bare limestone soil; **BS**: Bare soil; **W**: Wadis; **TW**: Turbid waters
Index **CVI**: Class Validation index; **CPI**: Class purity index; **CE**: Commission error; **OE**: Omission error; **GS**: Global success index; **K**: Kappa index

satellite overpass. According to Congalton [29], K and GS above 60%, indicates a good classification good. The GS and K values are similar to those found by Bouchiha and Besbes [17], who obtained 73 and 68% (respectively) for a supervised classification and 77 and 72% for a hybrid classification. The latter is considered as the best classification by Bouchiha and Besbes [17].

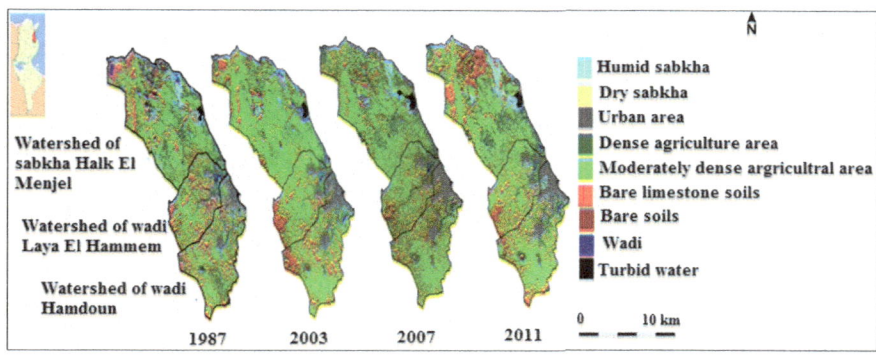

Fig. 4.5 Spatial and temporal variations of land use in Sousse region. *Source* Ben Salem, 2017

4.4.2 Spatio-Temporal Evolution of the Meskat System

Figure 4.5 shows the spatio-temporal variation in land use for the three studied watersheds. Visual analysis of spatial evolution revealed a high concentration of agricultural areas in the watershed of the Sabkha Halek El Menjel, representing the Meskat system.

The main changes are observed in Wadi Laya El Hammam and Wadi Hamdoun watersheds. The urban class, which is the most perceptible, mostly focuses in the watershed of Wadi Laya Hammam (downtown location). Other classes did not undergo significant spatial variation, except for the bare soil class which expanded. This is explained by the crop rotation practice, cultivated land alternates with fallows.

The comparison between four images illustrates the resemblance between years 2003, 2007 and 2011, whereas a change in land use can be observed in the image of 1987 (Fig. 4.5). Indeed, the temporal evolution of land use of Wadi Laya Hammam and Wadi Hamdoun watersheds reveals a remarkable extension of the areas occupied by urban class, while that of agricultural areas have declined. In 1987 the area was characterized by limestone bare soils (impluvium), and these bare limestone soils remarkably declined for the years 2003, 2007 and 2011. These areas are hilly and offer good-looking landscape and housing framework. They are currently occupied by the urban class.

4.4.3 Rate of Change of the Meskat System

Table 4.4 presents the surface variation of each land use class in the Sousse area. From this table, the temporal evolution of the surfaces urban class and that of the Meskat system were determined for the studied watersheds (Sabkha Halek El Menjel, Wadi Laya El Hammam and Wadi Hamdoun).

Table 4.4 Temporal variation in area covered by land-use classes (km^2)

Class	Year			
	1987	2003	2007	2011
HS	438	279	288	404
DS	133	161	165	145
UA	178	239	269	257
DAA	405	575	467	529
MDAA	1084	526	739	586
BLS	1118	396	533	527
BS	49	313	93	119
W	238	121	50	44
TW	138	6	12	6

For the period between 1987 and 2011, climate change has a remarkable effect on the wetland group (HS, DS, W, and TW) areas as well as those of the agricultural zone group (DAA, MDAA and BS). Indeed, the area of the latter decreased by 20% during the period of study. Regarding the dry Sabkhas (DS) class, the latter occupies an average area of 150 km^2 which has not varied over time. The analysis showed that the urban class (UA) increased by 30%.

An increase in urban areas was observed between 1987 and 2011, while the areas occupied by Meskats decreased. For urban areas, the rate of increase for the study period is 1, 8 and 4% for the Sabkha Halek El Menjel, Wadi Laya El Hammam and Wadi Hamdoun watersheds, respectively (Fig. 4.6). Therefore, the highest rate is observed in areas with high urban development (downtown Sousse and M'Saken). Thus, the urban population is distributed as follows: 32% (+2% per year) in the

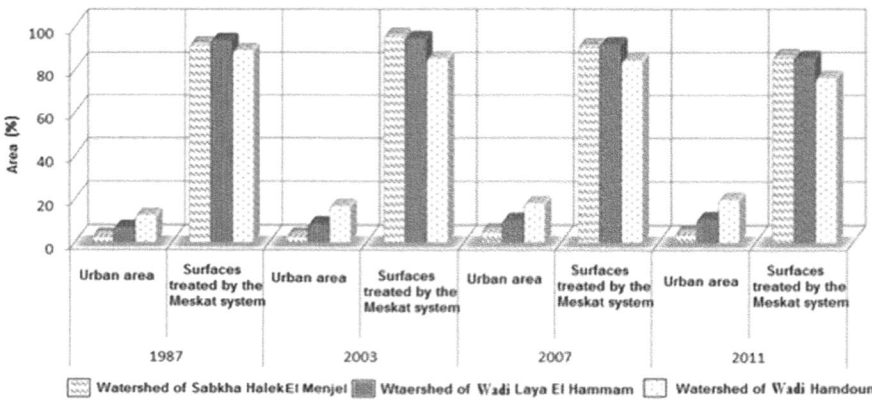

Fig. 4.6 Temporal evolution of the areas occupied by the urban and the Meskat system classes. *Source* Ben Salem, 2017

northern sector of the governorate, 23% (+0.6% per year) in the western sector and 45% (+7, 6 per year) in the southern sector. For the areas occupied by the Meskats, the rate of decline is of 6, 8, and 13% for the Sabkha Halek El Menjel, Wadi Laya El Hammam and Wadi Hamdoun watersheds, respectively.

Visual analysis of classified images showed the heterogeneity at the level of land use in the study area. The distribution analysis of the different classes revealed that the urban spaces dominate the landscape in coastal areas (downtown Sousse). In fact, the population was concentrated in the large cities maritime activities. This is explained by population growth (+4% growth rate). As an indication, the occupancy is 40% during the period 1995–2001 [30]. This phenomenon reflects also changes in population behavior and lifestyle. In continental regions (M'Saken region) agricultural areas are dominant. Thus, the most apparent classes are MDAA, DAA and BS. An alternation between bare limestone soil (bright areas) and agricultural areas (dark areas) was observed. These same observations were reported by Dallemand et al. [15], who mapped land use in Central Tunisia and characterized the Sousse region by low altitudes and corrugated landforms with more or less rugged slopes and cobbly upstream. At watershed downstream, the soil is heavy and characterized by the presence of arboriculture. On the southwest side, bare soil is the prevailing class. These areas are devoted to field crops (wheat, barley, etc.) and forage. In addition, Dallemand et al. [15] found that a strong reflectance is related to olive tree plantations, while zero reflectance corresponds to the bare soil. Class attribution of BLS and MDAA to the Impluvium (I) and the Mankaas (M), respectively, seems to be highly justified and is congruent with the results of [15, 31].

Urbanization is the main reason for the decline of Meskats. Indeed, the area corresponding to the extension of the urban class is the same that of that observed for the regression of Meskat systems. This confirms the findings of Houimli and Donadieu [32] in their descriptive study of the resistance and fragility factors influencing agriculture against urbanization in the Sousse region. The results do not simply illustrate urban sprawl and the decline of areas managed by the Meskats, but also, a productivity decrease over time. In this regard, the decrease in olive production is estimated around 30% which has been occurred in the past decades according to the local average. This is explained by weather conditions (lack of precipitation), lack of workforce and subsequent abandonment of olive groves, as well as urban sprawl through the construction of the buildings on the impluvium, which slows down rainwater downstream. Therefore, the olive trees present in the mankaas wilt, due to water shortage resulting reduction of their root zones. As an indication, the Ministry of Agriculture estimates that 15 millions olive trees have dried out under the effect of a water deficit in Tunisia.

The satellite images used in this study contributed to the classification and diachronic study of the Meskat system. However, differences were observed when comparing results with those of the field. For example, areas of some classes have been converted into other ones. As an indication, the reflectance of the vegetation hides that of water. Similarly, the shading effect of trees hides the reflectance of the soil. With respect to our study, limitations emerged for the identification of the Meskat system and its components. Despite these limitations, the resulting validation coefficients are acceptable and consistent with previous studies. The resolution of the images obtained from Landsat satellite (30 m × 30 m) is considered, by certain authors, as insufficient in smooth landscapes. In that regard, Bouaziz et al. [33] showed that the main limitation is the low resolution of the Landsat images as well as the resulting difficulty of recognizing small targets such as topographic details. Chartier and Servain [34] tried to improve urban spaces differentiation by performing a classification using dynamic clouds from Masked channels, i.e. by eliminating the pixels belonging to other themes. However, they noticed that in classified images, pixels located in the urban fabric merge with pixels located in farming plots or along quarries. They also found that their separation is impossible because their reflectance in the different channels is similar.

In order to follow the spatio-temporal evolution of areas managed by the Meskat system, a diachronic analysis of Landsat imagery allowed distinguishing nine classes that could be divided into three groups. The analysis of this classification helped to see the regional dominance of the agricultural sector and the high concentration of urban areas in the vicinity of the coast. The validation of the image classification with a confusion matrix revealed a global success index of 88, 90, 76 and 90%, for years 1987, 2003, 2007 and 2011, respectively. Similarly, the Kappa index was of 73, 94, 98 and 98%, for years 1987, 2003, 2007 and 2011, respectively. These results show that the adopted classification ranges from acceptable to good (indices ≥60%) and that is consistent with previous studies.

Interpretation of the images showed that the areas occupied by the Meskat system are located on the Sabkha Halek El Menjel, Wadi Laya El Hammam and Wadi Hamdoun watersheds. Urban areas increased between 1987 and 2011 of 1, 8 and 4% for the Sabkha Halek El Menjel, Wadi Laya El Hammam and Wadi Hamdoun watersheds, respectively. For areas occupied by Meskat systems, a decrease of 6, 8, and 13%, respectively, occurred for the same basins. The rate of reduction of areas managed by this system, due to urbanization, is very low during the study period (24 years). This leads us to believe that other factors were responsible for the decrease in areas managed by Meskat system.

4.5 Conclusion

In order to follow the spatio-temporal evolution of areas managed by the Meskat system, a diachronic analysis of Landsat imagery allowed distinguishing nine classes that could be divided into three groups. The analysis of this classification helped to see the regional dominance of the agricultural sector and the high concentration of urban areas in the vicinity of the coast. The validation of the image classification with a confusion matrix revealed a precision or global success index of 88, 90, 76 and 90%, for years 1987, 2003, 2007 and 2011, respectively. Similarly, the Kappa index was of 73, 94, 98 and 98%, for years 1987, 2003, 2007 and 2011, respectively. These results show that the adopted classification ranges from acceptable to good (indices ≥60%) and that is consistent with previous studies.

Interpretation of the images showed that the areas occupied by the Meskat system are located on the Sabkha Halek El Menjel, Wadi Laya El Hammam and Wadi Hamdoun watersheds. Urban areas increased between 1987 and 2011 of 1, 8 and 4% for the Sabkha Halek El Menjel, Wadi Laya El Hammam and Wadi Hamdoun watersheds, respectively. For areas occupied by Meskat systems, a decrease of 6, 8, and 13%, respectively, occurred for the same basins. The rate of reduction of areas managed by this system, due to urbanization, is very low during the study period (24 years). This leads us to believe that other factors were responsible for the decrease in areas managed by Meskat system.

4.6 Recommendations

The generated useful information in this study on spatio-temporal distribution of the Meskat system could be used for implementation of agricultural planning activities for sustainable agricultural development. However, in future investigations, the high resolution temporal data, as Modis would help to further improve the results accuracy and identify other factors that contribute to the decrease in the area managed by Meskat system.

Acknowledgements The authors would like to acknowledge the University of Sousse to subsidize the internship for the funding of this research with National Institute of Scientific Research of Quebec (INRS). The authors would also like to acknowledge the INRS to accept stage and to recognize the contribution of Jimmy Poulin and Ive Gauthier, researchers at the INRS, for their works on the first steps of Geomatica PCI. Finally, the authors would like to acknowledge the important and capital contribution of the reviewers to the final version of this paper.

References

1. Lei Y, Wang J, Yue Y, Yin Y, Sheng Z (2014) How adjustments in land use patterns contribute to drought risk adaptation in a changing climate. Land Use Policy 36:577–584
2. Ferrara A, Salvati L, Sateriano A, Nole A (2012) Performance evaluation and costs assessment of a key indicator system to monitor desertification vulnerability. Ecol Ind 23:123–129
3. Bakker MM, Govers G, Kosmas C, Vanacker V, Oost KV, Rounsevell M (2005) Soil erosion as a driver of land-use change. Agri Ecosyst Environ 105:467–481
4. Roose E, Sabir M, Arabi M, Morsli B, Mazour M (2012) Soixante années de recherches en coopération sur l'érosion hydrique et la lutte antiérosive au Maghreb. Physico-Geology 6:43–69
5. Khaledian Y, Kiani F, Ebrahimi S, Brevik E, Aitkenhead-Peterson J (2016) Assessment and monitoring of soil degradation during land use change using multivariate analysis. Land Degrad Dev J 28(1):128–141
6. El Garouani A, Chen H, Lewis L, Tribak A, Abahrour M (2008) Cartographie de l'utilisation du sol et de l'érosion nette à partir d'images satellitaires et du SIG IDRISI au nord-est du Maroc. Revue de la Télédetection 8:193–201
7. Provencher L, Dubois JM (2007) Méthodes de photo-interprétation et d'interprétation d'images. Précis de télédétection Volume 4. Presses de l'Université du Québec, pp 319–361

8. Bouchnak H, Felfoul M, Boussema MR, Snane MH (2004) Étude de l'évolution du ravinement en fonction de la lithologie dans les bassins versant des oueds el Hammam et Rmel en Tunisie Centrale. Revue de Télédétection 4(1):75–93

9. Anane, M., Masmoudi, M., Ben Mechilia, N., Oweis, T, (2008). Potential sites for hill reservoirs: An assessment methodology using GIS and Remote Sensing Tools. In: Proceeding of a Workshop Watershed Management in the Dry Area, Jerba, Tunisia, pp 109–116.

10. Abdelli F, Ouessar M, Khatteli H (2012) Méthodologie d'identification des ouvrages existants et des sites potentiels pour les jessours. Rev Sci Eau 25:237–254

11. Baccari N, Boussema MR, Lamachere JM, Nasri S (2007) Identification du risque de brèche des banquettes anti-érosives en région semi-aride Tunisienne à l'aide d'orth-images aérienne et de données multi-sources dans un système d'information géographique. Revue de la télédection. 7:405–417

12. Fourati M, Bouaziz R, El Amri A, Majdoub R (2015) Identification des anomalies de fonctionnement des ouvrages de conservation des eaux et du sol du bassin versant Sidi Salah. Int J Innovative Appl Stud 10(1):428–434

13. Majdoub R, Ben Salem A, Khlifi S, M'Sadak Y (2011) Aménagement antiérosif traditionnel (Meskat): exploitation des eaux de ruissellement et amélioration des caractéristiques du sol. Actes du Congrès Scientifique Euro Méditerranéen d'Ingénierie. Algesiras, Espagne, pp 159–165

14. Boujnah Mahjoub D (2004) Révision des densités de plantation en oléiculture pluviale en régions semi arides à arides Tunisiennes. Institut d'Olivier. Sousse Oliv 30:1–9

15. Dallemand JF, Lichtengger J, Reichert P, Soussi A, Rebillard P (1991). Cartographie de l'utilisation du sol en Tunisie centrale à partir de données microondes et optiques. Revue de la Télédétection 207–212

16. Houimli E (2008) Les facteurs de résistance et de fragilité de l'agriculture littorale faceà l'urbanisation: Le cas de la région de Sousse Nord en Tunisie. Thèse de Doctorat. Ecole Nationale Supérieure du Paysage de Versailles, pp 239–263

17. Bouchiha R, Besbes K (2008) Apport de deux méthodes de suivi d'évolution de la zone urbaine par imagerie Landsat, application à la région du Sahel Tunisien. Rev Franç Photog Téléd 190:40–48

18. Majdoub R, Khlifi S, Ben Salem A, M'Sadak Y (2013) Impacts of the Meskat water harvesting system on soil horizon thickness, organic matter, and canopy volume of olive tree in Tunisia. Desalin Water Treat 52:2157–2164

19. Snane MH, Toumi M, Chaabouni Z (1991) Un modèle d'optimisation des pluies pour les oliveraies. Sécheresse 1:12–16

20. Ben Salem A, Majdoub R, M'Sadak Y, Khlifi S (2013) Importance of the Meskat system and its landscape insertion through the olive groves of Sousse Region (Tunisian Sahel). Int J Innovation Appl. Stud 4:396–400

21. Irons JR, Dwyer JL, Barsi JA (2012) The next Landsat satellite: The Landsat data continuity mission. Remote Sens Environ 122:11–21

22. Chokmani K, Dever K, Bernier M, Gauthier Y, Paquet LM (2010) Adaptation of the SNOWMAP algorithm for snow mapping over eastern Canada using Landsat-TM imagery. Hydrol Sci J 55:649–660

23. Vanonckelen S, Lhermitte S, Van Rompaey A (2013) The effect of atmospheric and topographic correction methods on land cover classification accuracy. Int. J. Appl. Earth Obs 24:9–21

24. San Emeterio JL, Lacaze B, Mering C (2011) Détection des changements de la couverture végétale au sahel durant la période 1982-2002 à partir des données NDVI et précipitation. Revue de la Télédetection. 10:135–143

25. Rouse JW, Haas RH, Schell JA, Deering DW (1973) Monitoring vegetation systems in the great plains with ERTS. In: Third ERTS symposium, NASA SP, 351, pp 309–317.

26. Huete AR (1998) A Soil Adjusted Vegetation Index (SAVI). Remote Sens Environ 25:295–309

27. Caloz, R., Collet, C., (2002). Précis de Télédétection: Traitement numériques d'image de télédétection. Presses de l'Université du Québec/AUPELF, Sainte-Foy, 3.

28. Du Plessis WP (1999) Linear regression relationships between NDVI, vegetation and rainfall in Etosha National Park, Namibia. J. Arid Environ 42:235–260
29. Congalton RG (1991) A review of assessing the accuracy of classifications of remotely sensed data. Remote Sens Environ 37:35–46
30. INS: Institut National de Statistique (2014) http://dataportal.ins.tn/fr/
31. El Amami S (1983) Une nouvelle conception des aménagements hydrauliques en Tunisie. Livre Impact: Science et Société. Presse universitaire de France, Vendome, pp 61–68
32. Houimli E, Donadieu P (2005). Le Meskat, un système hydraulique de production oléicole menacé par l'étalement urbain ; Le cas de la région de Sousse Nord (Tunisie). Actes du séminaire « Étapes de recherches en paysage », n°: 7, École Nationale Supérieure du Paysage de Versailles, pp 31–43
33. Bouaziz R, Daoud A, Dahech S, Beltrando G (2005) Apports et limites de l'imagerie spatiale à l'étude des littoraux sableux: cas du littoral de Chaffar, Gouvernorat de Sfax. Photo Interpretation 3:1–17
34. Chartier M, Servain S (1996). Apports de la télédétection satellitaire à l'étude de l'environnement de sites archéologiques en milieu tempéré, le cas de la vallée de l'Aisne. Actes des Journées de Télédétection en Sciences humaines, Rev. Europ. Géog. https://cyb ergeo.revues.org/22419#entries

Chapter 5
Spatial Distribution of *Solanum elaeagnifolium* in the Tunisian Arid Region Using GIS Tools

Najla Sayari, Faiza Khebour Allouche, Amal Laarif, and Mounir Mekki

Abstract Alien plant invasions are getting more widespread and serious at an alarming rate around the world. In Tunisia, many plant species introduced accidently and deliberately have become invasive and continue to engender serious environmental, economic and sanitary problems. *Solanum elaeagnifolium* Cav. is considered as one of the most widespread invasive weeds in Tunisia, invading a wide range of habitats and generate considerable negatives impacts. Given the potential of its harmful impacts, *Solanum elaeagnifolium* management has become a high priority for the conservation of ecosystems specially crop fields. Weed mapping is considered the foundation for the development of a strategic long-term management plan to protect agro-biodiversity. In this context, our research consists to provide mapping surveys in a Tunisian arid region to estimate the actual extend of *Solanum elaeagnifolium* using geographic information system technology in the attempt to generate an accessible geodatabase and to develop a standardized mapping method to track and update weed population dynamics over time.

Keywords Geographic information system · Mapping · Spatial distribution · *Solanum elaeagnifolium* · Invasive alien plants

N. Sayari (✉) · F. Khebour Allouche · M. Mekki
Laboratory of Phytopharmacy and Weed Science, Higher Institute of Agronomy-Chott Meriem, University of Sousse, ISA CM BP 47, 4070 Sousse, Tunisia
e-mail: nagla_nd@yahoo.fr

F. Khebour Allouche
GREEN-TEAM Laboratory (LR17AGR01), Higher Institute of Agronomic Sciences-Chott Meriem, University of Sousse, Sousse, Tunisia

A. Laarif
Higher Institute of Agronomy-Chott Meriem, University of Sousse, ISA CM BP 47, 4070 Sousse, Tunisia

© Springer Nature Switzerland AG 2021
F. Khebour Allouche et al. (eds.), *Environmental Remote Sensing and GIS in Tunisia*, Springer Water, https://doi.org/10.1007/978-3-030-63668-5_5

5.1   Introduction

Biological invasions are considered as the second biggest threat to biodiversity after habitat destruction. Alien plant invasions are getting more widespread and serious at an alarming rate around the world [1]. Many more species are becoming invasive and cause enormous and often irreversible harm to biodiversity; they can compete with and displace native plants, cause the extinction or decline of many species, alter ecosystem functions, threat to biodiversity and also lead to economic consequences and human welfare impacts [2–4].

In Tunisia, many plant species from other parts of the world have been introduced for a range of purposes as garden ornamentals, as crop species, for stabilizing sand dunes and as barrier and hedge plants. Many of these alien species have become naturalized, surviving in disturbed and natural areas without needing to be tended, and some of these naturalized species have become invasive and may have serious implications for the invaded habitats [5]. Given the potential of harmful impacts, invasive alien plants (IAP) management has become an important challenge and a high priority for the conservation of ecosystems and human well being [6]. Important first step in invasive plants management is identifying species occurrence and estimating the invasion extent in the invaded habitat [7]. Thus, knowing the scale of an infestation, and where it is located, can set management priorities and feasibility [8]. This important first step should be achieved by the IAP species distribution prediction and mapping. In fact, mapping invasive plant populations is crucial for strategic management and monitoring. It is also essential for effective early detection [9]. Knowing where a plant currently grows is the foundation for knowing where to intervene and to survey for new occurrences [10]. Distribution data and mapping tools are needed to document the location and expansion of *Solanum elaeagnifolium*, a noxious invasive weed of the agro-ecosystems in Tunisia [11], threatening to thousands of hectares of irrigated fields in arid and semi-arid regions and generating considerable crop yield losses [12]. In this chapter we review briefly the biological invasions phenomenon and we focus on the urgent need to manage IAP, particularly *Solanum elaeagnifolium* by providing mapping surveys using GIS technology in a Tunisian arid region to estimate its actual extend and to highlight the importance of monitoring weed population dynamics for the implementation of a local and national management plans.

5.2   Introducing Concepts and Terms

5.2.1   Definitions

The understanding and management of invasive alien species (IAS) is an emerging science and its terminology continues to evolve and change [13].There is currently a variety of contentious terms used to describe species introduced outside their native ranges. Terms such as alien, invasive, weed, introduced, exotic and more are some-times used interchangeably or as synonyms, which can be confusing [14, 15]. Some basic definitions are important, and for the purposes of this chapter, we will use the following:

- **Alien plant**: (synonyms: allocthonous, introduced, non-indigenous, exotic, xeno-phytes) Plant taxa in a given area whose presence is due to intentional or unin-tentional human involvement [16]. Plant introduction is the movement of a plant outside its native range towards a new ecosystem, due to human activities. Two kinds of introductions are defined: (i) the intentional introduction which is an importation of a plant in a natural environment deliberately carried out by Man to satisfy or enrich his various activities (agriculture, breeding, pharmaceutical industry, ornamentation) and (ii) the non-voluntary or accidental introduction which is an introduction due to human activities not deliberately directed in order to introduce a plant [17]. The intensification of trade, with the development of human activities, is thus responsible for numerous involuntary introductions of plants [18].
- **Naturalized plant**: (synonym: established) Alien plants that sustain self-replacing populations without the direct intervention of people, through the recruitment of seeds or ramets capable of independent growth [16].
- **Invasive plant**: A subset of naturalized plants whose establishment and spread threaten ecosystems, habitats or species [19].
  Generally invasive species are alien, but native species may also become inva-sive, usually under changed environmental circumstances which affect vegetative communities' dynamics. The term "invasive" often includes a specific reference to environmental, economic or human health impacts. International agreements such as the Convention on Biological Diversity (CBD) define invasive alien species (IAS) as species whose introduction and/or spread threatens biological diversity [20].
- **Weed**: Species that has a perceived negative ecological or economic effect on agricultural or natural systems [21]. The words pests, weeds or noxious species are commonly used to refer to organisms that can negatively impact primary sectors of the economy such as agriculture and forestry.

5.2.2 Biological Invasion Process

According to [22] and [23] invasion process can be defined as the sequence of three principal phases:

(i) **Introduction of the species**: Species coming from another place must survive during and after the introduction. Many species fail to survive unless they are cared for. However, almost all invasive plants spread as seeds which do not require special care while being transported.

(ii) **Establishment and reproduction of the introduced species**: The survivors must persist and reproduce successfully until they establish a self-sustaining population.

(iii) **Spread**: In certain cases, established populations will multiply rapidly and spread across the landscape. This is the explosion phase which may only happen after a considerable lag phase, period during which an IAP occurs at low densities and its impacts are not noticeable.

5.3 Mapping Is an Effective Approach to Manage Invasive Alien Plants

Human activities are changing natural ecosystems and making them more susceptible to a great diversity of IAP. The rapid growth in trade, travel and transport is causing the introductions number increase [24, 25]. In fact, alien plant species enter countries through both legal and illegal routes, and introductions can be both intentional and unintentional. Becoming invasive, these vegetal alien species can have significant negative impacts on biodiversity, the economy and public health and their impacts on natural ecosystems are usually irreversible [26]. Facing to this global problem, which affects every region in the world, IAP management have become an important challenge and a high priority for the conservation of ecosystems and human well being [27]. Mapping IAP distribution is considered the foundation for the development of a long-term strategic management plan to protect ecosystems biodiversity and prevent invasion of other alien plant species [28]. Remote sensing (RS) and geographical information system (GIS) techniques have received considerable interest in the field of biological invasions in the recent years. They have already successfully been applied to map the distribution of several plant species, their ecosystems, abundance and distribution [29]. Mapping is needed to locate the spatial extension of IAP [30]. It presents many advantages by facilitating early detection and rapid response implementation, detecting and estimating the risk, identifying the infested areas and prioritizes management actions, allowing records update, showing treatment outcomes and infestations evolution over time [31].

Recently, different mapping resources are available online such:

– **BIOS**: Biogeographic Information and Observation System: a statewide California database and mapping system designed to enable the management, visualization, and analysis of biogeographic data (https://www.wildlife.ca.gov/Data/BIOS/).
– **EDDMapS**: Early Detection and Distribution Mapping System: providing distribution data on invasive species across the U.S. (www.eddmaps.org).
– **iMap Invasives**: Online tool for invasive species reporting and data management nationwide (www.imapinvasives.org/).
– **NAISMA**: North American Invasive Species Management Association: providing national data standards for mapping invasive plants (www.naisma.org/).

5.4 *Solanum elaeagnifolium* Mapping Using GIS

Solanum elaeagnifolium Cav. (#SOLEL) (Fig. 5.1) is native to Northeast Mexico and Southwest USA [32, 33]. Recorded as a highly invasive alien plant species in many regions, SOLEL is now considered as one of the most widespread invasive weeds in the world [34]. It is listed as a noxious weed in 21 USA states [35] and figures on the EPPO A2 list of species recommended for regulation to EPPO member countries [36]. In North Africa, the earliest introductions of SOLEL date back to 1949 in Morocco, and it is now considered the nation's most noxious weed especially in irrigated fields [37]. Later, it was detected in 1956 in Egypt, in 1985 in Algeria and in 2014 in Libya. In Tunisia, since 1985, SOLEL started to become a noxious weed [38]. The invaded area is increasing and this weed is becoming a potential threat to thousands of hectares of irrigated fields in arid and semiarid regions [39]. The most infested habitats and land-uses are roadsides, wastelands and summer crops. These infestations generate considerable crop yield losses [11].

Giving the SOLEL negative impacts in the agro-ecosystems, Tunisian researchers and authorities have increased their interest in its management. Several research programs were conducted to support the implementation of an appropriate control approach against SOLEL [40]. In our case, we propose a mapping method using the GIS technique that can be integrated into a national plan dedicated to the management of SOLEL in Tunisia.

Fig. 5.1 *Solanum elaeagnifolium* Cav. (Najla Sayari, 2016; personal photo)

5.4.1 Methodological Approach

5.4.1.1 Study Area

The study area, called El Alam, is located in the center of Tunisia, in the north of Kairouan governorate and belongs the Sbikha commune (Fig. 5.2). The areas likely to be irrigated in Kairouan represent 14% of the total surface area at the national level, hence the importance of the Public Irrigated Perimeter.

The climate is semi-arid with mild rainy winters and hot, dry summers. Annual rainfall ranges from 300 to 400 mm (Table 5.1). In 2015, the total annual rainfall was 338 mm, this year was characterized by two rainy periods in spring and autumn. The average annual rainfall of the last five years (2011–2015) is marked by strong annual and seasonal irregularities.

El Alam has sandy soils and poor humus. These are less heavy soils. The hydrographic network is organized around the big wadi with irregular regime but violent and devastating during floods: It is the Nebhana wadi which takes its birth with the trays of Kessra, it crosses the Jebel Oueslat then the plains of Sbikha to finally spread out in El Alam. The irrigated land of El Alem covers 2600 ha. The choice of this study site was based on land accessibility, which plays a crucial role in mapping alien plants, and makes mapping easier and geometrically correct. The study area is

Fig. 5.2 Study site's geographical location (adapted and modified from www.wikipedia.org)

Table 5.1 Average monthly rainfall (mm) at El Alem (2011–2015)

Year	January	February	March	April	May	June	July	August	September	October	November	December	Total
2011	29	31	29	22	65	33	0	4	16	50	108	11	398
2012	24	28	42	29	0	0	0	30	37	22	0	5	217
2013	8	4	24	42	11	0	16	37	2	78	10	75	307
2014	20	10	92	9	37	29	6	12	21	48	4	18	306
2015	5	44	49	0	0	0	0	105	85	29	21	0	338

Source National Institute of Meteorology, Sousse (2016)

also representative of the main irrigated crops that are usually affected by SOLEL in Tunisia [38].

5.4.1.2 Data Collection

Considering the fact that SOLEL is a highly invasive weed especially in irrigated summer crops, prospections were conducted during May–October 2015, period during which SOLEL have an active vegetative growth and is easy to be identified. To assess SOLEL weed populations over time, four steps have been used: (1) collection of GPS field data with land cover identified, (2) detection and visual assessment of SOLEL, (3) download of GPS data and creation of GIS geodatabase, (4) analysis of weed population using spatial identification and geodatabase tables. In the first step, the spatial delimitation of the irrigated perimeter is carried out using Garmin 64S GPS. The field work consists to take four GPS points in the corners of each parcel and a point in the center of the plot after crossing the diagonal. In the same time the type of land use and the estimation of the degree of infestation according to the guide are carried out in technical sheets (Fig. 5.3).

The third step is done in the laboratory. All GPS points and data are downloaded and imported from .kml to shapefile using QGIS. After the installation of 'OpenLayer Plugin' who helps to add the Google layer as base maps to the QGIS Map canvas.

Numéro:…………

Coordonnés GPS : ……………

Type de propriété : ☐ Privé ☐ Public

Longueur : ………………....... Largeur : …………………… Surface : ………………….

Biotope : ☐ Zone d'habitation ☐ C.M ☐ Olivier ☐ Jachère ☐ ……………

Flore adventice :

1	2	3	4	5	6	7	8	9	10	11	12	13	14	15	16	17	18	19	20

Patron de distribution de SOEL :

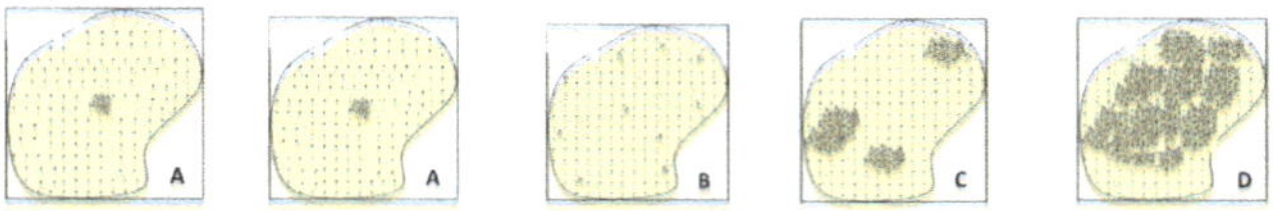

☐ Indemne ☐ PI ☐ PD ☐ TD ☐ DH

PI : plants isolés ; PD : plants dispersés ; TD : tâches dispersées ; DH : distribution homogène

Couverture de SOEL :

PI : ☐ < 1 % PD : ☐ 2-5 % ☐ 6-10 % TD : ☐ 11-20 % ☐ 21-40 %

DH : ☐ 41-60 % ☐ > 61 %

Fig. 5.3 A guide for the visual assessment of weed infestation as a percentage of groundcover

Table 5.2 Description of El Alam GIS data attributes

The attributes for the "GPS" point layer

Column name	Column type	Format	Description	Example
GPS	Text	5 characters	GPS number	« 547 »
X	Decimal number Precision = 2	6 characters	Latitude	35° 52′ 39.06″N
Y	Decimal number Precision = 2	6 characters	Longitude	10° 4′ 14.68″E

Attributes for the border line layer

Parcel	Texte	Characters	Parcel name	Exp: «F14»
Type	Text	1 characters	Border type	«P»: Track
Continuity	Nombre entier	1 characters	Continuity of plants along the border	«2»: batch (11–50% de L)
Density	Text	1 characters	Density of plants along the border	«F»: (<20 plants.m$^{-2)}$

Attributes for the polygon layer "Land cover"

Column name	Column type	Format	Description	Example
Parcelle	Text	5 characters	Parcel name	«F1»
2008/2009	Text	20 characters	Type of LULC	«Corn»
2009/2010				«Alfalfa»
2010/2011				«Tomato»
2011/2012				«Bersim/Oats»
2012/2013				«wheat»
2013/2014				«Fallow»
2014/2015				«oat»
Distribution	Text	2 characters	Type of distribution	«I»: No infestation
Cover	Text	6 characters	Percentage of SOEL coverage	«41–60%»
Superficie	Integer number	6 characters	Area of the plot	«270533»

These layers can be used for visualization and as well as base map for digitization (Table 5.2). The spatial rectification of each plot limits is done then all information collected were implemented in attributes per each shapefile. In total, 128 plots are studied. At least, the analysis of SOLEL density is based on the use of GIS spatial analyst tools.

5.4.1.3   Mapping Techniques

QGIS2.18 is a software that is designed to modify, display, query and analyze the weed mapping data collected using a GPS and can be used to generate maps and reports [41]. Across the irrigated perimeter, weed invasion is mapped as occurrences of populations. The prospections for the SOLEL took place at a very favorable period for its observation. The visits spread between May and October made it possible to carry out a complete inventory of this species. The accessibility of the plots made it possible to detect the species relatively correctly. The inventory can therefore be considered relatively satisfactory, even if it is not exhaustive.

5.5   Results and Discussion

5.5.1   EL Alam Land Cover Mapping in 2014–2015

The total area of the study site is 2600 ha. The average plot size is 20 ha. This area is subdivided into seven sectors. Figure 5.4 shows that in the North side extend five sectors with areas which vary between 185 and 474 ha. Sector B2 contains 11 parcels and sector A contains 20 parcels. Further the south, the areas are bigger. Sector E contains 27 parcels and occupies 513 ha and Sector F contains 24 parcels occupying 475 ha.

Figure 5.5 shows the variation in land use rates by sector. The distribution of crop types is disproportionate among sectors. However, wheat, sorghum, oats are crops almost present in all sectors and follow plots are present in each sector. The areas

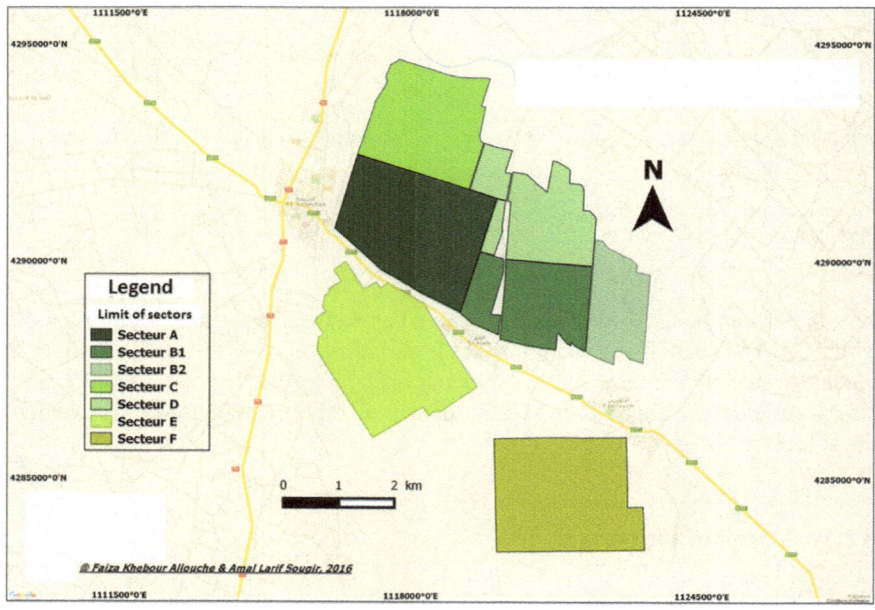

Fig. 5.4 Spatial delimitation of El Alam sectors (*Faiza Khebour Allouche, Amal Laarif, 2016*)

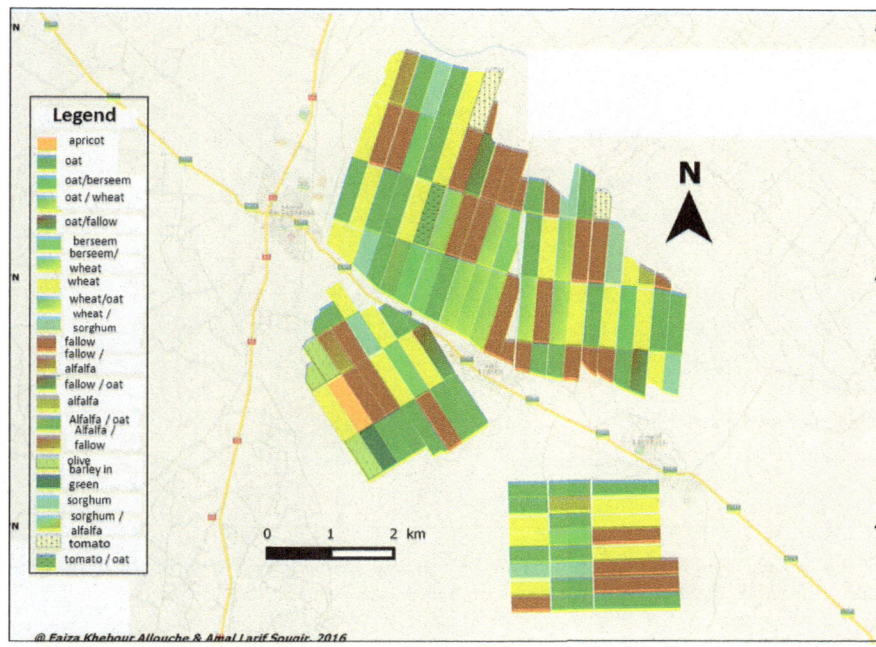

Fig. 5.5 Change in land cover type by parcel at OTD-El Alam in 2014–2015 (*Faiza Khebour Allouche, Amal Laarif, 2016*)

of sectors A, B2, E and F are more occupied by oats (more than 100 ha per sector); sectors B1 and D, are not cultivated (fallow) and sector C is occupied by wheat (92 ha). 4% and 7% in sectors D and C are occupied by tomato. 9% of the area is cultivated with olive trees in sector E.

5.5.2 Solanum elaeagnifolium *Spatial Distribution*

5.5.2.1 Spatial Delimitation of Infested Plots by *Solanum elaeagnifolium*

Figure 5.6 shows that more than 60% of the irrigated perimeter of Al Alam was infested by *Solanum elaeagnifolium*. In fact, it was present in 112 plots (over the 128 considered plots) which show the considerable spread of the weed in this region.

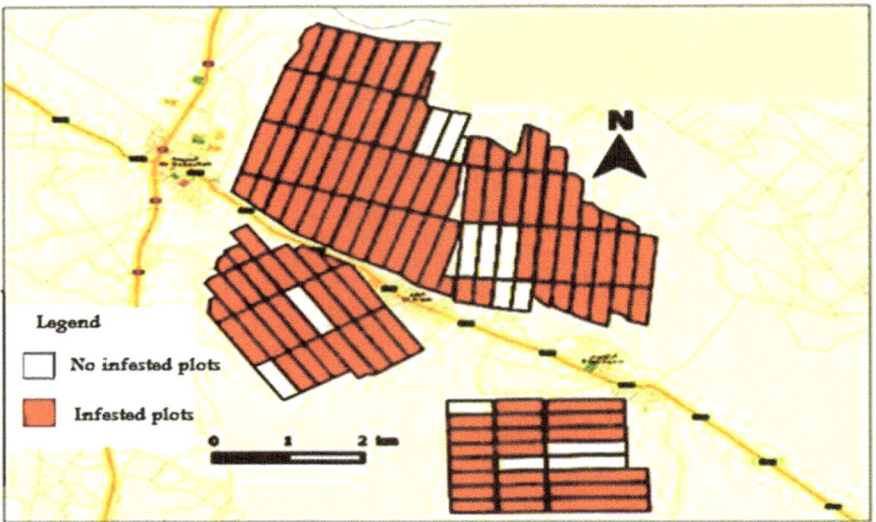

Fig. 5.6 Spatial delimitation of infested plots by *Solanum elaeagnifolium* in the irrigated perimeter of Al Alem (May–October 2015) (*Faiza Khebour Allouche, Amal Laarif, 2016*)

5.5.2.2   *Solanum elaeagnifolium* Distribution Patterns

The infested surface was 2314 ha which represent about 90% of the total studied area (2600 ha). This result confirms that SOLEL is becoming a serious problem in the irrigated crop fields and justify the urgent need to control infestations and avoid the weed spread. In fact, predicting species ranges and surveying their spatial distribution are important steps for early detection and active management of invasive weeds [42]. It allows characterizing the pattern of distribution of these plants, predicting their rate of spread and evaluating the relationship between their spatial extent and abundance [28]. On the other hand, SOLEL distribution pattern characterization reveals that the patchy pattern (Scattered spots) was the most common distribution type of the weed (Fig. 5.7), it represent 54% (60 plots) of the total number of infested plots (112 infested plots). This further confirms the wide expansion of this species in the studied area.

5.5.2.3   Ground Cover Spatial Distribution of *Solanum elaeagnifolium*

Figure 5.8 shows that the SOLEL ground cover in the infested plots was in most cases more in 40% (40–60% and >60%). This result explains the high abundance of SOLEL and its rapid infestation extend, it also gives an idea about its residence time in the region. Furthermore, high abundance may be explained by a long and continuing colonization process without any effective control program. These results confirm the ongoing invasion process of SOLEL in agriculture lands and considering

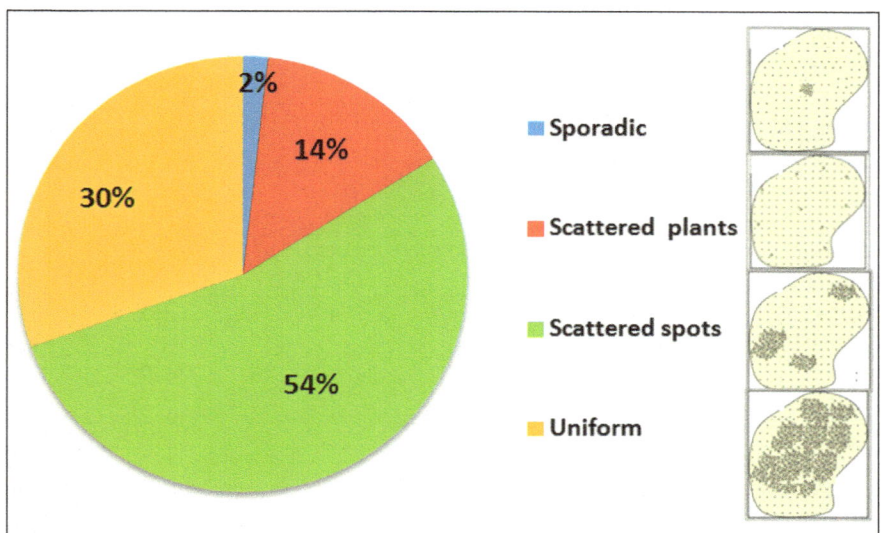

Fig. 5.7 Proportion of *Solanum elaeagnifolium* distribution patterns in the irrigated perimeter of Al Alem (May–October 2015)

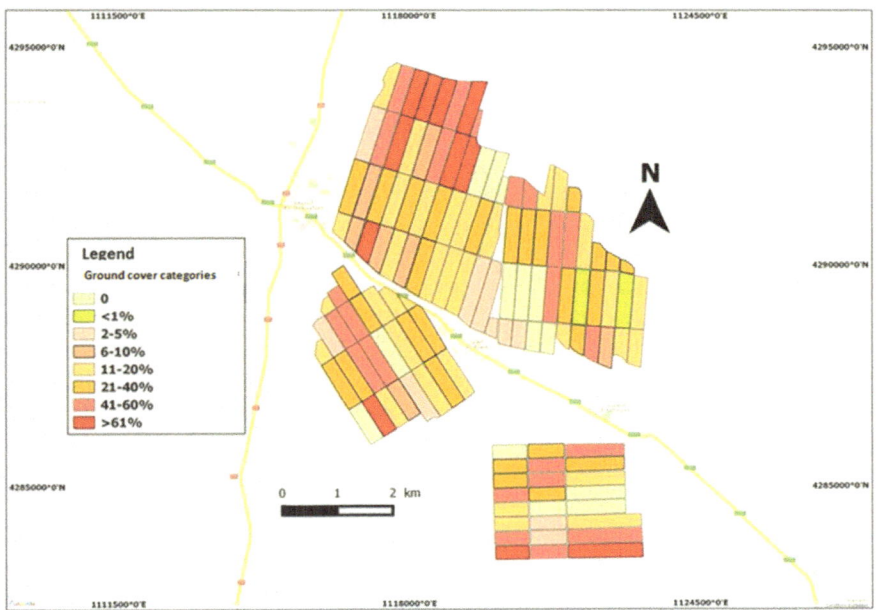

Fig. 5.8 *Solanum elaeagnifolium* ground cover in the irrigated perimeter of Al Alem (May–October 2015) (*Faiza Khebour Allouche, Amal Laarif, 2016*)

Table 5.3 Surface variation in ha of the spatial distribution of SOLEL in El Alam by sector

Superface (ha)	Sector A	Sector B1	Sector B2	Sector C	Sector D	Sector E	Sector F	Total
No infested	0	103.15	0	0	52.72	45.11	84.36	285
Sporadic	0	25.35	25.21	0	0	0	0	51
Scattered plants	79.99	95.77	17.71	52.69	0	62.26	32.13	341
Scattered spots	376.85	46.19	126.90	42.67	165.93	295.46	176.20	1230
Uniform distribution	17.79	24.83	15.48	266.50	75.40	110.18	182.58	693
Total infested	475	192	185	362	241	468	391	2314

its high invasiveness potential, this alien weed can also invade adjacent natural or semi-natural areas. Therefore, we can expect that SOLEL infestations extension will be advanced in the following years without any management strategy or specific action plan. Face to this alarming situation, we must actively seek solutions to control or, if possible, eradicate this species and even stop its reintroduction and protect non infested areas (Table 5.3).

5.6   Conclusions

Invasive alien weeds are a serious threat to native species, communities, and ecosystems. Prior to species establishment, early detection quickly followed by control and eradication is the most effective management process to stop IAP spread. Despite the increased interest given to biological invasions management, there is no yet a specific management program dedicated to IAP in Tunisia. Some works have been initiated but they are not well coordinated and currently, there are no specific methodologies of detecting and mapping IAP in the Tunisian agro-ecosystems. Estimates have been made of the spatial extent of alien plant invasions. Alas, each study used a different method or concentrated on particular species or areas. For these reasons, and because surveys were done at different times, these results cannot easily be merged to produce a national overview.

Therefore, our task was to develop a reproducible approach for mapping invasive plant species using GIS data that can be applied to other study areas and study species. In fact, when baseline mapping is successful, the results for each species can be used for a detailed analysis of their distribution patterns in order to support management decisions. The case study of SOLEL management in Tunisia confirms the ongoing colonization process of this invasive weed and shows that its control as a conventional weed is inappropriate. Thus, a national action plan to contain SOLEL spread is urgently required.

5.7 Recommendations

Knowing the scale of an infestation, and where it is located, can help identifying management priorities. This is an especially important step when resources are limited. The level of detail required to map the IAP distribution depend on the size of the site/area. For large sites, the use of a GPS device and GIS system is recommended. We can also make some effort to estimate the species density or ground cover which will take to successfully bring the plant under control and allow to measure the success, or otherwise, of the control program adopted.

To be useful for the implementation of an IAP management a distribution mapping system should:

- allow for data to be easily searched, queried, and downloaded in a variety of formats for use in management plans, reports, or GIS projects,
- offer Web-based mapping of invasive species distribution to help fill gaps and identify "leading edge" ranges,
- facilitate implementation of Early Detection and Rapid Response programs through online data-entry forms,
- combine local and national data in one database,
- be able to update records to show treatment outcomes and changes in infestations over time.

Finally, mapping is considered the first important step to early detect and manage IAP. In fact, identifying infestations in their early stages, give a chance to eradicate them completely. Finding and controlling invasive species early can result in huge savings in time and money when compared to the amounts spent in management of species that already cover large portions of our landscapes.

References

1. Powell KI, Chase JM, Knight TM (2011) A synthesis of plant invasion effects on biodiversity across spatial scales. Am J Bot 98(3):539–548
2. Cossu AT, Camarda I, Brundu G (2014) A catalogue of non-native weeds in irrigated crops in Sardinia (Italy). Webbia 69:145–156
3. Paini DR, Shepparda AW, Cookc DC, De Barroe PJ, Wornerf SP, Thomas MB (2016) Global threat to agriculture from invasive species. PNAS 113:7575–7579
4. Richter R, Berger UE, Dullinger S, Essl F, Leitner M, Smith M, Vogl G (2013) Spread of invasive ragweed: climate change, management and how to reduce allergy costs. J Appl Ecol 50:1422–1430
5. Sayari N, Mekki M (2017) Comportement de *Solanum elaeagnifolium* Cav. dans la région semi-aride de Chott-Meriem-Tunisie. 28ème Forum des Sciences Biologiques et de Biotechnologie, 21–24 Mars; 2017, Hammamet - Tunisie
6. Te Beest M, Esler KJ, Richardson DM (2014) Linking functional traits to impacts of invasive plant species: a case study. Plant Ecol 216(2):293–305
7. Essl F, Bacher S, Blackburn T, Booy O, Brundu G, Brunel S (2015) Crossing frontiers in tackling pathways of biological invasions. Bioscience 65:769–782

8. Cullors VM (2013) A geographic information system for invasive species: Sahara Mustard Weed. Master's thesis, University of Redlands, 54 p. Available on https://inspire.redlands.edu/gis_gradproj/205

9. Brundu G, Lozano V, Manca M, Celesti-Grapow L, Sulas L (2015) *Arctotheca calendula* (L.) Levyns: An emerging invasive species in Italy. Plant Biosyst 149(6): 954–957

10. Pengra BW, Johnston C, Loveland TR (2007) Mapping an invasive plant, *Phragmites australis*, in coastal wetlands using the EO-1 Hyperion hyperspectral sensor. Remote Sens Environ 108:74–81. https://doi.org/10.1016/j.rse.2006.11.002

11. Sayari N, Brundu G, Mekki M (2016) Mapping and monitoring an invasive alien plant in Tunisia: Silverleaf nightshade (*Solanum elaeagnifolium*) a noxious weed of agricultural areas. Tunisian J. Plant Prot. 11:219–227

12. Mekki M (2006) Potential threat of *Solanum elaeagnifolium* Cav. to the Tunisian fields. Invasive Plants in Mediterranean Type Regions of the World. In: Proceedings of the first international symposium on invasive plants in Mediterranean type regions of the world. Environmental Encounters Series no. 59: 165–170. Council of Europe Publishing, pp 235–242

13. Richardson DM, Pysek P, Carlton JT (2011) A compendium of essential concepts and terminology in biological invasions. In: Fifty Years of Invasion Ecology: The Legacy of Charles Elton. Richardson DM. Edition, Wiley-Blackwell, pp. 409–420

14. Blackburn TM, Pysek P, Bacher S, Carlton JT, Duncan RP, Jarosik V, Wilson JRU, Richardson DM (2011) A proposed unified framework for biological invasions. Trends Ecol Evol 26(7):333–339

15. Colautti RI, MacIsaac HJ (2004) A neutral terminology to define 'invasive' species. Divers Distrib 10:135–141

16. Pysek P, Hulme PE, Nentwig W (2009) Glossary of the main technical terms used in the handbook. In: Handbook of alien species in Europe. DAISIE Edition, Berlin, pp 375–378

17. Hulme PE, Bacher S, Kenis M (2008) Grasping at the routes of biological invasions: a framework for integrating pathways into policy. J Appl Ecol 45:403–414

18. Goudard A (2007) Fonctionnement des écosystèmes et invasions biologiques: importance de la biodiversité et des interactions interspécifiques. Thèse de Doctorat, Université Pierre et Marie Curie, Paris VI—France. 216 p

19. Richardson DM, Pysek P, Carlton JT (2011) A compendium of essential concepts and terminology in biological invasions. In Fifty Years of Invasion Ecology: The Legacy of Charles Elton. Richardson DM. Edition, Wiley-Blackwell, pp 409–420

20. SCBD (2017) What are Invasive Alien Species? Available on https://www.cbd.int/invasive/WhatareIAS.shtml

21. Booth BD, Murphy SD, Swanton CJ (2003) Weed ecology in natural and agricultural systems. CABI Publishing, Wallingford, p 299

22. Williamson M (1996) Biological invasions. Chapman & Hall, New York, 256 p

23. Richardson DM, Pysek P, Rejmanek M, Barbour MG, Panetta FD, West CJ (2000) Naturalization and invasion of alien plants: concepts and definitions. Divers Distrib 6:93–107

24. Costello C, McAusland C (2003) Protectionism, trade, and measures of damage from exotic species introductions. Am J Agr Econ 85(4):964–975

25. Levine JM, D'Antonio CM (2003) Forecasting biological invasions with increasing international trade. Conserv Biol 17(1):322–326

26. Kumar A, Prasad S (2015) Threats of invasive alien species. Int. Res. J. Manage. Sci. Technol. 4:605–624

27. McNeely JA, Mooney HA, Neville LE, Schei P, Waage JK (2001) A global strategy on invasive alien species. IUCN Gland, Switzerland and Cambridge, UK, 50 p

28. Trueman M, Standish RJ, Orellana D, Cabrera W (2014) Mapping the extent and spread of multiple plant invasions can help prioritize management in Galapagos National Park. NeoBiota 23:1–16

29. Joshi C, de Leeuwa J, van Duren IC (2004). Remote sensing and GIS applications for mapping and spatial modeling of invasive species. Comm. VII. In: Proceedings of the XXth ISPRS congress: geo-imagery bridging continents, 12–23 July 2004, Istanbul, Turkey, pp 669–677

30. Turbelin AJ, Malamud BD, Francis RA (2016) Mapping the global state of invasive alien species: patterns of invasion and policy responses. Glob Ecol Biogeogr 26(1):78–92
31. Rawlins KA, Griffin JE, Moorhead DJ, Bargeron CT, Evans CW (2011) EDDMapS: Invasive Plant Mapping Handbook. The University of Georgia. Center for Invasive Species and Ecosystem Health, Tifton GA. BW-2011-02. 32 p
32. Brunel S (2011) Pest risk analysis for *Solanum elaeagnifolium* and international management measures proposed. EPPO Bull. 41:232–242
33. Mekki M (2007) Biology, distribution and impacts of Silverleaf Nightshade (*Solanum elaeagnifolium* Cav.). EPPO Bull. 37(1):114–118
34. CABI (2018) *Solanum elaeagnifolium* (silverleaf nightshade). Invasive Species Compendium. Available on: http://www.cabi.org/isc/datasheet/50516
35. Roche C (1991) Silverleaf Nightshade (*Solanum elaeagnifolium* Cav.) In: Pacific Northwest Extension Publication, no. 365. Washington State University, Washington, USA, 2 p
36. EPPO (2018) EPPO lists of invasive alien plants. Available on http://www.eppo.int/INVASIVE_PLANTS/ias_lists
37. Taleb A, Bouhache M (2005) Etat actuel de nos connaissances sur les plantes envahissantes au Maroc. In: Proceedings of the International Workshop on Invasive Plants in Mediterranean Type Regions of the World, May 25–27, 2005, Mèze, France, pp 99–107
38. Chalghaf E, Aissa M, Mellassi H, Mekki M (2007) Control of the spread of *Solanum elaeagnifolium* Cav. in the governorate of Kairouan (Tunisia). EPPO Bull. 37:132–136
39. Mekki M (2006) Potential threat of *Solanum elaeagnifolium* Cav. to the Tunisian fields. Invasive Plants in Mediterranean Type Regions of the World. In: Proceedings of the first international symposium on invasive plants in Mediterranean type regions of the world. Environmental Encounters Series no. 59: 165-170. Council of Europe Publishing, pp 235–242
40. Mekki M (2011) Distinction between weed control and invasive alien plant management approaches: case study of *Solanum elaeagnifolium* management in North african countries. In: Proceedings of the International symposium on system intensification towards food and environmental security, organized by the Crop and Weed Science Society and Bidhan Chandra Krishi Viswavidyalaya, 24–27 February 2011, Turkey, pp 16–18
41. Brown K, Bettink K, Paczkowska G, Cullity J, Region S, French S (2011) DEC nature conservation service biodiversity standard operating procedure. Techniques for mapping weed distribution and cover in bushland and wetlands SOP No: 22.1
42. Hulme PE (2003) Biological invasions: winning the science battles but losing the conservation war? Oryx 37:178–193

Chapter 6
PROMETHEE and Geospatial Analysis to Rank Suitable Sites for Grombalia Aquifer Recharge with Reclaimed Water

Makram Anane, Rahma Souissi, Hanèn Faïdi, Rim Mehdaoui, and Khadija Gdoura

Abstract This work aims to rank the suitable sites for Grombalia aquifer recharge with reclaimed water using PROMETHEE II multicriteria analysis method coupled with Geographical Information System (GIS). The suitable sites were first defined and mapped using the disjunctive/conjunctive method. Then the suitable areas were ranked using the PROMETHEE II method. Several technical, environmental and economic criteria were identified, spatialized, weighted using the AHP method and normalized using the fuzzy functions and then aggregated using the PROMETHEE II method. The best site adapted for Grombalia aquifer recharge with reclaimed water of Bou Argoub wastewater treatment plant (WWTP) was selected. The suitable areas are about 1667 ha, exceeding by far the required area to absorb the amount of effluent produced by the wastewater treatment plant of Bou Argoub. The net flux φ obtained to rank the suitable areas varies between -0.33 and 0.32. The best site is located 3.6 km far from the WWTP, inside an agricultural area and having 9 m unsaturated zone thickness.

Keywords GIS · PROMETHEE II · Reclaimed water · Aquifer recharge · Grombalia

6.1 Introduction

The water sector is one of the most sensitive domains in Tunisia because of its scarcity and its economic and social importance. A strategic integrated water resources management at the national level was elaborated during the nineties in order to avoid a structural water scarcity situation. This policy considered the treated wastewater as an important, continuous and increasing source of water that could be used mainly in agriculture, which is the most water-consuming sector. Demographic growth, urbanization and the important sanitation development in Tunisia contribute to a continuous increase of this nonconventional source of water. The aquifer recharge with this

M. Anane (✉) · R. Souissi · H. Faïdi · R. Mehdaoui · K. Gdoura
Wastewater and Environment Laboratory (LR19CERTE03), Water Research and Technology Centre (CERTE), Borj cedria Technopark. BP 273, 8020 Soliman, Tunisia
e-mail: makram.anane@certe.rnrt.tn

© Springer Nature Switzerland AG 2021
F. Khebour Allouche et al. (eds.), *Environmental Remote Sensing and GIS in Tunisia*, Springer Water, https://doi.org/10.1007/978-3-030-63668-5_6

water constitutes an interesting solution of its storage and use for irrigation during the peak periods. Currently, this technique is under testing phase in two pilot sites, Oued Souhil and Korba at the governorate of Nabeul, pretending to the extent this technique all over the Tunisian territory. Nevertheless, the success of such practice requires the selection of an adequate site for aquifer recharge. This choice requires the consideration of criteria with different technical, economic, social and environmental aspects, minimizing the failure risks of the practice. Combining multicriteria analysis and GIS, through the geospatial analysis component, helps to contributing the delineation of the appropriate areas and select the best sites for treated wastewater reuse for artificial aquifer recharge. It is an essential step preceding the basins installation.

The geospatial analysis, through different operators, treats and combines data with different types and sources and generates new informational data. It contributes to manage the environmental phenomena and resolve ecological and natural issues. During the last decades, several works have been beneficed of the geospatial analysis to resolve issues with spatial components, using for instance interpolation, combinations of several spatial datasets and/or producing different scenarios. For instance, a set of spatial data was combined to map the aquifer vulnerability to anthropic pollution [1]. The aquifer water level was interpolated using fuzzy kriging approaches for hydrogeological studies, particularly as input data to groundwater models [2]. Different spatiotemporal data such as land use and mobile phones call records location were combined to relate the urban population activity with land use [3]. The salinity and nitrate content measured from a limited number of well were interpolated to the entire phreatic aquifer and related thresulting map with other spatial datasets such as water source irrigation and land use in order to define the source of aquifer pollution and make decisions to reduce pollution [4].

These capacities make the geospatial analysis powerful tools for planning and decision support. Nevertheless, it is notable on its own to carry out certain decisional tasks. In some cases, it is mandatory to combine their capacities with the multicriteria (MCA) analysis especially when the scope is to determine the best sites for a specific activity. MCA analysis put together several criteria with different economical, social, technical and environmental aspects and weights them with accordance to their influence to the objective, offering high possibilities in terms of decision making. The geospatial analysis manages and treats the geo-referenced data giving high capacities in terms of spatial data exploration. The consideration of this combination decreases the risk of technical failure and helps to resolve the socio-cultural conflicts that could submerge in this kind of projects.

Several methods of MCA analysis have been developed, namely ELECTRE, PROMETHEE, AHP, TOPSIS, AIM [5–7] and used to solve environmental and resources allocations [8–10].

Combination of MCA with the Geographic Information System (GIS) targeted different domains such as landfill [11, 12], airports [13], and marina construction [14]. In Tunisia, an important gap in using this kind of tools is observed. Only few studies focused on AHP and weighted sum methods were found [15–18]. The development of methodologies using these combinations becomes more and more mandatory for better land allocation and optimal management of certain resources, especially water.

This research work aims to delineate, rank and map the suitable sites for the Grombalia aquifer recharge with reclaimed water produced from Bou Argoub wastewater treatment plant (WWTP) using the PROMETHEE II (Preference Ranking Organisation Method for Enrichment Evaluations) MCA method linked to GIS.

6.2 Method

6.2.1 Description of the Study Area

The study area is located in the Northeast of Tunisia in the Cap Bon Peninsula (Fig. 6.1). It covers 720 km^2 and includes several cities such as Soliman, Bou Argoub, Grombalia and Menzel Bouzelfa. The climate is semi-arid with soft and humid winter and hot and dry summer. The mean annual rainfall is between 450 and 500 mm. About 80% of the amount is dropped between Septembre and March. The mean annual temperature is about 19 °C. The irrigated fruit trees cover the most important surface followed by cereals and vegetables. Some enclaves of halophytes are extended nearby of Soliman's sebkha. The area is located on a phreatic groundwater hosted in a sandy and sandstone quaternary aquifer with 50 m mean thickness. The exploitation of Grombalia aquifer begun during the 1950s. In 2005, the number of well exploiting the groundwater exceeded 11,000.

The study area contains four WWTPs; Menzel Bouzelfa, Grombalia, Soliman and Bou Argoub with a nominal capacity of respectively 5319, 3082, 2457 et 2735 m^3/d.

6.2.2 Methodology

To locate and rank suitable sites for recharging Grombalia aquifer with reclaimed water produced by Bou Argoub's WWTP a combination between MCA and GIS was applied. The procedure was carried out in two phases. The first phase consisted of the conceptualization of the different steps of MCA. The second phase consisted of bringing the first conceptual phase in a GIS environment through the application of a set of geospatial analysis operators. Figure 6.2 represents the methodology flowchart.

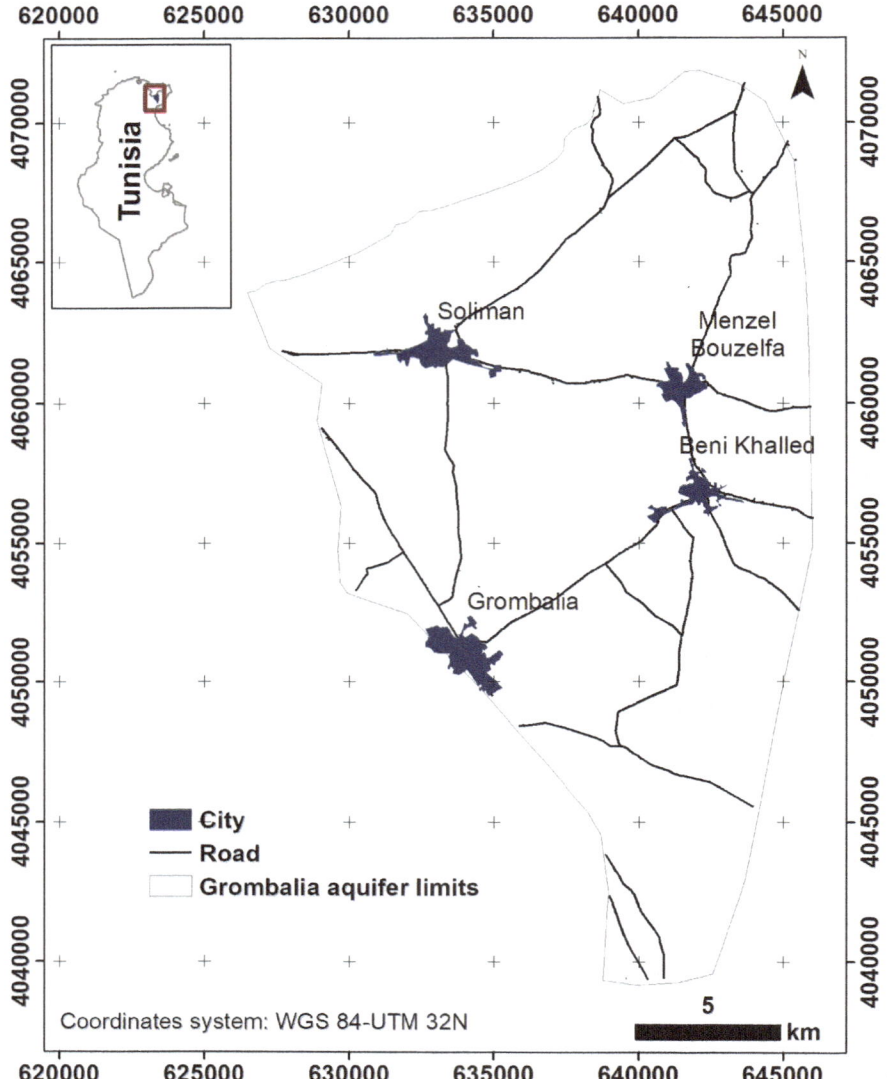

Fig. 6.1 Geographic location of the study area

6.2.2.1   Phase I: Multicriteria Analysis

Problem Identification and Objectives Definition

The first step of all multicriteria projects is the identification of the problem at hand and definition of the project objectives. In this study case the problem arises due to

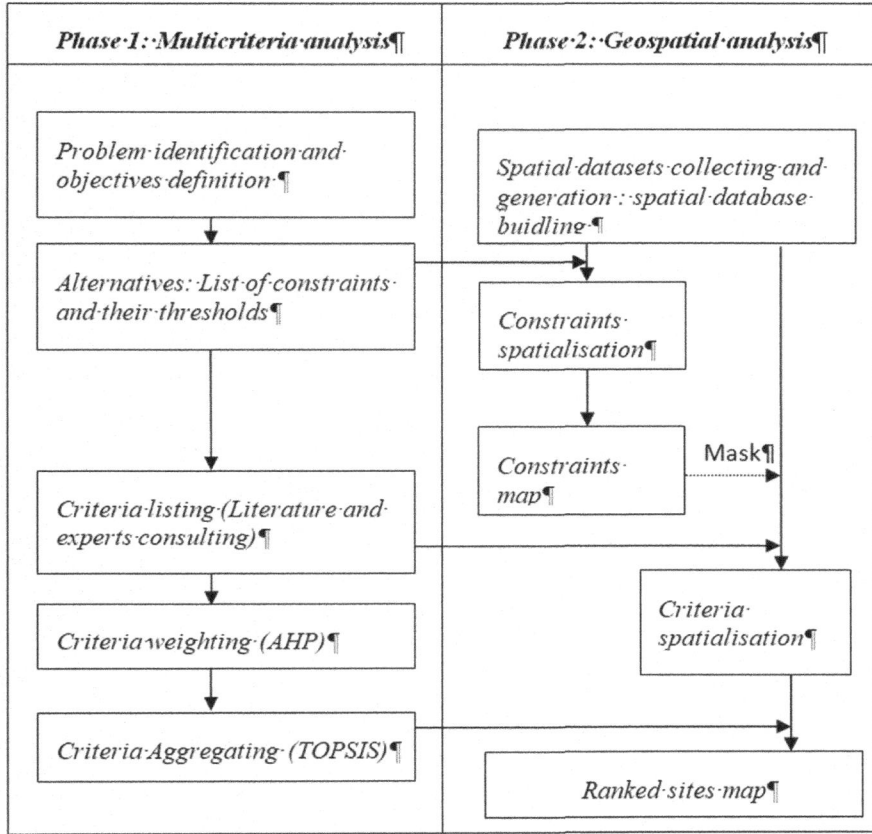

Fig. 6.2 The methodology flowchart

the presence in the region of a high amount of treated wastewater disposed of in the hydrographic network, flowing down to the sea. This wastewater pollutes Grombalia region surface water, producing odor nuisance and deteriorating groundwater quality. The rationale of this study is to suggest a way to use this water in order to increase the amount of available water for irrigation meanwhile saving the environment. The objective is the ranking of suitable sites for Grombalia phreatic aquifer recharge with reclaimed water of the Bou Argoub's WWTP.

6.2.3 Calculation of Surface Needed for Aquifer Recharge

To calculate the needed area for Grombalia aquifer recharge with Bou Argoub plant effluent we applied the Eq. 1 developed by the Environmental protection agency of United State [19].

$$A = (Q * P * 10^4)/(N * L) \tag{1}$$

where

A is the required area [ha].

Q is the WWTP outfow. For Bou Argoub effluent, it is equal to 2735 m^3/day.

P is the days number per season [day]. According to evapotranspiration data given by [20], we suppose that two seasons characterize the study area; summer and winter. The summer season lasts 214 days from march to september and the winter season lasts 151 days from october to february.

N is the cycles number per season [cycle]. It is calculated based on Eq. 2 [19]

$$N = days\ number\ per\ season/days\ number\ per\ cycle \tag{2}$$

A cycle is the alternance between the wet and dry periods. The wet period is when the basin is filled with reclaimed water and the dry period is when the basin is empty. The latter is needed to avoid basin clogging and guarantee reclaimed water infiltration during the dry period [18]. The wet period lasts 2 days for both seasons [18]. Consequently N is equal to 11 for the summer season and 24 for the winter season.

L is the hydraulic loading rate per cycle of reclaimed water [m/cycle]. It is 10% of the hydrauling loading of fresh water (L_{fw}) [19]. L_{ef} is obtained according to the Eq. 3 [19].

$$L_{fw} = Kv[cm/h] * 24[h/day] * 365[day/year]*[m^2]/100[cm/m] \tag{3}$$

where Kv is the soil hydraulic conductivity. It is calculated based on soil characteristics. For Grombalia region Kv is estimated to 981 cm/h. Accordingly, the mean hydrauling loading of fresh water is 5.54 m^3/day and of reclaimed water is 0.55 m^3/day.

Based on these data, the needed area to absorbe the total Bou Argoub effluent is 6 ha during the winter season and 4 ha during summer season. The highest area between both seasons is considered. An area of 30% is added for basins protection. Then the total required area for recharge the aquifer with Bou Argoub reclaimed water is of 8 ha.

6.2.3.1 Alternatives Identification

An alternative is all site considered suitable and able to be selected for aquifer recharge with reclaimed water. These sites were delineated through the selection and treatment of a list of parameters considered as constraints. For each constraint, a threshold of feasibility was identified. According to the literature and expert consulting [15–18], eleven constraints were identified. Table 6.1 presents the constraints and their feasibility thresholds.

Table 6.1 Constraints and their feasibility threshold to delineate the suitable sites for Grombalia phreatic aquifer recharge with reclaimed water of Bou Argoub WWTP

Constraints	Feasibility threshold
Slope	<12%
Soil texture	Clay <10%
Thickness of vadose zone	>5 m
Geology	Gravies and coarse sand
Quality of groundwater	Chlorate >16.95 mg/l; Calcium >15 mg/l; Sulfates >12.5 mg/l; Nitrates >0.8 mg/l
Land use	Sebkha, forest, urban areas and Dams
Distance to residential area	>200 m
Distance to roads	>50 m
Distance to dams	>500 m
Distance to rivers	>50 m
Faults	>50 m

To be feasible, a site should respond to the feasibility conditions of all the constraints simultaneously.

6.2.3.2 Criteria Identification

The suitable sites were ranked from the best to the worst based on thirteen criteria. The same criteria identified by [18, 21] were used in this work. These criteria were grouped into three dimensions; technical criteria cost criteria and environmental criteria.

The technical criteria depict the factors influencing the building and management of the basins used for the recharge. They are directly linked to the land characteristics. They are made up of six criteria: soil texture, geology, slope, the thickness of vadose zone and soil salinity. Soil texture and geology should be sufficiently permeable with a low rate of clay and high rate of sand to favor infiltration, but not highly permeable in order to favor natural purification of reclaimed water before reaching the aquifer. Thicker is the vadose zone better is the site for the aquifer recharge. More the slope is gentle better is the vertical infiltration and less costly is the construction of basins and then more suitable is the site for aquifer recharge. The soil salinity should be low to avoid groundwater quality deterioration by leaching.

The economic criteria depict the costs needed to transfer the treated wastewater from the WWTP to the site of recharge and the land cost of the site. It is composed by three criteria: (i) the elevation difference between the WWTP site and the recharge site (ii) the distance between the WWTP site and the recharge site and (iii) the land use. The elevation difference represents the cost of pumping to transfer reclaimed water from the WWTP to the site. Higher is this difference less suitable is the site for the aquifer recharge. The distance depicts the cost of pipes installing. Larger is this distance less suitable is the site for aquifer recharge. The Land use represents the cost of land for basins installation. More costly is the land less suitable is the site; the non irrigated agricultural land is cheaper than the irrigated land and the crop irrigated areas are cheaper than the fruit trees irrigated areas.

The environmental criteria are made up of two criteria which are the distance to the residential areas and the quality of groundwater. The recharge basins should be as far as possible from the residential areas to avoid the negative effect of odors and insects on population. The quality of groundwater is represented by four criteria which are salinity and chloride, sodium and nitrates concentrations. The three first parameters reflect the quality of groundwater for irrigation. More the concentration is low more the site is good for the recharge. The nitrates rate indicates the aquifer pollution due to the anthropologic activities, mainly from agricultural land fertilization. Treated wastewater is rich in nitrogen component so that the site is considered more suitable for aquifer recharge with reclaimed water above the groundwater less concentrated with nitrates.

6.2.3.3 Criteria Weighting

The criteria weighting was applied using AHP (Analytic Hierarchy Process) method. It is based on organizing the objects in a hierarchy structure [22]. The system was made up into four levels. The first presents the objective of the study. The second contains the group of criteria Technical, Economic and Environmental. The last ones present the criteria (Fig. 6.3).

From this structure, the elements sharing the same node were pairwise compared according to their importance to the objective. The comparison is based on the assignment of what it is called priority. Priority values vary from 1 to 9; 1 indicates that two compared elements have the same importance and 9 indicate that the first element has extreme importance compared to the second. These values were organized in a matrix. A set of the matrix was carried out exploring their eigenvectors and values. The resulting output is the local and global weights of each criterion used. The consistency of the obtained local weights was checked using the consistency index [22].

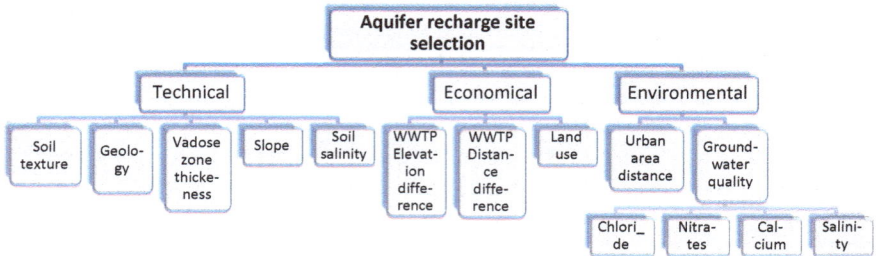

Fig. 6.3 The hierarchical structure of the decisional problem at hand

Table 6.2 Local and global weights obtained using the AHP method

Criteria Group	Local weight	Criteria	Local weight	Criteria	Local weight	Global weight (w_i)
Technique	0.33	Soil texture	0.29			0.096
		Geology	0.29			0.096
		Thickness of vadose zone	0.29			0.096
		Soil salinity	0.07			0.023
		Slope	0.05			0.017
Economic	0.33	Distance from WWTP	0.61			0.201
		Elevation difference	0.3			0.099
		Land use	0.09			0.03
Environmental	0.33	Distance to urban area	0.8	Salinity	0.73	0.264
		Groundwater quality	0.2	Nitrate	0.1	0.048
				Sodium	0.08	0.007
				Chloride	0.08	0.005
						0.005

If the index is less than 0.1 the attributed priorities are considered satisfactory and consistent, otherwise they should be revised.

The pairwise comparison is a particularity of AHP, which make easier the weighting process. Table 6.2 presents the local and global weights of the different criteria obtained using the AHP method.

6.2.3.4 Criteria Aggregation

Criteria aggregation was achieved using the PROMETHEE II outranking method. PROMETHEE II is based on the calculation of the net flux value. The alternative is better if the net flux is higher. The followed steps to carry out the PROMETHEE II are presented hereafter.

1. Step 1. Elaboration of the performance table

Once the alternatives and criteria had been defined, the next step consisted of the elaboration of the performance matrix (Eq. 1)

$$
D = \begin{array}{c} A_1 \\ A_2 \\ \vdots \\ A_i \\ \vdots \\ A_j \end{array} \left(\begin{array}{cccccc}
F_1 & F_2 & \ldots & F_i & \ldots & F_n \\
f_{11} & f_{21} & \ldots & f_{i1} & \ldots & f_{n1} \\
f_{12} & f_{22} & \ldots & f_{i2} & \ldots & f_{n2} \\
\vdots & \vdots & \vdots & \vdots & \vdots & \vdots \\
f_{1i} & f_{2i} & \ldots & f_{ii} & \ldots & f_{in} \\
f_{1j} & f_{2j} & \ldots & f_{ij} & \ldots & f_{mn}
\end{array} \right) \qquad (1)
$$

A_j are the alternatives corresponding to all the suitable sites for Grombalia aquifer recharge with reclaimed water. F_n are the criteria, n is equal to 13 in our case study. f_{mn} are the performance values of each alternative related to each criterion.

2. Step 2. Determination of preference function and preference and indifference thresholds

For each criterion a preference function P (a, b) was selected. This function reflects how the alternative a ourtranks the alternative b. There are six preference functions. For this work the linear function was chosen (Eq. 5) because it is the most used in projets related to environment.

$$
\begin{cases}
Pi(a,b) = 0 \; if \; fi(a) - fi(b) \leq qi \\
0 < Pi(a,b) < 1 \; if \; qi < fi(a) - fi(b) < pi \\
Pi(a,b) = 1 \; if \; fi(a) - fi(b) \geq pi
\end{cases}
\tag{5}
$$

The linear preference function requires the definition of two thresholds; preference (p_i) and indifference (q_i). These thresholds were defined by consulting experts in aquifer recharge with reclaimed water (Table 6.3).

3. Step 3. Global preference index calculation

The global preference index (π (a, b)) represents the preference intensity of the alternative a with regard to the alternative b. π (a, b) was calculated according to the Eq. 6.

$$
\pi(a, b) = \sum_{i=1}^{N} Pi(a,b)w_i
\tag{6}
$$

π (a, b) varies between 0 and 1. More π (a, b) is close to 1 more important is the preference of a compared to b.

4. Step 4. Positif, negative and net flows determination

To be able to get a complete ranking of the alternatives we proceeded to calculate first the preference positive (ω^+) and negative (ω^-) flows using the Eqs. 7 and 8.

$$
\omega^+ (a) = = \frac{1}{m-1} \sum_{x \in A} \pi(a, x)
\tag{7}
$$

$$
\omega^- (a) = = \frac{1}{m-1} \sum_{x \in A} \pi(x, a)
\tag{8}
$$

Then the net flow ω was calculated (Eq. 9).

$$
\omega(a) = \omega^+(a) - \omega^-(a)
\tag{9}
$$

Table 6.3 Indifference (q_i) and preference (p_i) thresholds of the 13 criteria

	Soil texture	Geology	Thickness of vadose zone	Soil salinity	Slope	Distance from WWTP	Elevation difference	Land use	Distance to urban area	Water salinity	Nitrate, sodium chloride
q_i	1	1	3	2	1	100	10	1	60	0.15	0.15
p_i	3	3	5	4	3	1000	100	3	200	0.25	0.25

This net flow ranks the alternatives from the best to the worst. The alternative a is better than b if $\omega(a) > \omega(b)$. So, the site to receive Bou Argoub's effluent for Grombalia aquifer recharge should have the highest ω value recorded.

6.2.3.5   Phase II: GIS and Geospatial Analysis

Applying the MCA methodology described previously requires the use of GIS and geospatial analysis to locate and rank the suitable sites for aquifer recharge and select the best site.

6.2.3.6   Constraints and Criteria Spatialisation

Several geospatial operators were used to derive from different raw spatial datasets the spatial dataset of constraints and criteria necessary to carry out the different steps of the MCA. The different operators applied for each constraint and criteria are presented hereafter:

The slope was obtained from the 30 m-resolution DEM raster dataset of "Shuttle Radar Topographic Mission" (SRTM) using *slope operator* in ArcGIS. The soil texture and salinity were obtained directly from the "Carte Agricole" official database of the Nabeul governorate. The geology spatial dataset was obtained by digitizing the scanned and georeferenced national geological maps of the region. The thickness of vadose zone and the groundwater salinity and nitrates, chloride and calcium contents spatial datasets were obtained by interpolating the contents of punctual dataset of 41 wells homogenously distributed over the study area. IDW, spline, and kriging interpolation methods were tested for each parameter, and the one with the least error was chosen. The land use was obtained through the application of supervised classification on five Sentinel 2 A and Lansat 8 images of the 2016/2017 agricultural period. Distance to residential areas and dams were obtained first by extracting from the land use map urban areas and dams and them applying on them the operator "*Euclidian distance*." Distances from roads and from rivers were obtained using the operator *Euclidian distance* to the road and rivers datasets of the Carte Agricole. The same operator was used to the punctual dataset of WWTPs to get the distance from Bou Argoub's WWTP. The difference of elevation was obtained subtracting the elevation of the Bou Argoub WWTP from the SRTM using "*Map calculator*" operator.

1. *Constraints*

From the resulting datasets, a Boolean logic analysis was used to delineate the suitable and unsuitable sites for each constraint on the base of its corresponding thresholds (Table 6.1). Twelve constraints spatial layers were obtained in which the suitable areas were delineated. An intersection operator was applied to the eleven constraints, and the resulting constraints layer was derived. This layer defines the limits of the suitable area for Grombalia aquifer recharge with reclaimed water.

2. *Criteria*

The thirteen criteria obtained over the entire Grombalia phreatic aquifer limit were clipped based on the suitable areas limits. Two operators were used "*extract by mask*" raster data and "*clip*" for vector data.

6.2.3.7   Criteria Aggregation Using PROMETHEE II

One of the PROMETHEE II strengths is the possibility to use vector or raster data. In our case, we chose the raster model for ease of use and precision. So, all vector criteria were converted to raster. Then all the criteria were normalized using fuzzy functions using the "*Fuzzy Membership.*" This function transforms the values/qualitative classes of all the criteria to the same interval of values [0, 1]; 0 represents the values less suitable for aquifer recharge with reclaimed water and 1 the most suitable.

Then the weighted normalized performance matrix for each criterion was obtained by multiplying each normalized criterion dataset with the corresponding global weight using the operator "*raster calculator*". Afterwards, the positive and negative ideal alternatives were calculated using the same operator "*raster calculator*". Finally, the distance to the positive and negative ideal alternatives and the relative proximity to the alternative C_j^* were obtained using again the operator "*raster calculator*". The final result is a raster dataset in which each pixel is assigned a value between 0 and 1 representing their ranking according to their aptitude to receive the reclaimed water for Grombalia aquifer recharge. The small values are the alternatives less suitable for aquifer recharge, and the higher values are the alternatives most suitable.

6.3   Results and Discussions

6.3.1   Spatial Datasets Used

Figures 6.4, 6.5 and 6.6 present the spatial datasets used to get the criteria and constraints layers: slope, soil texture and salinity, land use, thickness of vadose zone, geology, closeness to residential area, dams, roads, rivers and sebkhas, groundwater salinity and groundwater chloride, sulfate, nitrates and calcium contents.

The slope of the study area varies between 0 and 75% with the dominance of gentle slopes. More than 98% of the area presents slope between 2 and 5%. A small part of the study area located to the east at the Piedmont has an abrupt relief.

The soil texture is characterized by a large variability, from heavy to lightweight soils. However the most widespread are silt and silt-sandy soils, representing respectively 22 and 20% of the study area.

The soil salinity is variable, being the healthy soils the most widespread, covering 73% of the total area. The salty soils are very limited.

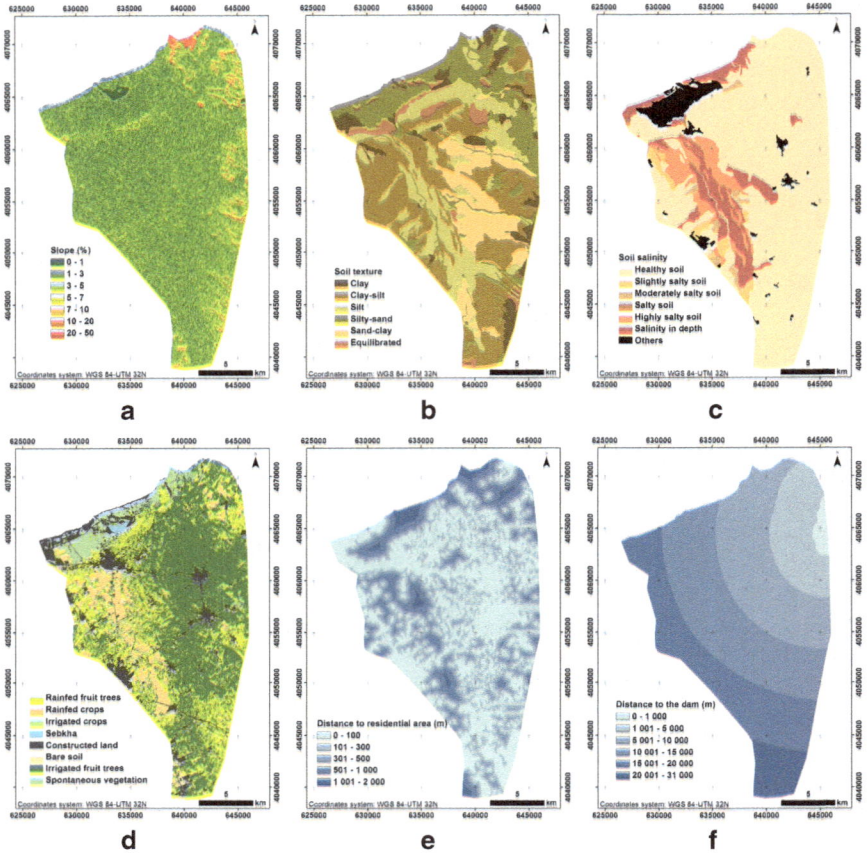

Fig. 6.4 Spatial datasets used to get the constraints and criteria spatial layers: **a** slope, **b** soil texture, **c** Soil salinity, **D** Land use, **e** closeness to the urban areas and **f** closeness to the dams

Different land uses are present in the study area. The urban area covers about 15% of the total area, the irrigated fruit trees occupy 39%, the crops cover 28% of the total area split between 22% rained and 6% irrigated. The olive trees cover 9% of the study area. The water surfaces cover 1037 ha spread between the Soliman's sebkha and Bzigh's dam. The farthest distance from the sebkha is 25,263 m and from the dam is 25,940 m.

The distance to the urban areas varies from 0 to 2300 m. The farthest areas are located on the coast line and small enclaves in the center of the study area.

The road network 107 km long divided in 3 km highways, 23 km national roads, 76 km regional roads and 5 km local roads. The farthest distance from the roads is 4116 m.

Concerning the hydrographic network, the farthest distance is 3944 m.

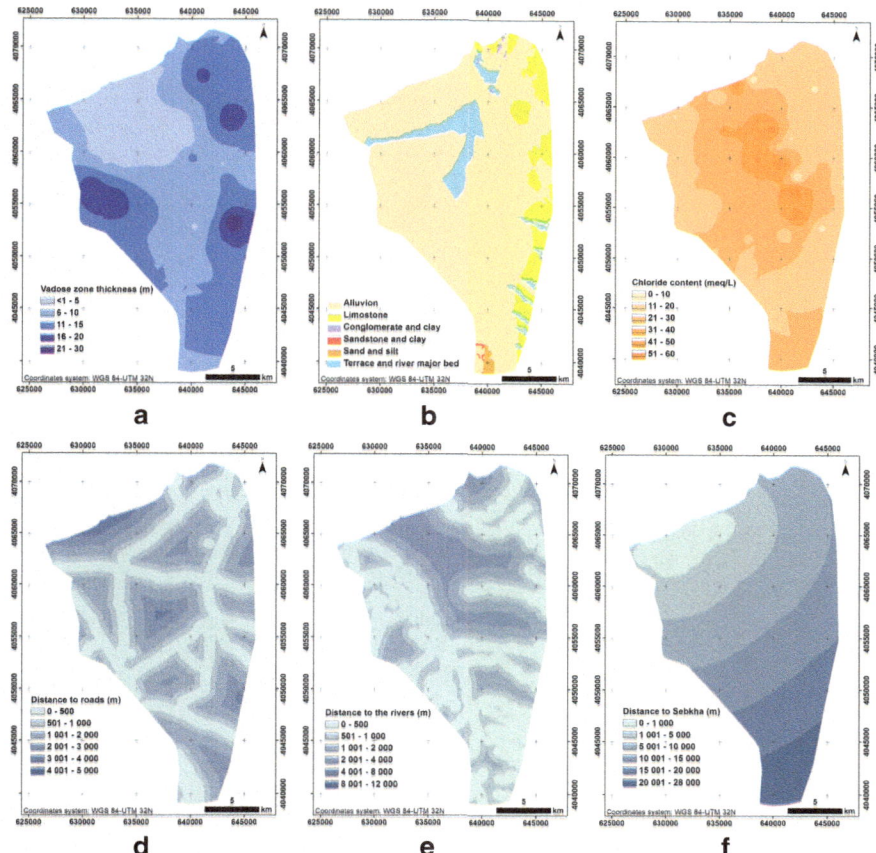

Fig. 6.5 Spatial datasets used to get the constraints and criteria spatial layers: **a** Vadone zone thickness, **b** geology, **c** chloride content, **d** distance to the roads, **e** distance to the rivers and **f** Distance to the sebkhas

The thickness of vadose zone varies from centimeters near the coast and in the Bou Charray irrigated area to twenties meters in the Piedmont in the east. The geological formations are very variables; from the permeable formation (alluviums), moderately permeable (limestones and sandstones) and impermeable formations (clay). Two faults of 1800 m length are located in the extreme northeast of the study area.

Concerning the groundwater quality, the salinity varies from 2 to 6 g/l. The calcium and sulfates contents vary respectively from 9 to 17 meq/l and from 0 to 21 meq/l. The nitrates and chloride concentration vary respectively from 0 to 43 meq/l and from 0 to 55 meq/l. The highest contents are observed in the south and/or the center of the area and the lowest in the north.

The highest elevation difference between the site and the Bou Argoub's WWTP is 232 m. The farthest distance from the WWTP is 25364 m.

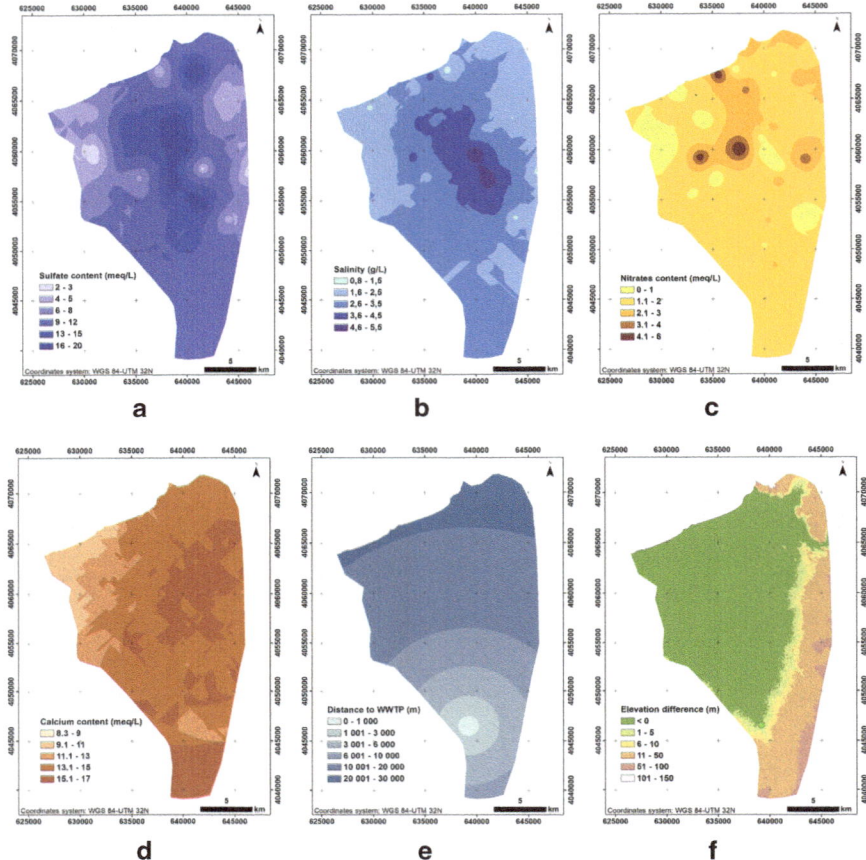

Fig. 6.6 Spatial datasets used to get the constraints and criteria spatial layers: **a** sulfates content, **b** groundwater salinity, **c** nitrates content, **d** calcium content, **e** Distances to Bou Argoub's WWTP and **f** elevation difference with Bou Argoub's WWTP

6.3.2   Constraints Layer

The resulting constraints layer obtained by intersecting the eleven layers of constraints is presented in Fig. 6.7. The total suitable area is 1667 ha, exceeding by far the needed area to absorb the total effluent produced by the Bou Argoub's WWTP. The most suitable area 37,202 ha is given by the fault constraint layer and the lowest suitable area is the distance to a residential area, which is 13,859 ha.

Fig. 6.7 Resulting
constraints layer

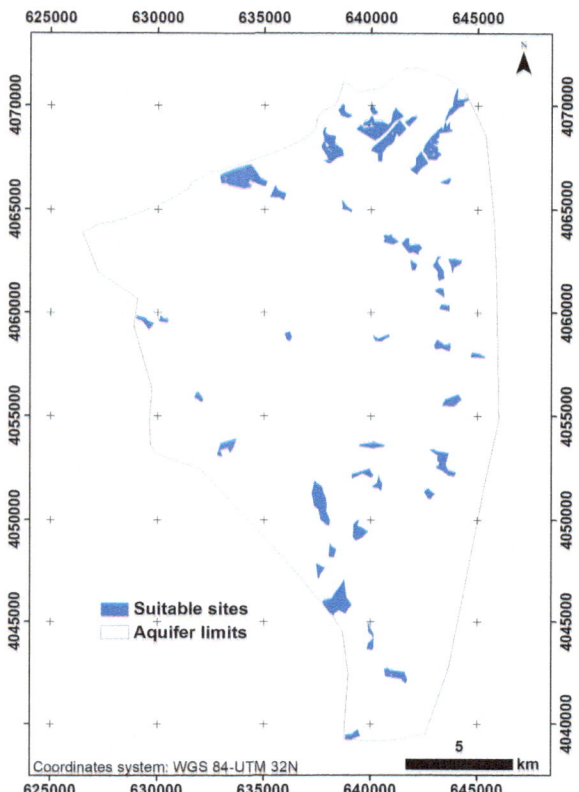

6.3.3   Suitable Sites Ranking

The net flux φ of PROMETHEE II method obtained to rank suitable sites for aquifer recharge with reclaimed water of Bou Argoub varies between −0.33 and 0.32 (Fig. 6.8). The less suitable areas have the lowest net flux and the most suitable have the highest net flux.

The best site for the Grombalia phreatic aquifer recharge with Bou Argoub WWTP has the highest net flux φ recorded, equal to 0.32. This site is located in an agricultural area, 3.6 km from the WWTP and about 500 m far from the residential areas. The vadose zone thickness of this site is about 9 m.

These results show the capacity of integrating the multicriteria analysis into a GIS environment in helping decision makers to allocate reclaimed water for aquifer recharge. However among plenty of multicriteria analysis methods we tested only PROMETHEE **II**. It is known that the results could change from one method to another so it is recommended to test many, such as weighted sum, ELECTRE and TOPSIS. The convergence between these methods enforces the results obtained.

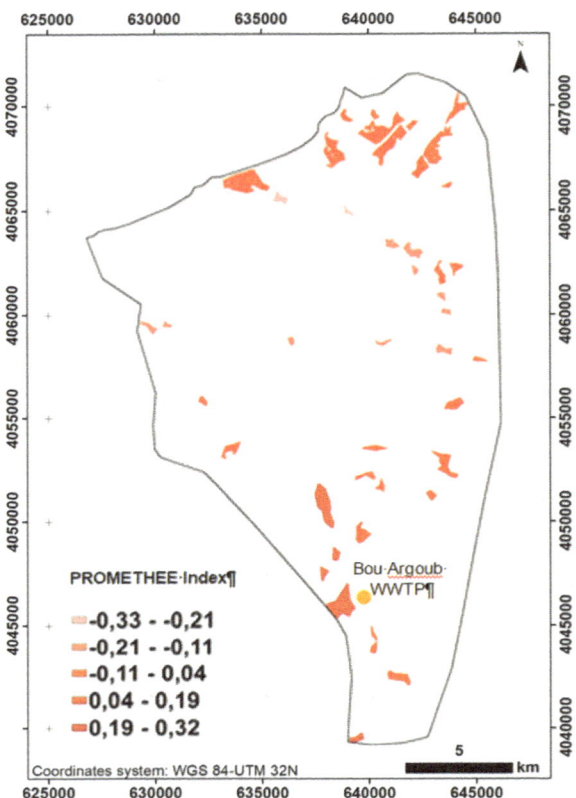

Fig. 6.8 Ranking of suitable sites for Grombalia phreatic aquifer recharge with Bou Argoub WWTP reclaimed water using the PROMETHEE II method

When divergence is observed it is possible to apply a vote method such as "Borda" or "Copeland" to decide.

Uptaking the recharged reclaimed water for its use in irrigation requires prior understanding of its dynamic in the aquifer. So before basins installation it is appropriate to simulate the impact of this recharge on the aquifer dynamic, in terms of amount and solutants, through hydrogeological modeling. The modeling helps to know how, when and where the uptake should be done and then optimizes the use by constructing different management scenarios.

6.4   Conclusions and Recommendations

The eleven constraints identified for delineating the suitable sites for Grombalia phreatic aquifer recharge with Bou Argoub WWTP reclaimed water are Slope, Soil texture, Thickness of vadose zone, Geology, Quality of groundwater, Land use, Distance to a residential area, Distance to roads, Distance to dams Distance to rivers and Faults. These constraints show that the suitable area for Grombalia aquifer recharge with reclaimed water is 1667 ha.

The weighting of the thirteen criteria to rank the suitable sites using AHP method states that the distance to the residential areas and to the WWTP are the two most influential criteria for site selection decision. The first is an environmental criterion and the second is an economic criterion.

The combination of GIS and the multicriteria PROMETHEE II method offer an efficient tool to rank the suitable sites for Grombalia aquifer recharge with reclaimed water. The decisional net flux obtained using the PROMETHEE II method varies between -0.33 and 0.32. The most suitable sites are located around the Bou Argoub's WWTP and the less suitable sites are 20 km far from the WWTP.

The methodology carried out is a useful decision support tool that could help the decision maker to better choose the suitable sites for phreatic aquifer recharge with reclaimed water in the different region of Tunisia. However it is important to test different aggregation methods and in case of divergence a vote method (i.e. Borda or Copeland) could be used to decide which site should be selected.

Before the execution of the project and basins construction, it is appropriate to apply the hydrogeology modeling in order to simulate the impact of the aquifer recharge in the selected site on the dynamic of the aquifer in terms of water amount and pollutant transfer.

Acknowledgements We appreciate financial support from WaterFARMING project (Grant Agreement No: 689271) and Tunisian Ministry of Higher Education and Scientific Research.

References

1. Anane M, Abidi B, Lachaal F, Limam A, Jellali S (2013) GIS-based DRASTIC, Pesticide DRASTIC and the Susceptibility Index (SI): comparative study for evaluation of pollution potential in the Nabeul-Hammamet shallow aquifer, Tunisia. Hydrogeol J 21:715–731
2. Peeters L, Fasbender D, Batelaan O, Dassargues A (2010) Bayesian data fusion for water table interpolation: incorporating hydrogeological conceptual model in kriging. Water Resour Res 46:1–11
3. Widhalm P, Yang Y, Ulm M, Athavale S, Gonzalez M, Marta C (2015) Discovering urban activity patterns in cell phone data. Transportation 42:597–623
4. Anane M, Selmi Y, Limam A, Jedidi N, Jellali S (2014) Does irrigation with reclaimed water significantly pollute shallow aquifer with nitrate and salinity? An assay in a per-urban area in North Tunisia. Environ Monit Assess 186(7):4367–4390
5. Malczewski J (2006) GIS-based multicriteria decision analysis: a survey of the literature. Int J Geogr Inf Sci 20:703–726
6. Rahman MA, Uddin SB, Rusteberg M, Lutz A, Abu Saada M, Sauter M (2013) An integrated study of spatial multicriteria analysis and mathematical modeling for managed aquifer recharge site suitability mapping and site ranking at Northern Gaza coastal aquifer. J Environ Manage 124:25–39
7. Ghoseiri K, Lessan J (2014) Waste disposal site selection using an analytic hierarchal pairwise comparison and ELECTRE approaches under fuzzy environment. J. Intell. Fuzzy Syst 26(2):693–704

8. Pascal O, Xie Q, Ling X, Eugene O (2017) A fuzzy TOPSIS model framework for ranking sustainable water supply alternatives. Water Resour Manag 31(9):2579–2593

9. Mahboubeh G, Tooraj H, Mohammad R (2018) A hybrid TOPSIS-agent-based framework for reducing the water demand requested by stakeholders with considering the agents' characteristics and optimization of cropping pattern. Agric Water Manag 199:71–85

10. Elleuch MA, Anane M, Euchi J, Frikha A (2019) Hybrid fuzzy multi-criteria decision making to solve the irrigation water allocation problem in the Tunisian case. Agric Syst 176:1–13

11. Barzehkar M, Dinan NM, Mazaheri S, Tayebi RM, Ian Brodie G (2019) Landfill site selection using GIS-based multi-criteria evaluation (case study: SaharKhiz Region located in Gilan Province in Iran). SN Appl Sci 1:1082

12. Abba AH, Noor ZZ, Yusuf RO, Din MFMD, Abu Hassan MA (2013) Assessing environmental impacts of municipal solid waste of Johor by analytical hierarchy process. Resour Conserv Recycl 73:188–196

13. Erkan TE, Elsharida WM (2019) Overview of airport location selection methods. Int J Appl Eng Res 14:1613–1618

14. Gumussay MU, Koseoglu G, Bakirman T (2016) An assessment of site suitability for marina construction in Istanbul, Turkey, using GIS and AHP multicriteria decision analysis. Environ Monit Assess 188(1):676–690

15. Anane M, Kallali H, Jellali S, Ouessar M (2007) Soil Aquifer Treatment areas in Tunisia: Jerba Island. En: Wastewater Reuse–Risk Assessment, Decision-Making and Environmental Security. Book series: NATO Security through Science Series. Editeur: Springer Netherlands. ISBN. 978-1-4020-6026-7, pp 65–72

16. Kallali H, Anane M, Jellali S, Tarhouni JN (2007) GIS-based multi-criteria analysis for potential wastewater aquifer recharge sites in Hammamet-Nabeul Aquifer (Tunisia). Desalination 215:111–119

17. Anane M, Bouziri L, Limam A, Jellali S (2012) Ranking suitable sites for irrigation with reclaimed water in the Nabeul-Hammamet region (Tunisia) using GIS and AHP-multicriteria decision analysis. Resour Conserv Recycl 65:36–46

18. Gdoura K, Anane M, Jellali S (2015) Geospatial and AHP-multicriteria analyses to locate and rank suitable sites for groundwater recharge with reclaimed water. Resour Conserv Recycl 104:19–30

19. EPA (1984) U.S. Environmental protection agency center for environmental research information Cincinnati, process design manual for land treatment of municipal wastewater supplement on rapid infiltration and overland flow. EPA, Cincinnati, Ohio

20. Ben Hamouda FM (2008) Hydrochemical and isotopical approaches of Cab Bon coast aquifers systems: Cote Oriental and Haouaria aquifers case study – Tunisia. PhD Thesis

21. Anane M, Kallali H, Jellali S, Ouessar M (2008) Ranking suitable sites for SAT in Jerba island (Tunisia) using GIS, remote sensing and AHP-multicriteria decision analysis. Int J Water 4:121–135

22. Saaty TL (1980) The analytic hierarchy process. McGraw-Hill, New York, NY

Chapter 7
Using RS and GIS to Mapping Land Cover of the Cap Bon (Tunisia)

Monaem Nasr, Hedi Zenati, and Mohsen Dhieb

Abstract The Peninsula of Cap Bon (Tunisia) is an area containing a much-contrasted land use. It is divided into large natural subareas next to the mountains and "artificialized" spaces scattered now along the coast. Yet only a few land use maps have been elaborated on the topic. Within this context, the design and implementation of one specific GIS called LUIS (Land Use Information System) issued from one database may lead to a systemic and "multiscalar" approach to study and understand the complex occupancy of this unique Tunisian region. "Multisource" and "multi-scalar" remote sensing images represent certainly the potential source of data and knowledge in this work. The main goal of this research is to develop a methodology that can identify, locate and map, at different scales and by using various modes of presentation, the land use as an overall information system. It contains three phases bringing together a useful synergy between GIS and remote sensing: The design of the LUIS using the HBDS method; Remote sensing image processing. The proper development and implementation of the LUISystem. The data issued from multiple land use layers helps to represent the finest classes of land uses that can be superimposed altogether by creating more or less homogenous subareas. Thus, the various themes which are analyzed, including the social, economic and even historical aspects, ranging from the local to the regional levels. Of course, the LUIS system allows storing, updating, querying and synthesizing heterogeneous data. It also helps to monitor and understand the landscape dynamics, the population changes and the multiple violations of the land use regulations.

M. Nasr (✉) · H. Zenati · M. Dhieb
GEOMAGE Team-Laboratory "SYFACTE", Faculty of Arts and Humanities of Sfax, University of Sfax, 1168, 3000 Sfax, Tunisia

H. Zenati
e-mail: hzenati@kku.edu.sa

H. Zenati
Department of Geography, Faculty of Humanities, King Khalid University, Abha, Saudi Arabia

M. Dhieb
Department of Geography and Geographic Information Systems, Faculty of Arts and Humanities, King Abdulaziz University, 80202, 21589 Jeddah, Saudi Arabia

© Springer Nature Switzerland AG 2021
F. Khebour Allouche et al. (eds.), *Environmental Remote Sensing and GIS in Tunisia*, Springer Water, https://doi.org/10.1007/978-3-030-63668-5_7

117

Keywords GIS · Remote sensing · Mapping · Land cover/land use · LUIS · Modeling · Geodatabase · Cap Bon

7.1   Introduction

One of the various reasons that originated the remarkable development of the environmental phenomena, the Geographic Information Systems (GIS) is the complexity and diversity of data within the geographic area, especially those related to the land cover and land use. These latter originated the development of computer systems to be capable of meeting the needs of the collection, the analysis and the layout of data. Despite their relative newness, these systems became essential tools for understanding and monitoring dynamic facts and processes, particularly the evolution of the land cover and land use. "It is a dynamic variable because it reflects the interaction between socio-economic activities and regional environmental changes, and for this reason mapping land cover/land use is necessary to be updated frequently" [1, 2].

Therefore, the implementation of a GIS can be seen as an innovative and interesting solution to modeling complex portions of spaces. It may eventually constitute a valuable diagnostic tool for the management and analysis of land cover/use regardless of their level of perception and can be taken into account in the various planning schemes.

Nowadays, the establishment of such systems has become a necessity given the relative immensity of the studied area (2822 km^2) of the "Cap Bon" and its proximity to the national metropolis "Tunis". The implementation of one GIS for the land cover and land use in the region of Cap Bon contributes indeed to the management of natural resources in this multifunctional space. We assume that this system may facilitate the conservation, the management and the exploitation of a large amount of data on land cover/use.

The first part of this work gives a brief description of the geographical context: presentation of the peninsula of Cap Bon, setting the objectives and the methodology, citation of the documents used. Then we will describe the essential steps of the classification of the image, the design and construction of a GIS, namely the development of the Conceptual Data Model (CDM) and the physical creation of the Geodatabase. Finally, we discuss the whole mapping process: a multi-scalar mapping of land use and land cover in the Cap Bon with the creation of the derived spatial and thematic information [3, 4].

7.2   Objectives of the Present Study

The overall goal of this work is to create a Geographic Information System for the land cover/use in the Cap Bon.

The main objectives are:

- The design of a Conceptual, Logical and Physical Data Model using the HBDS (Hypergraph Based Data Structure) method, respecting the rules of the graphics semiology (especially when dealing with colors and color-values);
- The conception of a Geodatabase, coming at the end of the work, fitting the described CDM and using Arc Catalog software;
- The image processing and creation of datasets derived (supervised classification) from remotely sensed imagery and visual image interpretation approach;
- The creation of a geographic database of the land uses area (BADOS), multi-source, multiscalar and updatable, using the software ArcGIS 9.X or 10.X;
- The conduction of a multiscalar mapping procedure figuring the land cover and land use (urban, agricultural land, natural areas, wetlands, water bodies…) for the Cap Bon.

7.3 Methodology

To achieve these objectives, the current paper presents the different steps of the construction of this GIS, dedicated to the land cover/use in the Cap Bon in Tunisia. These tasks required acquiring a large amount of heterogeneous data, multi-sources, multi-scales and multi dates; thus, the implementation and updating of this amount of data becomes quite complex and difficult to implement.

In order to accomplish these tasks, we chose the ESRI ArcGIS package which allows the creation, the integration, the structuring and the analysis of a wide range of geographic data of multiple uses.

This software has an integrated set of GIS software and remote sensing can manage complex Geographic Information Systems. It also provides a scalable structure for the implementation of GIS for a single or many users on desktop systems and connected to servers for use in intranet, internet or directly on the ground.

Our choice of one valuable method to process and analyze spatial phenomena was based on approaches inspired by hyper graphic theory, known as HBDS (Hypergraph Based Data Structure). Created by F. BOUILLE in 1977 [5], this theory is summarized in a Conceptual Data Model (CDM), structuring the spatial and thematic information used to represent by graphs and hypergraphs, classes and superclasses, links and hyperlinks, the internal structure of simple phenomena. When linked to each other, these latter contribute to the understanding of complex phenomena, and the build the CDM.

For the implementation of geographic data, several steps were taken: data collection, georeferencing, conversion of data on a digital medium, analysis and visualization of useful information.

7.4 Documents Used

Many applications implemented using GIS such Land Use Land Cover depend on datasets derived from remotely sensed imagery or make use of the imagery directly as a background such Base map from ArcGIS Online in graphic displays, the digital remotely sensed data could be directly imported into a GIS and the photographic image can also be used if scanned and rectified to create GIS digital imagery [4].

Improved GIS tools and software packages will make the integration of remote sensing information and GIS possible and easy (Fig. 7.1).

The developed GIS was based on:

- Data base maps of the OTC (Office of Topography and Cadastre) of Tunisia;
- 4 ASTER scenes (14 m resolution) acquired from the TERRA satellite, May 30, 2001 and 2009. The images were delivered ortho-rectified, format ".hdr" and coded on 8 bits;
- 2 Landsat satellite scenes of 30 m spatial resolution;
- Databases from Regional Commissary for Agricultural Development (CRDA) of Nabeul. These latter have the advantage of integrating the details of crops grown in 2001, the date of acquisition of ASTER images used;
- Structure plans of the cities of Grombalia, Soliman, Menzel Bouzelfa, and Beni Khalled etc. … (2005).
- Base map from ArcGIS Online (these basemaps are map services that require an internet connection for them to draw and to edit in many map or application according to ESRI… Fig. 7.2). Recalling that ArcGIS is not an open source platform.

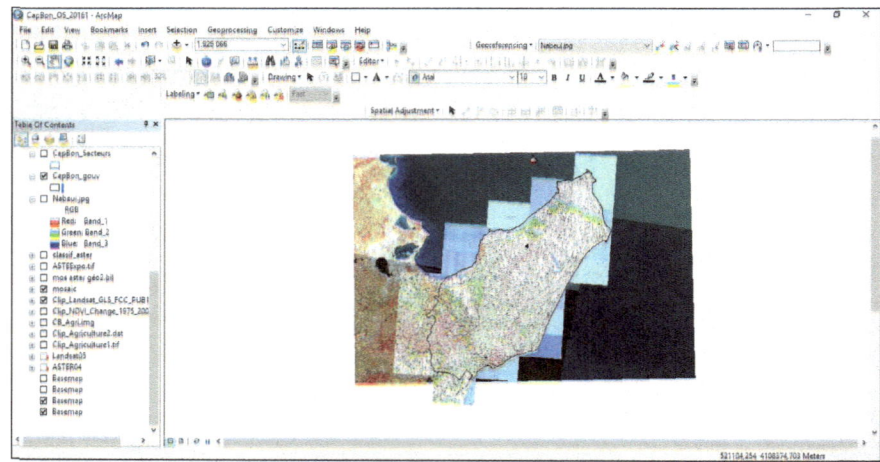

Fig. 7.1 Georeferencing, mosaicking and conversion of data used on a digital interface in ArcGIS from ESRI

Fig. 7.2 Base map from ArcGIS Online interface from ESRI

7.5 The Cap Bon as an Application Field

7.5.1 Overview of the Study Area

The peninsula of Cap Bon which coincides with the administrative "governorate of Nèbil" (Fig. 7.3), appears as a peninsula-wide open sea, bounded by two major gulfs: The Gulf of Tunis in the north and the Gulf of Hammamet in the south; it should be considered as a full part of the changing metropolitan area of Tunis.

With an area extending over 2822 km^2 or 1.8% of the total land area, and a population of over 787,920 inhabitants [6] or 7% of the Tunisian population, and

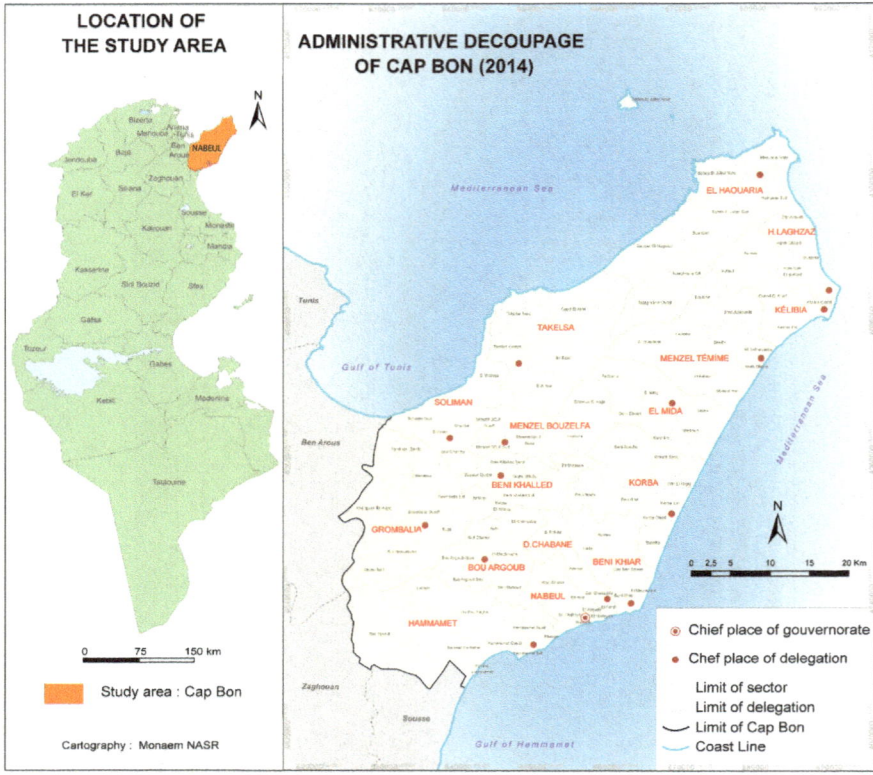

Fig. 7.3 Administrative map of the Peninsula of Cap Bon

with only 4% of agricultural area in Tunisia, the Cap Bon does participate for 14.3% of the national agricultural production and is considered as the leading producer of citrus fruits (over 80% of national production in 2014).

Natural resources of the governorate are quite rich and varied, however these resources have been, all along the recent history of Tunisia, suffering from the significant human pressures that are beginning to cause certain types of damage which certainly are not yet irreversible, but that do need very special attention in the future.

7.5.2   Model Space for a Better Understanding

In this work, we will focus more on the modeling and the organization of the database within a Geographic Information System (GIS). The first step in this process is to modulate the spatial objects of the research. The modeling is intended to structure the information space (geographical) and/or aspatial (theme) to acquire (logical data modeling), store (physical modeling data) and to manage (Computer Science), to process, to manipulate, to create, to recreate it and then be able to interrogate, analyze, and make simulations to create scenarios to verify and test hypotheses issued at the beginning (spatial information theory) [7].

Data modeling makes it possible to express the internal structure of the phenomena that shape the real world, as well as the links between and within them. During the development of a GIS, this important step will define the structure of the geographic database in order to represent with certain accuracy the reality on the ground.

We chose to model the data that form the basis of data on land use in the Cap Bon by using the method HBDS (hypergraph Based Data Structure). This modeling approach offers the advantage of being graphically simple, so the data patterns are very easy to read. The HBDS method also allows to better identifying all data (both spatial and thematic) to collect or create, and will then be organized according to this model [8]. Moreover, in view of the construction of a Geodatabase in ArcCatalog, HBDS method seems to be most suitable.

By representing the structure of simple phenomena like classes, graphs and directional links, the method can reproduce HBDS conceptual data models (CDM) which are easily understood by unskilled people to geomatics.

7.5.3 The Conceptual, Logical and Physical Data Model

Conceptual, logical and physical models are three different ways of modeling data in a domain of land cover/use. The proposed Conceptual Data Model is the simplest model among all. It is intended to be reproduced. It aims to apply to other areas of Tunisian and extended to other similar geographical areas. However, it must take into account the specificities of the Cap Bon. Thus, the presence of a long coastline (approximately 200 km) requires consideration alongside the classic components of the natural environment of physical criteria specific to the natural environment, including the geological, geomorphological and soil aspects as well as the other historical variables, social, economic and legal issues.

The conceptual data model is made up of 8 interconnected hypergraphs encompassing both the natural and anthropogenic environment as well as the exogenous factors and elements ensuring governance and territorial management of Cap Bon. By following the HBDS methodology, the MCD took into account simultaneously the composition of the real world, linked to the land use at Cap Bon, and the mode of operation of these components. It is indeed a model allowing a systemic and systematic approach to the real world. The relational dimension in this model is ensured not only by the links appearing in red in Fig. 7.4 but also by the graphic nesting illustrated in the same figure connecting the two hypergraphs (The territorial division and The main operations) between them and to the other hypergraphs.

The developed model may be read as follows.

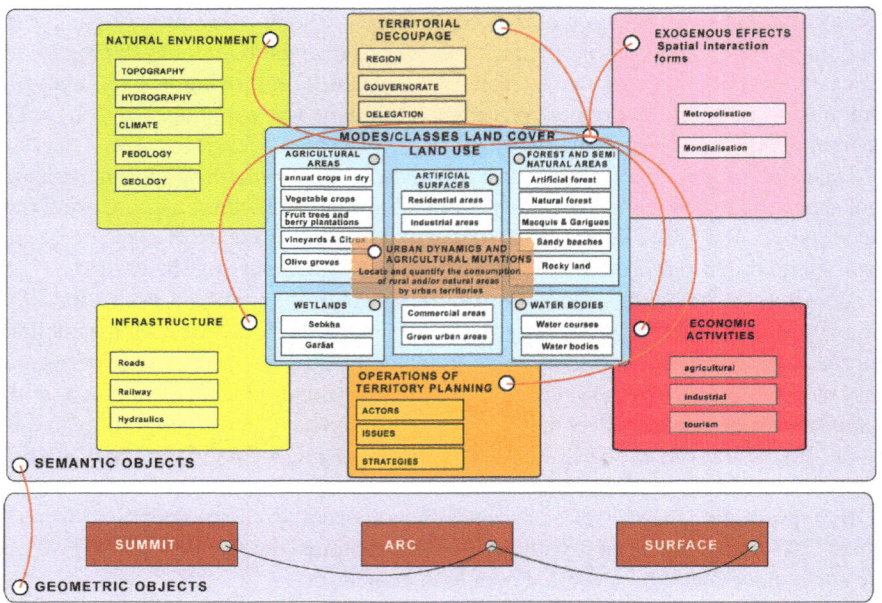

Fig. 7.4 Conceptual Data Model using the HBDS method depending on the land use and land cover of the Peninsula of Cap Bon

Figure 7.5 shows how the semantic objects correspond to the particular division of space and some non-spatial data. They include complex objects including the hyperclasses "Land use", "Territorial Division", "Natural Environment", "Infrastructure", "Economic activities", "Operations Management", "Historical process" and "Exogenous effects". The natural environment combines the physical and natural determinants that shape and influence the land. Similarly, the territorial division and infrastructure determine and influence the types of land use. Economic activities, operations management and exogenous effects modify the various modes of land use and spatial dynamics involved in agricultural and mutations. The land meets the classes "urban areas, agricultural areas, forests and semi-natural wetlands and water bodies."

All these complex objects are represented by geometric objects corresponding to points, lines or polygons.

The software implementation phase consists of making a representation of the data according to the data model of the DBMS, the core of the GIS, namely the relational model. Conceptual, logical (Fig. 7.5) and physical (Fig. 7.6) model are three different ways of modeling data in a domain of Land Cover and Land Use in The Cap Bon.

Although the development of CDM is independent of the technologies proposed by the GIS software, creating a Physical Data Model (Fig. 7.6) requires a platform that represents the physical medium [9].

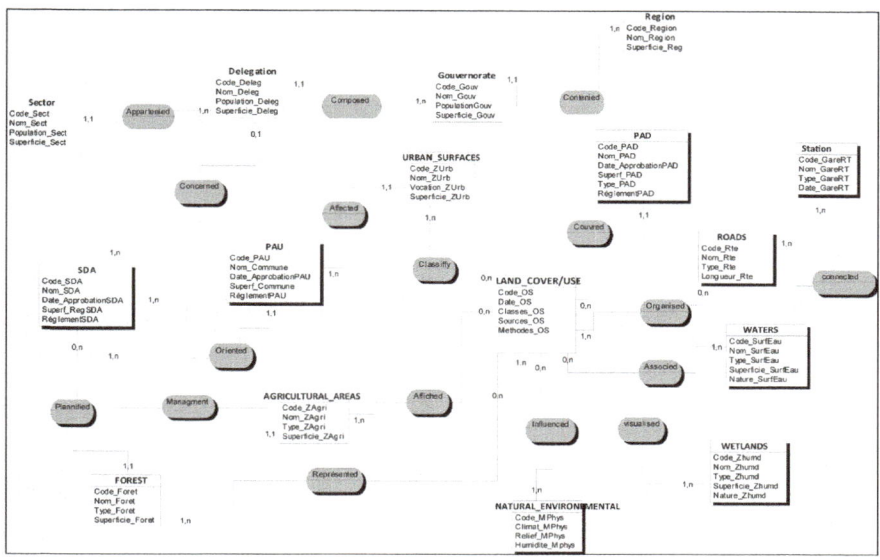

Fig. 7.5 Logical Data Model using the HBDS method depending on the land use and land cover of the Peninsula of Cap Bon

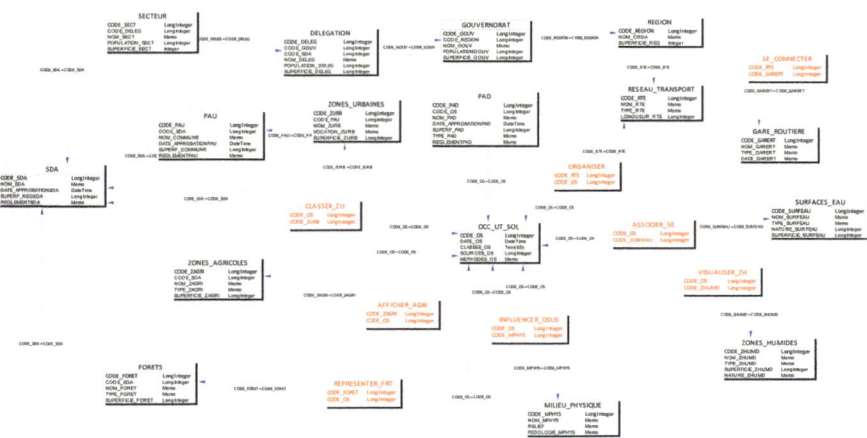

Fig. 7.6 Physical Data Model using the HBDS method depending on the land use and land cover of the Peninsula of Cap Bon

However, the CDM that we developed previously will be implemented physically in an ergonomic computing environment and allowing interconnections and crossings at multiple scales of analysis as much as the components of the Land Use and Land Cover system of the Peninsula of Cap Bon are variable. This is the simplest model among all.

Terminology HBDS	ArcGIS (ESRI)
Hypergraph	Geodatabase
Hyperclass	Feature class set
Hyperclass	Subtype
Class	Feature class
Link, topologic relation	Association
HBDS subject	Subject
Topology	Topology
Domain	Field (range of values taken by attributes)
Valuation	Field

Table 7.1 Correspondence between HBDS concepts and ESRI geodatabase [7, 8] (translated by the authors)

Source www.esri.com

Finally, the MCD enables us to transit a logical model to a physical model and with the transition relationship maintained about the creation of the Geodatabase.

7.6   Creation of the Geodatabase

7.6.1   The Structure of the Geodatabase

The CDM using the method HBDS configures the Geodatabase to be created in Arc Catalog: it is the Physical Data Model (PDM). A parallel can be drawn between the concepts of this hypergraphic method and concepts developed in ArcGIS, where each cell of a class diagram corresponds to a class (or subclass) of CDM [10]. The correspondence is explained in the following Table 7.1.

The geodatabase can be defined as "a common platform for storing and managing data. It brings in all the GIS layers of information available. It is like the skeleton of the GIS. It follows the representation of the CDM, bringing together in each sub-file segments of the same theme" [9] (translated by the authors). A geodatabase can store data tables, layers (points, lines and polygons) and raster images (Fig. 7.7).

7.6.2   The Image Processing and Loading of Datasets

The Geodatabase environment is favorable for the integration of remote sensing images directly into a GIS.

Remote sensing is the science and art of obtaining information about an object, area or phenomena through the analysis of data acquired by a device that is not in contact with the object, area, or phenomena under investigation [11]. The data analysis process involves

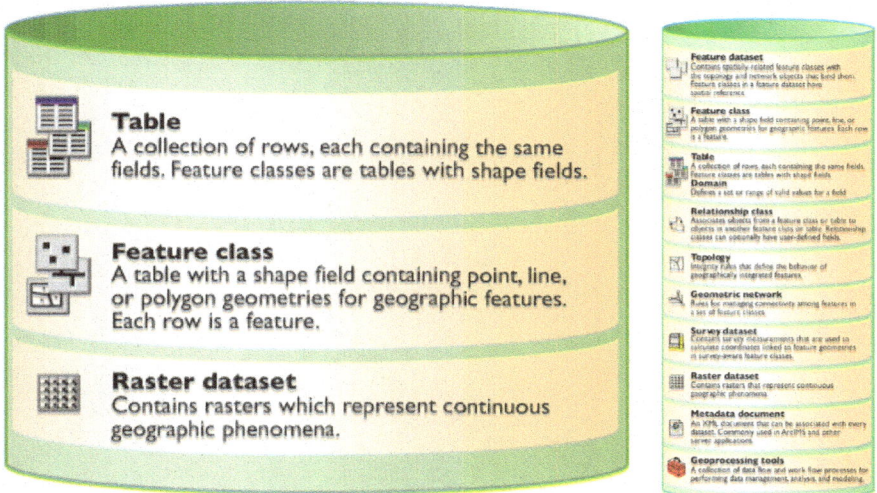

Fig. 7.7 The various data stored in a geodatabase

examining the data using various image processing techniques by a digital computer. Its application in the field of environmental management is of great prominence.

Land cover features were obtained using supervised classification. It is the main method of extraction of information from images. It represents the most important part since it will lead to the map of land cover sought. This classification relies heavily on the visual interpretation of the images…

The purpose of this classification, in this work, is to make it easy to process and interpret the raw image "ASTER or LANDSAT" [12], to match the radiometric reality with the thematic reality because starting from the radiometry we can distinct the land use types in the Cap Bon Peninsula. The numerical classification of this image uses the spectral information contained in the values of the three visible and near-infrared spectral bands to classify each pixel individually [13, 14].

The supervised method comprises two phases: the training phase and the classification phase (Fig. 7.8).

- The first is to select areas of interest which we know the real Land cover/use of the ground. These zones must be relatively homogeneous, that is to say representative of the classes created and it takes at least 9 pixels per region or area of interest.
- In the second phase, we classify the pixels of the whole image by comparing each pixel with the known domains. The procedure assigns each pixel of the image one of the land cover/use categories. This generalization over the entire image is done using an algorithm.

This method has several advantages. From a conceptual point of view: The classification is hierarchical. Also, this classification is designed to map land cover/use at all possible scales. The objective is to develop a georeferenced land-use and geographic

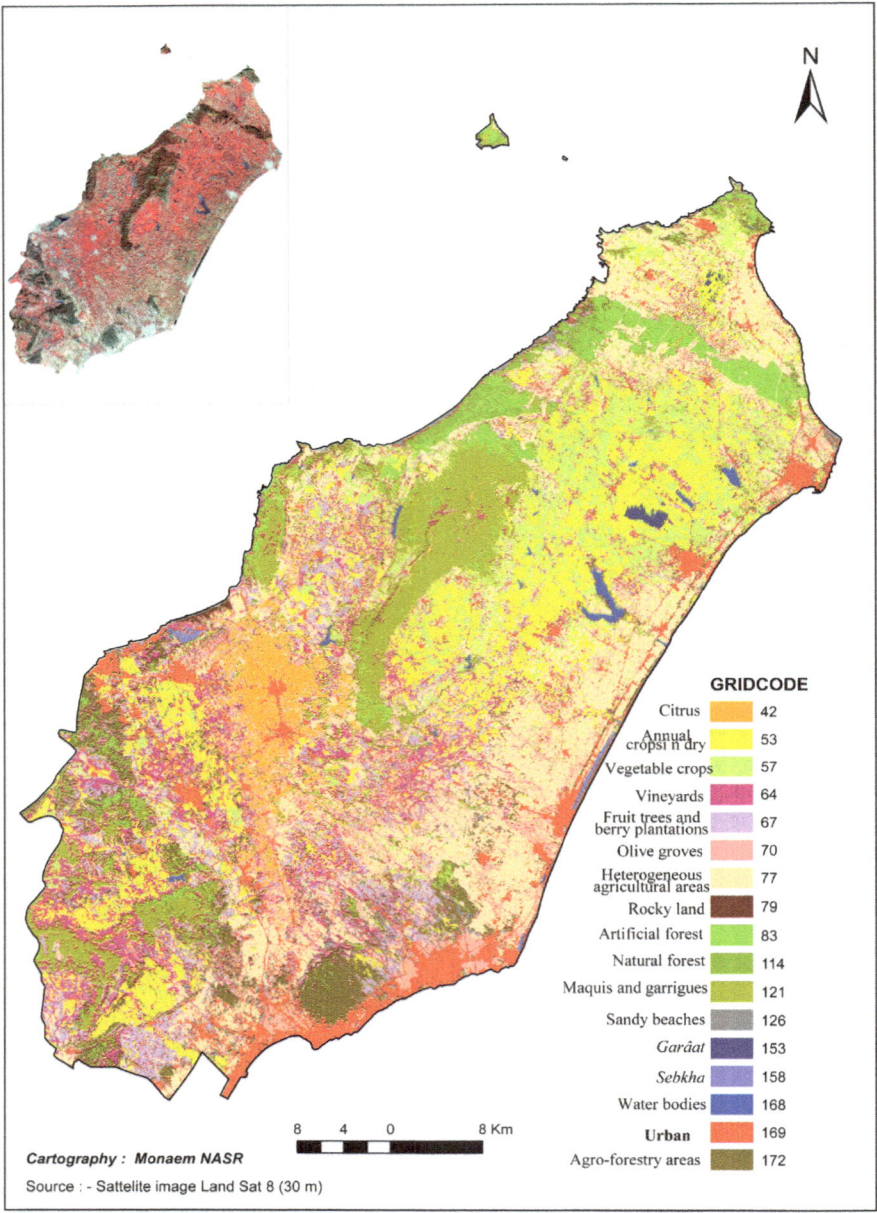

Fig. 7.8 Supervised classification results using Landsat 8 (2014): land use and land cover in the Peninsula of Cap Bon

reference database at a scale of 1:200,000 to 1:25,000 for certain priority and generalized areas, updated, homogeneous and comparable from a thematic and geographical point of view throughout the Cap Bon territory.

This classification can be used as a reference, for two main reasons: the classification contains a large number of classes; the importance is also focused on a set of criteria rather than a name.

From a practical point of view, its specific conception allows it to integrate easily into a GIS and a database. Because the way classes are built facilitates the overlay procedures. From an interpretation point of view, we note great flexibility in relation to available information and time.

This streamlines the field data collection, which allows the interpretation work to be done independently of the field data collection work. In addition to that, this method offers to facilitate, standardizing and interpreting by contributing to its homogeneity.

7.6.3   The Implementation of the Geodatabase

It can be realized in several steps:

***Step 1: Creating the Geodatabase "OCSOL_CAPBON" empty using ArcGIS*:**
The structure of the geodatabase is directly taken from the CDM.

The Geodatabase is built based on the architecture of the Conceptual Data Model previously realized.

– **Creation of hyperclass (Feature dataset)**

First, we must enter a name for the first hyperclass, then set the spatial reference of the hyperclass and import the existing spatial reference (WGS 84 Zone 32 N) from an existing file. The reference system is set automatically on the latter: UTM Cartage Zone 32 N. All classes, which will then be created in this hyperclass, automatically take the above reference system.

We created 5 "Feature Dataset" corresponding to level 1 of the nomenclature for the theme of land cover/use within this geodatabase:

– Urban territories;
– Agricultural areas;
– Forest and semi-natural areas;
– Wetlands;
– Water bodies.

To improve the geodatabase (Fig. 7.9), we have also created other "Feature Dataset" and also "tables" including:

– Territorial decoupage (division);
– Natural environment;

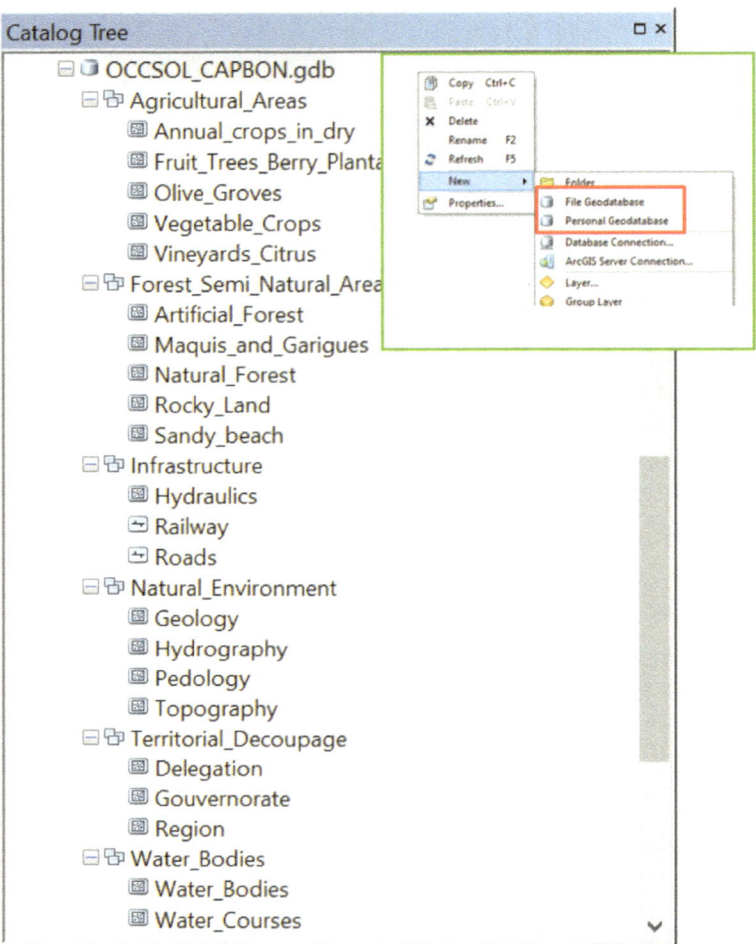

Fig. 7.9 Extract from the geodatabase developed in Arc Catalog

- Infrastructure;
- Economic activities;
- Exogenous effects;
- Operations of territory planning (management);
- Historical process.

Similarly, we have created the "Raster Dataset" in the geodatabase including ASTER images from 2001 and 2009, Landsat images from 1987 and 2014 and topographic maps at 1:25,000 and 1:200,000.

– **Creation of classes (Feature class)**

Each of these "Feature Dataset", previously created, is itself divided into one or more "Feature Class" operational.

As with "hyperclass", we enter the name of the class. Information "Geometry" in Data type defines the type of spatial location of the class. In our case, the objects of class "delegations" for example will be represented by polygons. Then we verify that the class "delegations" is in the right frame of reference as thee created empty fields in the class "delegations" which corresponds to those of the cover "Territorial division". This procedure based on F. Pirot work preserves both spatial information and thematic loading [10].

For each field, it is necessary to indicate the type of data (text, integer …).

Step 2: Loading data into classes:
The only missing thing to constitute a solid database is to make this geodatabase functional and active. In coverage "Territorial division," we took the data source representing polygons (Polygon Feature Class). This information is loaded into the class Delegations.

Step 3: Creating links in Arc Catalog:
The last step is the creation of links between classes. We have to select the classes that will be affected by this relationship, indicating the class of origin and the destination class.

7.7   Multiscalar Mapping of Land Cover and Land Use

The Geodatabase allows us to create the physical model of the geographic information system of the land for the study area whereas geovisualization allows us to display geographic information. It covers all the work on the geographic and thematic information. This includes GIS mapping of different levels of the land as required by the planner or decision maker. The land with its different levels, its implications on the lifestyle of urban and even rural areas, its use of space and the governing relations are needed to define precisely the different class of land cover land use in the Cap Bon.

The multiple land use data categories help us to visually represent the different levels of land use map in order to achieve the finest classes that can be superimposed by creating more or less homogeneous areas, but also to undertake a knowledge and approach of the major social, economic and even historical study area.

Thus, the Table 7.2 contain a list of levels of the land according to the themes selected from the regional to local. It is inspirited from CORINE LandCover nomenclature [15, 16].

Table 7.2 Nomenclature of land cover land use for the Cap Bon

Level 1: Cap Bon	Level 2: Plaine de Grombalia	Level 3: test Sites: Solimane
1-Urban Territories (Artificial Surfaces)	11-Urban fabric	111-Continuous urban fabric
		112-Discontinuous urban fabric
	12-Industrial and commercial areas	121-Industrial areas
		122-Commercial areas
		123-Port areas
	13-Careers	131-Careers
	14-Artificial, non-agricultural vegetated areas	141-Green urban areas
		142-Sport and leisure facilities
2-Agricultural Territories (Agricultural Areas)	21-Arable land	211-Annual crops in dry
		212-Vegetable crops
	22-Permanent crops	221-Vineyards
		222-Citrus
		223-Fruit trees and berry plantations
		224-Olive groves
	23-Heterogeneous agricultural areas	231-Heterogeneous agricultural areas
	14-Agro-forestry areas	241-Agro-forestry areas
3-Forest And Semi Natural Areas	31-Forests	311-Artificial forest
		312-Natural forest
	32-Sclerophyllous vegetation	321-Maquis and garrigues
	33-Open spaces with little or no vegetation	331-Sandy beaches
		332-Rocky land
4-Wetlands	41-Inland wetlands	411-*Garâat*
	42-Maritime wetlands	421-*Sebkha*
5-Water Bodies	51-Inland waters	511-Water courses
		512-Water bodies
	52-Marine waters	521-Sea

7.7.1 Level 1 Mapping of the Land Cover and Land Use: Across the Cap Bon

Graphically, the land territory of the Peninsula of Cap Bon is distributed, according to the level 1 classification, into five major types of land uses as listed in the

Table 7.2. This distribution gives us a broader view and more space (Fig. 7.10 and 7.11 respectively).

- The urban areas represent 6% of the surface of the Peninsula, about 15,300 ha concentrated at the eastern coastal fringe as well as the hinterland being much more rural, with small urban centers.
- The agricultural land traditionally occupies various plains (Grombalia, Haouaria, east coast). Agricultural lands comprise around 188,000 ha are 67% of the Peninsula. They are overrepresented in annual crops in dry and vegetable crops.
- The natural areas occupy 26% of the territory of the Peninsula, more than 74,000 ha. They are dominated by forests, "maquis and garrigues" and are over-represented in the area of Jebel Sidi Abderrahmane, Jebel Korbous, the hills of Nabeul-Hammamet hinterland.
- Wetlands and water bodies in various parts of the Peninsula of Cap Bon comprise around 4500 ha.

7.7.2   Mapping Level 2 of the Land Cover and Land Use: Across the Plain of Grombalia

At a finer scale, the plain of Grombalia (Fig. 7.12), mapping the land use according to the level 2 helps us to identify and represent the subclasses resulting from the breakup of hierarchical classes of Level 1, and we found, for example:

- For urban areas: four subclasses: (urbanized areas, Industrial areas, urban green spaces, careers);
- For agricultural land: four subclasses: (arable lands, permanent crops, Heterogeneous agriculture, Agro-forestry zone);
- For forests and semi-natural: three subclasses: (forests, sclerophyllous vegetation, open spaces);
- For wetlands and water bodies: no breakdown for these two modes in the plain.

These subclasses give us an idea about the distribution of different types of land use in the plain of Grombalia including the over-representation and the dominance of permanent crops (including citrus fruits, vineyards, olive groves and fruit trees …).

7.7.3   Mapping Level 3 of the Land Cover and Land Use: Local Delegation of Soliman

Level "3" gives us an idea about the distribution of different modes of land cover land use on a local scale, that is to say, in a test site. The towns of Soliman (Fig. 7.13) were chosen according to the available documents databases such as topographic

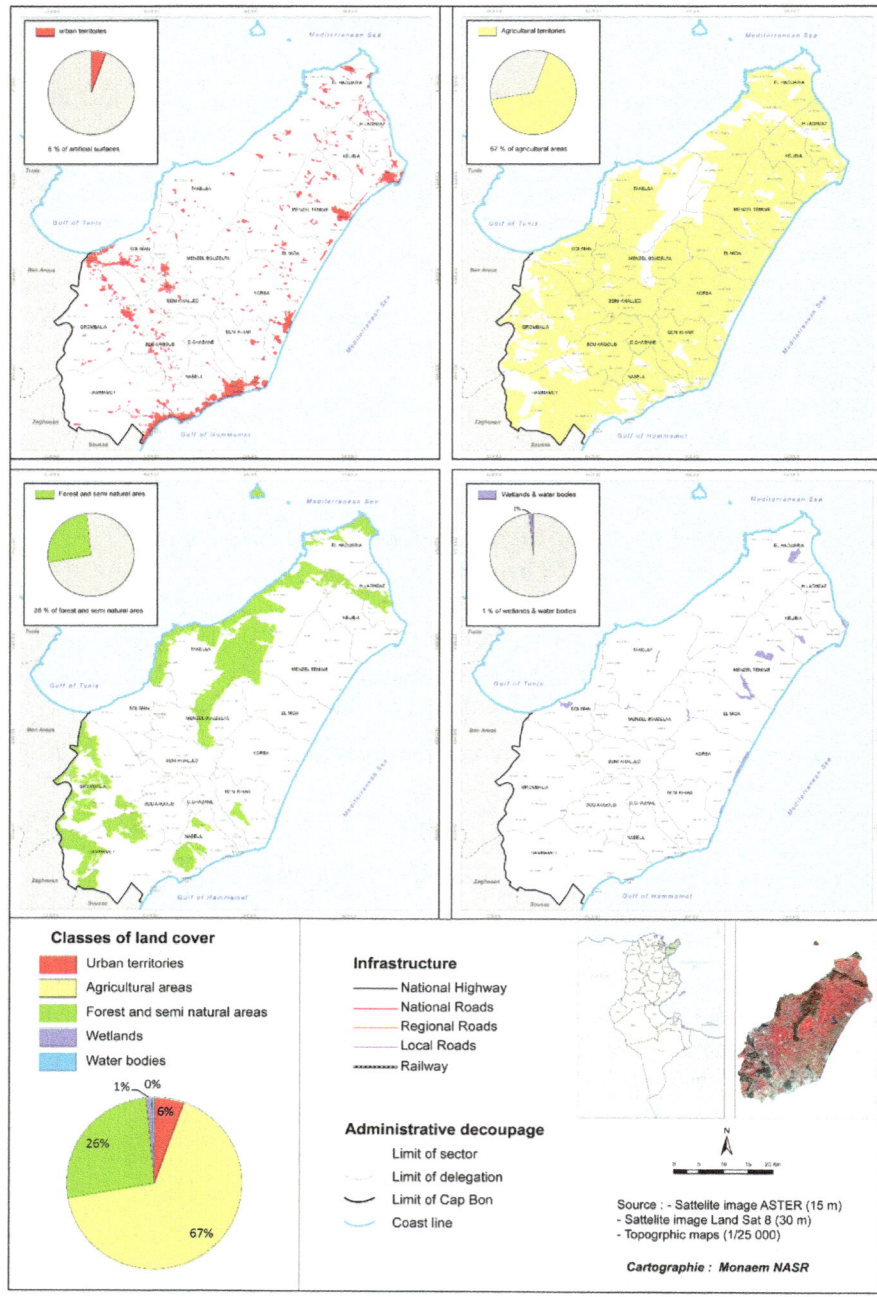

Fig. 7.10 Extraction and visualization of 5 classes of land cover: Cap Bon

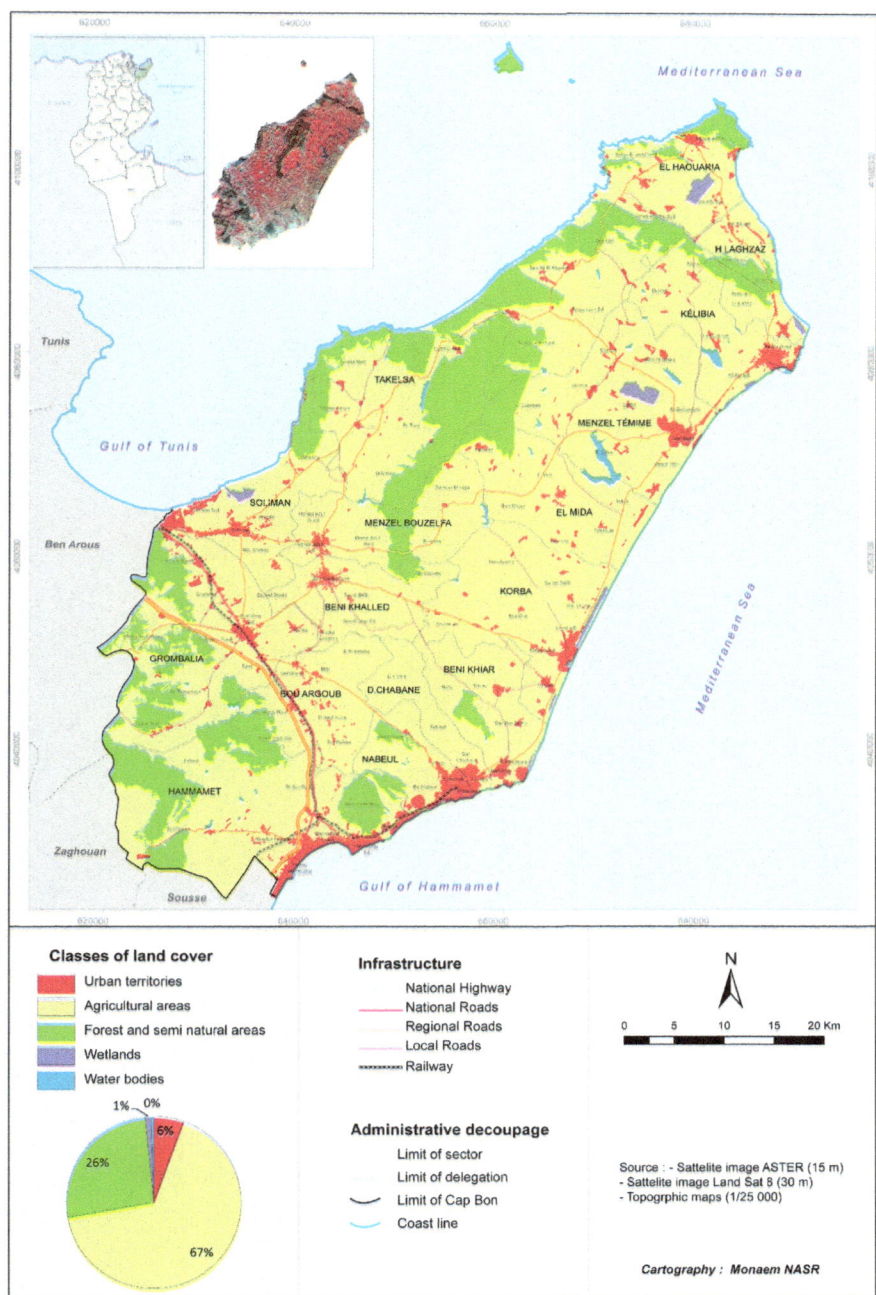

Fig. 7.11 Mapping of the Level 1 of the land cover in 2014: Peninsula of Cap Bon

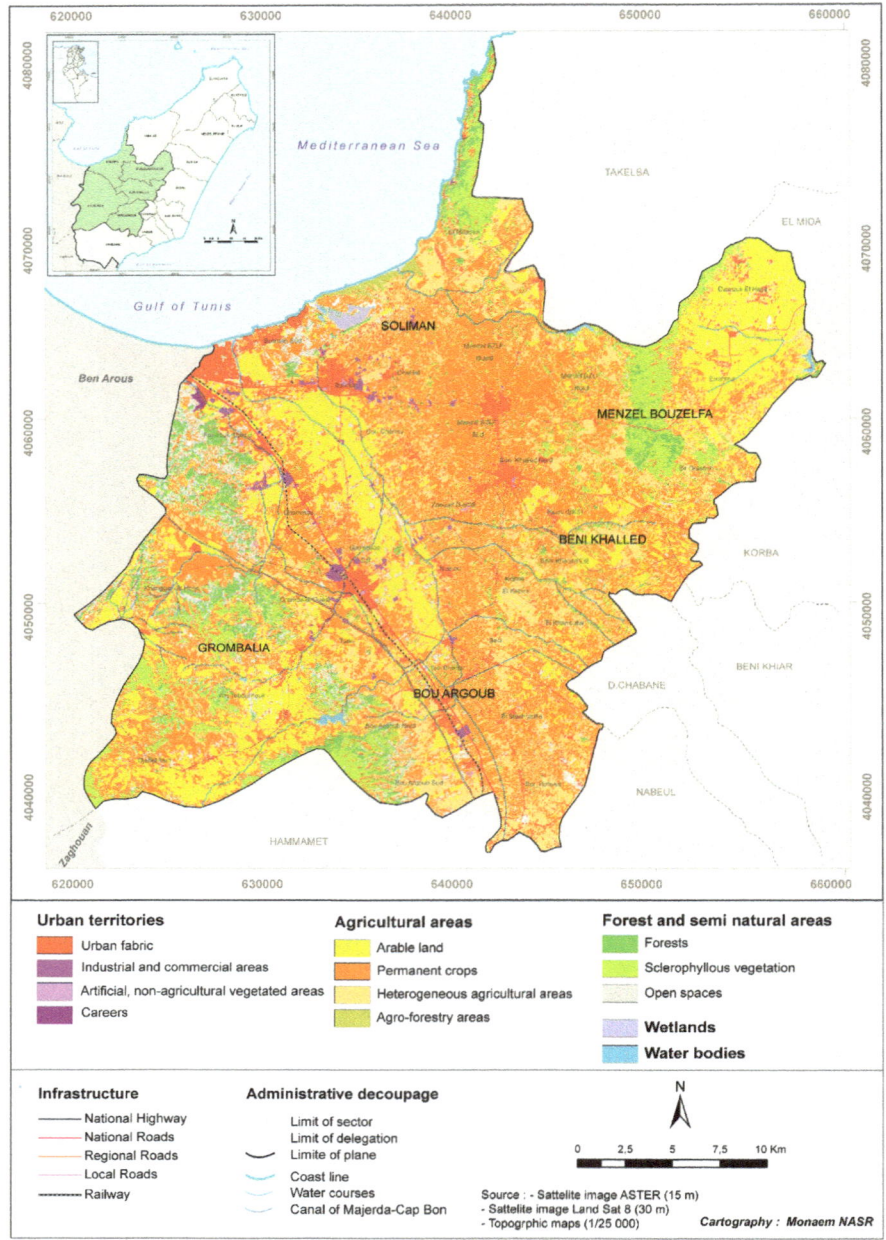

Fig. 7.12 Mapping of the Level 2 land cover/use: plain of Grombalia 2014

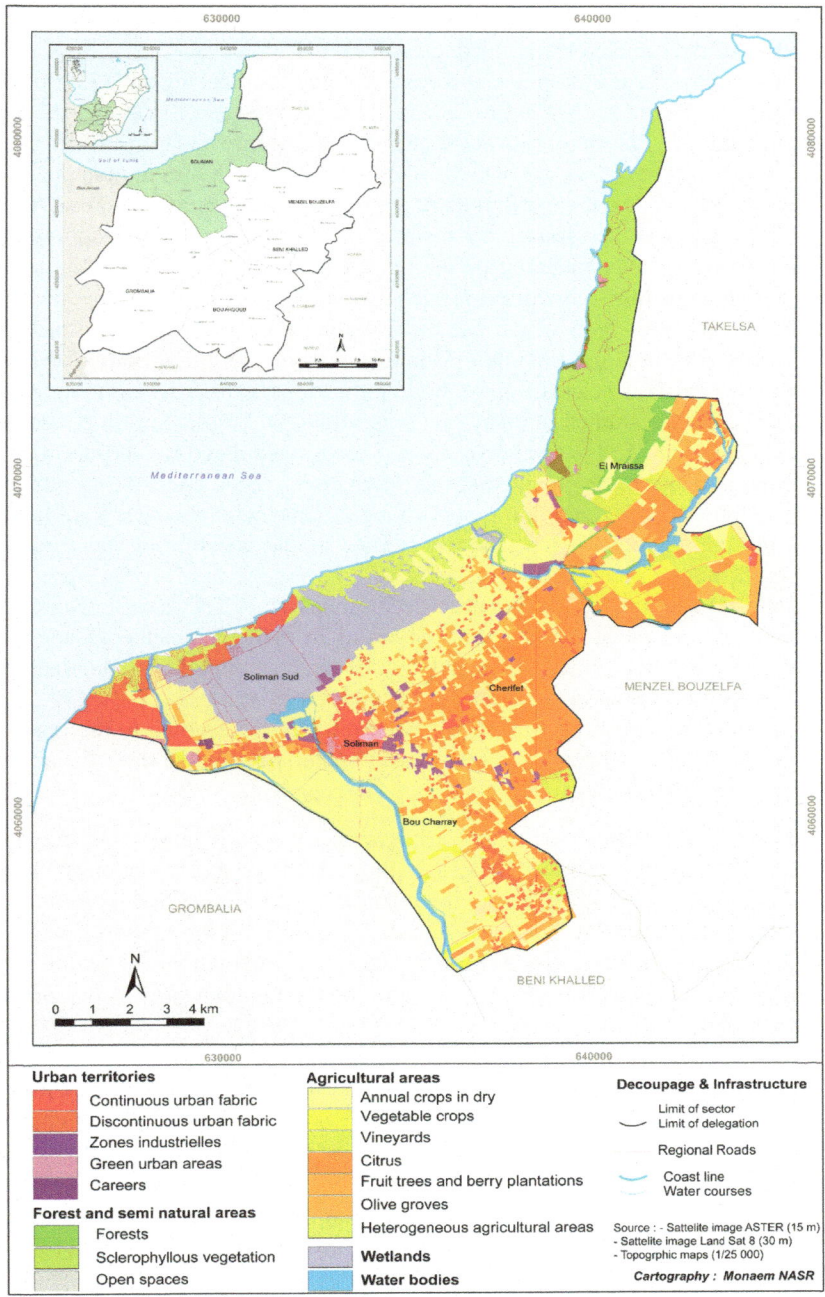

Fig. 7.13 Mapping of the Level 3 of land cover/use: Soliman delegation

maps and plans of urban development by visual image interpretation approach. This is the result of the satisfied hierarchical class level "1" and level "2".

In cartography, the use of new technologies of Remote Sensing and GIS applied to the Cap Bon Peninsula has led to new methods and issues of mapping the land cover and land use. The study shows the accuracy, flexibility and cost-effectiveness of remote sensing, as well as the benefits of combining it with GIS to map the inner structures and dynamics of the different modes of the land cover and land use. It also raised new paradigms and issues, for instance with the multi-scalar mapping that needs different legends depending on the map scale and data accuracy. Numerous works and previous research witnessed such aspects.

However, the land cover which has to be distinguished from the land use, has a major interest in land use planning. Monitoring the evolution of both land cover and land use reveals very useful in several applications, especially in urban areas. This aspect has been studied in many previous studies and researches in geography and other disciplines and several projects and programs related to land-use and its monitoring have been implemented at the international level, constituting a fundamental component of research on global change. Some scholars have gone farther considering that land cover and land use studies should constituting the core of a new discipline.

In Tunisia, little interest has been accorded to such issues. This is sufficient reason that led us to investigate and to discuss it departing from the Peninsula of Cape Bon. This region reveals appropriate to abroad the cartographic methods used in land use because of its great landscapes' diversity and their rapid changes.

But our methodological approach has been refined and adapted to the regional and local needs of this complex region. In particular we used several distinct but complementary approaches to apprehend the land use issue:

– First, we used a multidisciplinary cartographic approach integrating data and aspects of human and physical geography, sociology and economics. These various domains help in analyzing the organization and functioning of the territory, the changes on the land uses and their impacts on the land-use patterns, the new land use classes, and the new management and planning constraints. Of course, these changes revealed by the cartographic methods need new planning methods and tools.

– Second, the authors used a multi-scalar approach, that uses different scales to analyze the land cover and land use of the soil evolving under the joint action of the natural and anthropic factors. These changes are part of spatially differentiated and often nested scales: from the smallest (the plot) to the biggest features (the region as a whole territory and its integration into the globalized market economy) and all the possible intermediate levels whether thematic or territorial: farms, neighborhoods, villages, regions, metropolitan area, state ….etc. The multi-scalar approach involved the exploitation of a variety of tools and data combining direct or survey observation, supervised classification and photo-interpretation.

- Third, the multi-source approach that has been used involved the crossing of several data sources (satellite images, topographic maps, thematic maps, management plans, ArcGIS 10.3 Base Map, GPS data …).
- Fourth, the authors used simultaneously an analytical and a systemic approach in a complementary way. The first seeks to bring back useful details of each classes of land cover of the ground apart. The second considers land use modes in their totality, their complexity and their dynamics, all what constitutes the overall landscape of the region.

The main result of this study based on the combination of such approaches, thoroughly discussed, conducted to the production of a prototype land cover and land use map based on the strict rules of graphic semiology and deduced from that of the project and the "CORINE Land Cover" nomenclature.

One benefit of this multi-theme, multi-source, multi-scale and multi-date approach is that it is likely to take into account all the data exploited by a Geographic Information System such as the ArcGIS 10.3 program. In addition, the prototype produced is not rigid: new themes from exogenous data can be added, new features can be incorporated when necessary, and even new data obtained by combining layers or processing certain elements can be created. We think that this developed prototype will have a real utility on territory planning and development when exploited by local the actors participating in the management and the development of their space.

On the other hand, the main objectives of this study, namely the correct diagnosis of land-use types using a Multi-scalar mapping approach, have been achieved at the level of the final cartographic product developed. Using data from different sources, we have been able to map land use and land use on the Cap Bon Peninsula at different scales. The documents used were processed and analyzed for the realization of this new way of mapping the land use. The multi-spectral characteristics of the pixels of the satellite images matched with the reality of the ground highlighted the characteristics of the surface of the ground. It appears that the method of photo-interpretation and the numerical methods of supervised processing of satellite images make a very effective contribution to the cartography of land cover and land use at different scales in the Peninsula of Cap Bon. Indeed, the use of earth observation satellites for the inventory and cartography of agro-natural and artificial resources finds its effectiveness by the establishment of real Geographic Information Systems.

The remote sensing data used in this work proved to be well suited to the problem addressed. Together with remote sensing, Geographic Information Systems (GIS) as emerging effective tools make available objective territorial-based decisions to spatial planning managers. The conceptual model (CDM) developed on the Cap Bon represents the links and the interlocking that can exist between the different levels of analysis, the different objects and the interactions between these different objects. It can be easily transferred to other geographical areas to check and track the changes in land use.

On the contrary, the description of the steps should not obscure the difficulties encountered during the various manipulations. We may cite in particular:

- The huge data size which requires long and heavy tasks in processing
- The data standardization of the images that need to ensure a total matching when superposing the pixels on images.
- Topological problems revealed when territorial limits and boundaries are changed between two dates.

All these difficulties may be successfully overcome using adequate and appropriate hardware and dedicated software, a full access to the data sources and for GIS and RS users a full training.

7.8 Conclusion

The present study proposes an alternative methodology for mapping land cover land use in Cap Bon by integrating GIS and Remote sensing images. Today, land cover/land use mapping has a great significance in scientific research, in planning and management.

It also shows the territorial reality of the Peninsula of Cap Bon by establishing first a georeferenced database in ArcGIS. However, the production of relevant information is always dependent on the basic data. As in most developing countries such as Tunisia, the data must be taken with a lot of accuracy and care.

The method of formalization HBDS seems to be of little use among GIS users. Nevertheless, in recent years it has evolved among geographers, urbanists, archaeologists and historians (F. Pirot, T. Saint-Gérand, J.C. Ernould, F. Fournet, 2002; E. Gauthier, 2005; T. Tran, 2006; X.L. Saligny and Rodier, 2007; B. Lefebvre, 2008; Labarthe H. and F. Pirot, 2008; H. Labarthe, 2009).

However, a number of difficulties have occurred in the design and management of the GIS dedicated to the Cap Bon region. The biggest challenges are related to:

- The difficulty of identifying certain types of land cover and land use;
- The differently spatial resolution remote sensing imagery;
- The reliability of some data (badly gotten, not updated, etc.);
- The errors in data entry in the database;
- The errors in handling GIS operations;
- The withholding of information by certain structures.

Finally, Remote Sensing and GIS technology have been a new approach to the spatializing of the information to help with the decision, particularly in land cover land use for planning and territory development. This is because a well-informed land use decision will enhance sustainable development. The research established remote sensing and GIS technology as a viable tool for natural resource management in the developing countries.

7.9 Recommendations

Due to recent technological developments in GIS and other geospatial technologies, including remote sensing, the scope of land use mapping has been expanded to include the diffusion and sharing of data via the new platform ArcGIS Online. This really leads to the release of the power of GIS to provide spatial intelligence in various areas including land cover and land use. In this sense, we recommend the implementation of a web application in the form of a real "organization" allowing the various potential actors of the territory of Cap Bon and the different users to access this geographic information through the Web browser. Also, this organization needs to be as easy to edit and manage as it is, at the same time, richer in terms of processing and data volume considering all the components of the initial GIS, that of the Geodatabase. And finally, we can affirm that "GIS Web remains very promising in the coming years in all fields" and this according to GIS specialists.

References

1. Rujoiu-Mare MR, Mihai BA (2016) Mapping land cover using remote sensing data and GIS techniques: a case study of Prahova Subcarpathians. Procedia Environ Sci 32:244–255
2. Vimla S, Alok D (2012) Land use mapping using remote sensing & GIS Techniques in Naina—Gorma Basin, Part of Rewa District, M.P., India. Int J Emerg Technol Adv Eng 2(11). Website: www.ijetae.com (ISSN 2250-2459, November)
3. NASR M (2007) Elaboration of a prototype land cover map for Grand Sfax. Master's thesis, Faculty of Letters, Arts and Humanities of Manouba, 149 p
4. NASR M (2017) Mapping of recently evolution of land cover and land use in Cap Bon. Thesis, Faculty of Letters, Arts sciences humans of Sfax, 435 p
5. Bouillé F (1977) A universal model for database simultaneously shareable, portable and distributed. PhD thesis State of Science (speciality: mathematics, mention: IT). Université Pierre et Marie Curie-Paris VI, 1977, p. 447
6. National Institute of statistics (2014) Population and housing general census (Tunisia 2014)
7. Pirot F, Saint-Gérand Th (2004) From concept to HBDS geodatabase topology: 25 years apart, Francophone ESRI Conference, Issy-les-Moulineaux, 6 and 7 October 2004. http://www.esrifrance.fr/sig2004/communications/pirot/pirot.htm
8. Arab R, Minelli F, Pirot F (2005) From modeling to implementation: a proposed methodology for the census of ponds in the Nord-Pas-de-Calais, Francophone ESRI Conference, Issy-les-Moulineaux, 5 and 6 October 2005. http://www.esrifrance.fr/sig2005/communications2005/pirot1/pirot1.htm
9. Saint-Gérand Th (2005) GIS understand to measure ... or measure to understand? HBDS: for a conceptual approach to real-world geographic modeling. GEOSYSCOM University of Caen FRE CNRS IDEES 2795
10. Pirot F, Saint-Gérand Th (2005) La Geodatabase in ArcGIS, the conceptual basis for software implementation. GIS Expert, February–March, No. 41/42, pp. 62–66
11. Sivakumar R (2010) Image interpretation of remote sensing data, By Geospatial World, in https://www.geospatialworld.net/article/image-interpretation-of-remote-sensing-data/
12. Corbane C, Baghdadi N, Hosford S, Somma J, Chevrel S (2004) Application of an object-oriented classification method for land cover mapping: results on Aster and Landsat etm, French Review of Photogrammetry and Remote Sensing No. 175 (2004-3)

13. Miu M, Zhang X, Dewan MAA Wang J (2017) Aggregation and visualization of spatial data with application to classification of land use and land cover. Geoinfor Geostat 5(4):2
14. Wentz EA, Nelson D, Rahman A, Stefanov WL, Sen Roy S (2008) Expert system classification of urban land use/cover for Delhi, India. Int J Remote Sens 29(15–16):4405–4427
15. INSPIRE Data Specification for the spatial data theme Land Cover and Land Use—Technical Guidelines. European Commission Joint Research Center; 2013. Available for download at http://inspire.ec.europa.eu/
16. Bartholomé E, Belward AS (2005) GLC2000: a new approach to global land cover mapping from Earth observation data. Int J Remote Sens 26(9):1959–1977

Chapter 8
A GIS Based DRASTIC, Pesticide DRASTIC and SI Methods to Assess Groundwater Vulnerability to Pollution: Case Study of Oued Laya (Central Tunisia)

Asma El Amrı, Makran Anane, Lotfi Drıdı, Manel Srasra, and Rajouen Majdoub

Abstract The delimitation of so-called sensitive areas is one of the means adopted to prevent contamination of groundwater. This study aims to assess the vulnerability to pollution of Oued Laya shallow aquifer (Central Tunisia) using three parametric approaches linked to Geographic Information System: Standard DRASTIC, Pesticide DRASTIC and Susceptibility Index (SI) methods. According to Standard DRASTIC, the low vulnerability represents the major part of the aquifer, 97%, and the moderate vulnerability, presents only 3% appearing in two small areas in the North-East and South-East. The SI and Pesticide DRASTIC methods however highlight higher susceptibility to pollution. The former assigned 15% to the moderate vulnerability detected in the North-East and South-East and the later allocated 35.5 and 0.5% to the moderate and high vulnerability, observed along the Eastern part and in the North-East part, respectively. Vulnerability methods comparison and spatial distribution of groundwater nitrate content tend to indicate that Pesticide DRASTIC better reflects the specific aquifer vulnerability than SI. Standard DRASTIC, as intrinsic vulnerability method, reflect worst the specific vulnerability to anthropic activities. For the three methods, the most vulnerable parts of the aquifer corresponds to areas subjected to the agricultural pollution from the irrigated lands and to urban contamination, which comes from the wastewater treatment plant (WWTP) of Kalaa Sghira and the uncontrolled landfill located in the Oued Laya river bank close to Akouda. The groundwater protection of Oued Laya aquifer against these pollutions requires removing the uncontrolled landfill and improving the wastewater treatment performance of Kalaa Sghira WWTP.

A. El Amrı (✉) · L. Drıdı · M. Srasra · R. Majdoub
Higher Institute of Agronomy-Chott Meriem, University of Sousse,
ISA CM BP 47, 4042 Sousse, Tunisia

M. Anane
Water Research and Technology Center, Technopole of Borj-Cédria, Tourist
Route of Soliman, BP 273, 8020 Soliman, Tunisia
e-mail: makram.anane@certe.rnrt.tn

© Springer Nature Switzerland AG 2021
F. Khebour Allouche et al. (eds.), *Environmental Remote Sensing and GIS in Tunisia*, Springer Water, https://doi.org/10.1007/978-3-030-63668-5_8

143

Keywords Drastic · Pesticide drastic · Susceptibility index · Land use · Nitrates · Central Tunisia

8.1 Introduction

Plenty of groundwater resources are subject to degradation consequence of over-exploitation and exposure to punctual and diffuse pollution coming from different sources such as urban and industrial treated wastewater plants, excessive use of chemical fertilizers, etc. [1]. Assessing the contamination risks of aquifers with the anthropologic pollution through mapping the inherent vulnerability improves understanding the reaction of the aquifer against the anthropogenic activities and contributes to the long-term planning protective measures. This would ensure sustainable management of water resources and protect the environment. This vulnerability is assessed through different methods [2] varying from complex models taking into account physical, chemical and biological processes to parametric methods aggregating a number of weighted criteria affecting the aquifers vulnerabilities. According to [3], the used methods are classified on hydrogeological complex and setting method (HCS), the point count system model (PCSM), rating system (RS) analogic relation (AR) and matrix system (MS). The parametric methods (PCSM) are by far the most used since they reflect better the conditions on field without using a large amount of data input [4]. The most known of these methods are Standard DRASTIC [5], Pesticide DRASTIC [6], GOD [7], SI [8], AVI [9], SYNTACS [3], EPIK [10], etc.

The DRASTIC method assesses the vertical intrinsic vulnerability to pollution. It was developed in 1987 by the United States Environmental Protection Agency (USEPA) and the American National Water Well Association (NWWA) to assess the groundwater contamination of the semi-arid areas of USA. DRASTIC is an acronym in which the seven letters correspond to seven hydrogeological, lithological and soil parameters intervening in contamination potential of groundwater [11]. The acronym is: Depth to water table, D; Recharge, R; Aquifer media, A; Soil media, S; Topography, T; Impact of vadose zone, I and hydraulic Conductivity, C. DRASTIC is the most used in North America and is more and more used all over the world [12]. Indeed, since its conception, the method was applied in Africa [13], in Sweden [14], in the USA [15] and in Russia [16]. In Tunisia, this method was applied by [17] for the Nabeul-Hammamet aquifer [18], for the alluvial coastal aquifer of Metline-Ras Jebel-raf-Raf (North-East of Tunisia) [19], for the aquifer of Sfax-Agareb and [20] for Chaffar aquifer (South-East of Tunisia), etc. Pesticide DRASTIC method is a modified version of the Standard DRASTIC method to assess the specific vulnerability, based on the likelihood of a nonadsorbed, nondegrading solute leaching to groundwater [21]. It uses the same parameters and highlights the pesticide behaviour as an organic pollutant by assigning different weights to each factor. The Pesticide DRASTIC index is often used to standardize the evaluation of groundwater pollution potential within various hydrogeological settings for intense agricultural activity

[22]. Susceptibility Index (SI) method assesses also the vulnerability to a specific pollution of a particular pollutant or a group of pollutants. It takes into consideration the properties of the pollutant and its relation with the intrinsic characteristics of the aquifer [23]. It was developed by [8] in Portugal. It includes fives parameters, four are included in DRASTIC (Depth to water table, Recharge, Topography, Aquifer media) and the fifth is the land use. Land use reflects the impact on the anthropogenic activities on groundwater such as agricultural, industrial, urban, tourism activities, etc. [24]. Indeed, many recent studies have shown that land use is a key issue that has to be considered when predicting the effect of anthropogenic activities on groundwater quality and the potential future hydrological responses [25, 26].

The assessment of the aquifer vulnerability is of extreme importance when it is subject to multiple contamination sources. This is the case of the Oued Laya aquifer, located in the Sousse region in the Central East of Tunisia. The groundwater is threatened by urban and agricultural pollution, especially the uncontrolled landfill of Akouda, the wastewater treatment plant (WWTP) of Kalaa Sghira discharging its waters in Oued Laya and the agricultural lands requiring an important amount of chemical fertilizers. This study aims to evaluate the Oued Laya aquifer vulnerability aquifer vulnerability to pollution through three parametric methods Standard DRASTIC, Pesticide DRASTIC and Susceptibility Index (SI), coupled with the Geographical Information System (GIS).

8.2   Assessment of Vulnerability of Shallow Aquifer to Pollution and Methodology

8.2.1   Study Area

The Oued Laya phreatic aquifer covers 177 Km^2. It is located in the Sahel region in the Centre East of Tunisia (Sousse Governorate), between the latitudes 35°44' and 35°51'N and the longitudes 10°23' and 10°35'E. It is spread over four delegations namely, Kalaa Sghira in the South-West, Kalaa Kbira and Akouda in the North and Hammam Sousse in the South-East (Fig. 8.1). The Oued Laya and its effluents cross all the study area and outlets in the Mediterranean Sea.

The climate of the area is Mediterranean semi-arid [27] with a mean annual rainfall of 409 mm and mean temperature of 21 °C. The rainy seasons are winter and autumn, accumulating equally 70% of the total annual rainfall. The Oued Laya phreatic aquifer media is composed of level of Pliocene with alternation of thin and coarse sand and clay. Its thickness is about 30 m. Overexploitation of Oued Laya groundwater mainly for irrigation purpose induced alarming declines in water levels and quality. The available amount of groundwater was 2.64 Mm^3/year in 1995 decreased to 2.6 Mm^3/an in 2000, and water salinity varies between 1.5 and 2.5 g/l [28]. The topography of the study area is gentle made mainly by low hills and plateaux and coastal plain [29]. Oued Laya watershed covers an area of 224 km^2, representing

90% of the study phreatic aquifer, and has a perimeter of 633 km. İts hydrographic network is 5849 m lenght, and the main Oueds are: Oued Sidi Bou Ali and its tributaries, Oued El Ghares and its tributaries, these last tributaries converge towards Ras El Oued which discharged directly into Oued Laya with Oued El Kharroub and its tributaries. There are also Oued El Mderij, Oued Gamgam and Oued Ghdir Ellila and their tributaries discharge their flows in Oued El kbir. The intersection of Oued El kbir with Oued Laya forms Oued El Hammam which discharges into the sea.

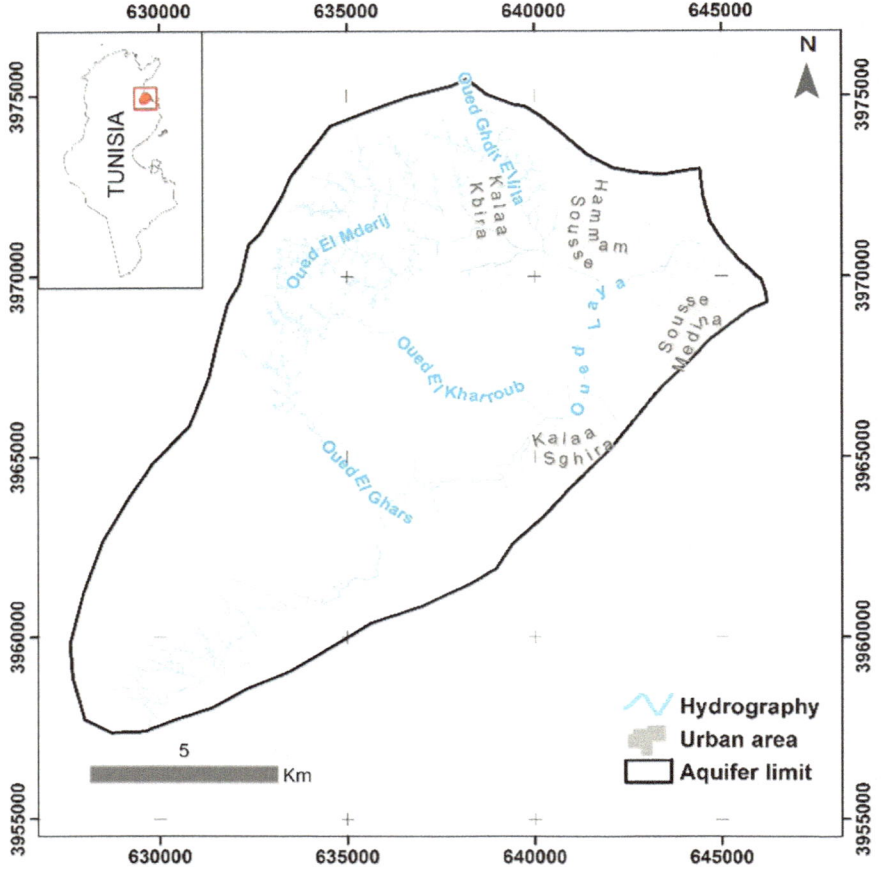

Fig. 8.1 Geographical location of the study area (Oued Laya phreatic aquifer)

8.2.2   *Methodology*

The methodology for Oued Laya aquifer pollution vulnerability analysis using Standard Drastic, Pesticide Drastic and SI models is presented in Fig. 8.2. The vulnerability assessment was carried out following these four steps: (i) the identification and gathering the needed data to derive the geospatial layer corresponding to each parameter, (ii) deriving the geospatial layers using various geospatial operators, (iii) aggregating the layers (seven for Standard Drastic and Pesticide Drastic and five for SI) and get the Standard Drastic, Pesticide and SI vulnerability indexes layers, (iv) classification of each vulnerability index layer into vulnerability categories (low, medium, high and very high). It is mandatory the use of GIS, considered an important tool for environmental studies and geospatial modelling of natural phenomena [30]. ArcGIS 10.2 was used to carry out the methodology.

8.2.2.1   Parameters Identification and Data Gathering

Table 8.1 presents the collected data and the sources from which they are obtained. The vadose zone depth was obtained from 259 wells 67 of which are collected from the CRDA "Regional Commissariat for Agricultural Development" of Sousse and

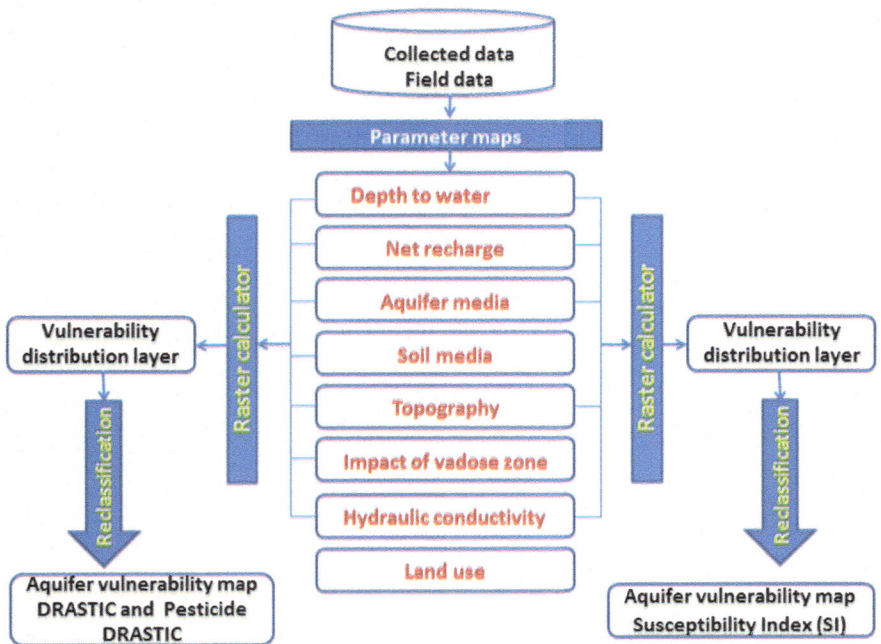

Fig. 8.2 Flowchart of methodology for groundwater pollution vulnerability analysis

Table 8.1 Geospatial data and sources used by the adopted three methods

Parameter	Method		Data	Source
	DRASTIC and Pesticide DRASTIC	SI		
Depth to water table (D)	x	x	Static level of the aquifer – 67 wells – 192 wells	CRDA of Sousse Agricultural map of Sousse
Net Recharge (Rn)	x	x	Annual rainfall and temperature from six stations meteorological stations (2010–2015) Soil depth and texture	INM Agricultural map of Sousse
Aquifer media (A)	x	x	Geological map of Sebkha Kalbia (1/50000) Geological map of Sousse (1/200000) Ten lithological logs for water drilling	ONM ONM CRDA of Sousse
Soil media (S)	x			Agricultural map of Sousse
Topography (T)	x	x	MNT	SRTM-DEM (90 m)
Impact of vadose zone (I)	x		Geological map of Sebkha Kalbia (1/50000) Geological map of Sousse (1/200000) Ten lithological logs for water drilling	ONM ONM CRDA of Sousse
Hydraulic conductivity (C)	x		Lithology of the aquifer	
Land use (LU)		x		Agricultural map of Sousse

CRDA: Regional Commissariat for Agricultural Development; INM: National Institute of Meteorology; ONM: National Office of Mines; SRTMDEM (Shuttle Radar Topography Mission Digital Elevation Model); MNT: Digital ground Model; DEM: Digital Elevation Model

192 from the official database "la Carte Agricole" (Table 8.1). These data were validated through the static level measurement in the field. The net recharge was estimated through two methods: the hydric balance method and the Williams and Kissel method. The hydric balance method applies the Eq. 8.1.

$$Rn = P - Er \qquad (8.1)$$

Where Rn: Net recharge of the aquifer (mm); P: annual rainfall (mm); Er: Actual evapotranspiration (mm). Er was calculated using the Eq. 8.2 of Turc [31].

$$Er = \frac{P}{\left(0.9 + (P^2/L^2)\right)^{1/2}} \qquad (8.2)$$

Where $L = 300 + 25T + 0.05T^3$ and T is the temperature (°C). All the climatic data needed to calculate the net recharge were obtained from the meteorological stations of Sousse Port, Akouda, Hammam Sousse, Kalaa Kbira, Kalaa Sghira and Msaken (Table 8.1). An interpolation operator was used to estimate the climatological data from the stations to the entire extend of the study area.

Williams and Kissel method, developed for the semi-arid regions of USA [32], considers the infiltration capacity of the soil. It classifies the study area in four hydrological groups according to the soils depth and texture. Then it estimates the Rn using the Eqs. 3–6 [33].

Hydrological soil group A:

$$Rn = (P - 10.28)^2 / (P + 15.43) \qquad (8.3)$$

Hydrological soil group B:

$$Rn = (P - 15.05)^2 / (P + 22.57) \qquad (8.4)$$

Hydrological soil group C:

$$Rn = (P - 19.53)^2 / (P + 29.29) \qquad (8.5)$$

Hydrological soil group D:

$$Rn = (P - 22.67)^2 / (P + 34.00) \qquad (8.6)$$

Where P is the annual rainfall in inches.

The geospatial layer of the net recharge using Williams and Kissel method was derived from the meteorological data to get the annual rainfall and from the soil layer of the "carte agricole" database to get the hydrological groups (Table 8.1).

The impact of vadose zone geospatial layer was obtained through the use of the available logs and the geological maps of Sousse and Sebkha El Kalbia after georeferencing and digitizing (Table 8.1). The hydraulic conductivity layer was obtained assigning to each lithological entity the corresponding hydraulic conductivity value [34].

8.2.2.2 Vulnerability Indexes

Each method, Standard Drastic, Pesticide DRASTIC and SI, combines the parameters in a weighted sum to obtain the vulnerability index according to Eqs. 8.7 and 8.8. For Pesticide Drastic, the vulnerability index is estimated by the same formula as that of

Table 8.2 Assigned weights to different parameters considered by the three adopted methods

Parameters	Weights		
	Standard DRASTIC	Pesticide DRASTIC	SI
Depth to water (D)	5	5	0.186
Net recharge (R)	4	4	0.212
Aquifer media (A)	3	3	0.259
Soil media (S)	2	5	–
Topography (T)	1	3	0.121
Vadoze zone (I)	5	4	–
Hydraulic conductivity (C)	3	2	–
Land use (LU)			0.222

the Standard DRASTIC (formula 8.7) by applying a simple change of the weights according to Table 8.2.

$$I_{DRASTIC} = (D_r D_w) + (R_r R_w) + (A_r A_w) + (S_r S_w)$$
$$+ (T_r T_w) + (I_r I_w) + (C_r C_w) \tag{8.7}$$

$$I_{SI} = (D_r D_w) + (R_r R_w) + (A_r A_w) + (T_r T_w) + (LU_r LU_w) \tag{8.8}$$

Where: $I_{DRASTIC}$ is the Drastic vulnerability index and I_{SI} is the SI vulnerability index. D, R, A, S, T, I, C and LU are the parameters rating and w the weight attributed to the parameter.

The rating parameter is a normalized value between 0 and 10 indicating the relative importance of a site compared to the other sites. The weight of each parameter reflects the importance of such parameter in the pollutants fate and transfer from the surface to the groundwater. The weights assigned to the parameters (Table 8.2) vary between 1 and 5 for Standard DRASTIC, between 2 and 5 for Pesticide Drastic and between 0.121 and 0.259 for SI [5, 8]. Higher is the weight more influential is the parameters in the groundwater pollution process.

The index rankings obtained for each method were then discretized in categories (Table 8.3) according to the thresholds given in [11 and 20].

8.2.2.3 Vulnerability Maps Assessment

The vulnerability maps obtained using DRASTIC, Pesticide DRASTIC and SI were evaluated and compared through groundwater nitrates content sampled in fifteen wells for which the main purpose of the extracted water was irrigation. The groundwater samples were taken in January 2017 and the coordinates of each well were

Table 8.3 Vulnerability categories for the three adopted methods

Vulnerability	$I_{DRASTIC}$ and $I_{Pesticide\ DRASTIC}$	I_{SI}
Very low	$I_{Drastic} < 80$	–
Low	$80 \le I_{Drastic} \le 120$	$I_{SI} < 45$
Moderate	$120 < I_{Drastic} \le 160$	$46 < I_{SI} \le 64$
High	$160 < I_{Drastic} \le 185$	$65 < I_{SI} \le 84$
Very high	$I_{Drastic} > 185$	$85 < I_{SI} < 100$

recorded using a handheld GPS. Nitrate content was obtained using Ionic Chromatography apparatus. The location and the data gotten were introduced in an excel sheet and converted to a geospatial dataset using ArcGIS software. The nitrate content data was grouped in three groups; low, medium and high content. The first group involves the contents less than 50 mg/L; the second between 50 and 200 mg/L and the third is higher than 200 mg/L. A symbol map of nitrate content groups was elaborated.

A Spatial join operator was applied using ArcGIS to relate the wells location containing the nitrate content groups with the two specific vulnerability maps obtained using Pesticide DRASTIC and SI methods. The evaluation was done through the determination of the number of high and very high actual nitrate content in the low vulnerable area. Higher is this number worse is the vulnerability map and vice versa.

8.3   Results and Discussion

8.3.1   Thematic Maps Presentation

The map of the depth of water table shows a progressive decrease from the North-West to the North-East of the aquifer towards the sea (Fig. 8.3a). This variation is depicted by seven ranges going from 1 to 10. The ratings between 1 and 3 illustrate a low vulnerability to pollution and represent 83.50% of the total area. However the ratings 7, 9 and 10 reflect a higher vulnerability and represent only 16.50% of the area. The low vulnerability is observed in sites where the vadose zone is thick and can even reach 100 m. The high vulnerability is recorded in areas where the water table comes close to the surface up to 2 m. This variability in the unsaturated zone thickness is in part explained by topography.

The map of the aquifer net recharge estimated according to water balance and Williams and Kissel methods reveals a unique type rating 1 (Fig. 8.3b). This rating proves that for all over the aquifer area, water quantity which may infiltrate into the soil is low and doesn't exceed 50 mm. This low net recharge makes difficult the vertical contaminant transport and reaching the water table. It should be noted that

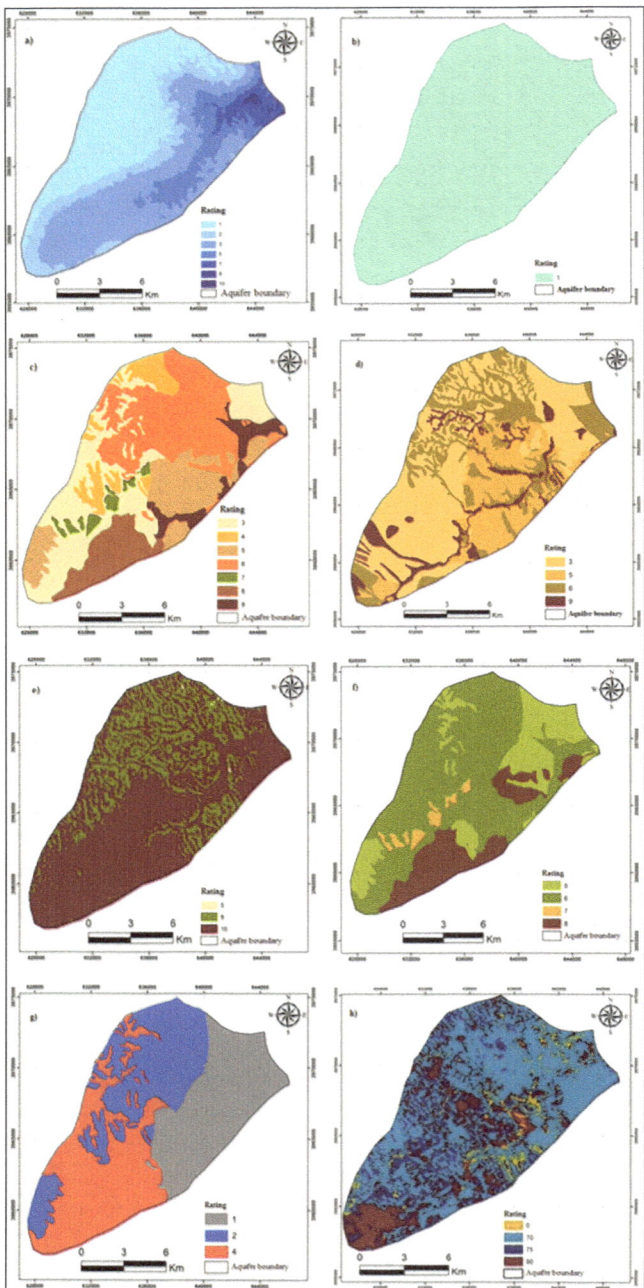

Fig. 8.3 Parameter's ratings according to the three adopted methods: (a) depth to water table, (b) net recharge (c) aquifer media, (d) soil media, (e) topography, (f) impact of vadose zone, (g) hydraulic conductivity of the aquifer, (h) land use

for this study's conditions (climate and soil) using one method or another doesn't affect the net recharge magnitude.

The map of saturated zone lithology displays seven ranges with ratings varying from 3 to 9 which respectively correspond to the moderate and high aquifer media vulnerabilities to pollution (Fig. 8.3c). The lowest ratings 3 and 6 characterized the moderate vulnerability. These occupy 81.60% of the total aquifer area and are assigned to the clay formations characterized by low infiltration rates. The ratings between 7 and 9, however apply to 18.40% allocated to the sandy formations promoting a rapid migration of pollutants and reflecting the high contamination risk of the aquifer.

Regarding soil media, four ranges are obtained for which the ratings are 3, 5, 6 and 9 (Fig. 8.3d). Ratings from 3 to 6 apply to a great part 90.50% of the total study area and indicate a moderate vulnerability. These ratings are mainly observed in the silt and clay soils. The highest rating, 9 apply to only 9.50% and confer a high vulnerability assigned to sandy soils which allows the groundwater pollutants transfer.

Three ranges are displayed by the slope map, corresponding to the ratings 5, 9 and 10 (Fig. 8.3e). The highest ratings 9 and 10 show the lowest slopes varying from 0 to 6% for which the water infiltration rate can highly promote transfer of contaminants to the aquifer. This risk of infiltration decreases towards the rating 5 characterizing steep slopes from 6 to 12%. Ratings 9 and 10 are observed for 99.9% of the total area, reflecting the low susceptibility to pollution. Less than 1% of the study area presents a low vulnerability with a rating of 5.

The map of the vadose zone (Fig. 8.3f) shows a ratings ranging from 5 to 8. The ratings 5 and 6 occupy 73.84% of the total area, conferring a moderate vulnerability. These ratings are assigned to clay texture (bright red sandy clay, brown clay, siliceous limestone blue, red silts, etc.). Ratings 7 and 8 concern 26.15% of the study area and correspond to high vulnerability, since they are assigned to the sandy soils offering a very low attenuation capacity.

Concerning the aquifer hydraulic conductivity (Fig. 8.3g), three ranges are obtained and correspond to the ratings 1; 2 and 4. The hydraulic conductivity is increasing from the North-East side (next to the sea) towards the South-West side. The lowest rating 1 occupies 32.74% of the total area, the rating 2 represents 30.81% and the rating 4 assigned to highest susceptibility to pollution occupies the highest part, 36.45%. In this zone, hydraulic conductivity reaches 17.28 m/day because of the lithological composition of the saturated zone in this part. Indeed it consists on coarse sand and silty sand that promote the pollutant migration.

The map of land use reveals four ranges (Fig. 8.3h). The low vulnerability part with a rating of zero is occupied by forests and the high vulnerability parts with ratings of 70; 75 and 90 correspond to the agricultural zone and urban area. Agricultural zone occupy the major part 88.85% of the study area. It contains 43.18% of olive trees, 25.21% of cereals, 12.36% course and 0.75% of vegetable crops which requires substantial quantities of fertilizer that can pollute the groundwater by infiltration. The urban area, which represents only 11.50% of the area, is also a potential source of pollution from the swage network and the uncontrolled industrial landfill located in the North-East of the aquifer.

8.4 Standard DRASTIC Vulnerability Assessment

The intrinsic vulnerability assessment of Oued Laya aquifer identified two categories of vulnerability: low and moderate, characterized by DRASTIC indexes between 0 and 120; and between 120 and 160, respectively (Fig. 8.4). Low vulnerability represents 97% of the total area of the aquifer and moderate vulnerability occupies

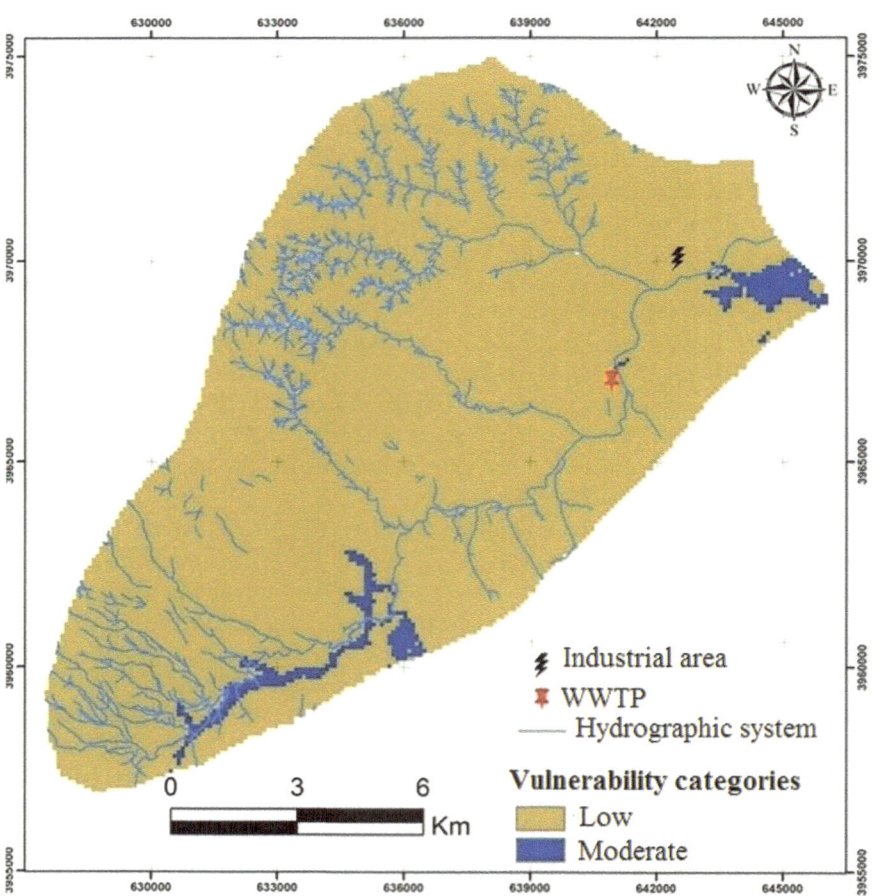

Fig. 8.4 Vulnerability mapping of Oued Laya aquifer (DRASTIC method)

the remaining 3% of the area. The moderate vulnerability is observed in the North-East and South-East parts of the study area. The North-East part covers Sousse city and a part of Kalaa Sghira and Akouda. The South-East part includes Sidi El Hani, Msaken and a part of Kalaa Sghira.

The low vulnerability of the majority of the study area is mainly due to the very high attenuation effect imposed jointly by the large thickness of the vadose zone and the low net recharge. Indeed, when the recharge is more important the extent of contamination increases [12]. For the other parameters such as lithology of the saturated and unsaturated zone, soil media and slope, their effect differs from one zone to another. In some areas these parameters contribute to the attenuation of the contamination and in others they promote the transfer of pollution, but not enough to move the vulnerability to a higher category. The moderate vulnerability located at the Northeastern part is mainly explained by a thin vadose zone (<3 m) and a gentle slope (<2%) facilitating the infiltration of the pollutants to the groundwater and a moderate attenuation capacity of the soil and the saturated and unsaturated zone. A gentle slope reduces runoff and favors the vertical infiltration which increases the risk of the pollutants to reach the aquifer [12, 35]. In regards to the South-East part, the moderate vulnerability is interpreted mainly by the low attenuation capacities of the soil (sandy), saturated and non-saturated zone and the slope associated with moderate vulnerability of the depth to water table (between 15 and 27 m). Indeed, the sandy soils have a high intrinsic permeability which accelerates the vertical migration of pollutant to the groundwater, by the physicochemical process, adsorption, ionic exchange, oxidation, biodegradation [20]. The hydraulic conductivity in the saturated zone in South-East part, estimated at 17.28 m/d increases the transfer of contaminants and water [36].

8.5   Pesticide DRASTIC Vulnerability Assessment

The specific vulnerability assessment according to pesticide DRASTIC reveals three vulnerability categories to pollution: low, moderate and high (Fig. 8.5) for which the index ranges within the intervals 0–120; 120–160 and 160–200. The low vulnerability where the pesticide DRASTIC index is in the range 0–120 is always predominant. It covers 64% of the total aquifer area, extended almost in the entire Western part of the study area. This low susceptibility to pollution in more than the half of the total area reflects the high reduction effect imposed jointly by the large thickness of the vadose zone (between 24 and 100 m) and the low net recharge. The moderate vulnerable areas for which the index ranges between 120 and 160 occupy 35.50% of the total area and extend from the North-East to the South-West parts. This vulnerability is due to a modearte thickness of the vadose zone (between 10 and 32 m), a low net recharge (<38 mm/ year), a low pollution attenuation capacity of the soil and saturated and non-saturated zones and a gentle slope (<2%). Areas of high vulnerability, are confined in the Northeastern part and represent only 0.50% of the total area. They are located in Sousse Medina, Sousse Jawhra and Kalaa Sghira, 356 m from the WWTP. This high vulnerability depicts characteristics such as shallow depth to groundwater (<6 m), flat topography (slope <2%) as well as low attenuation capacity of the saturated, unsaturated zone and soil, natures is coarse sand. Although this high vulnerability class represents less than 1% of the total area, the risk is considered very high since it's located near the industrial landfills, the anarchic discharge of akouda and the discharges from the WWTP of Kalaa Sghira.

8.6   SI Vulnerability Assessment

The specific vulnerability assessment of Oued Laya aquifer (Fig. 8.6) reveals two categories of vulnerability, low and moderate characterized by an SI vulnerability

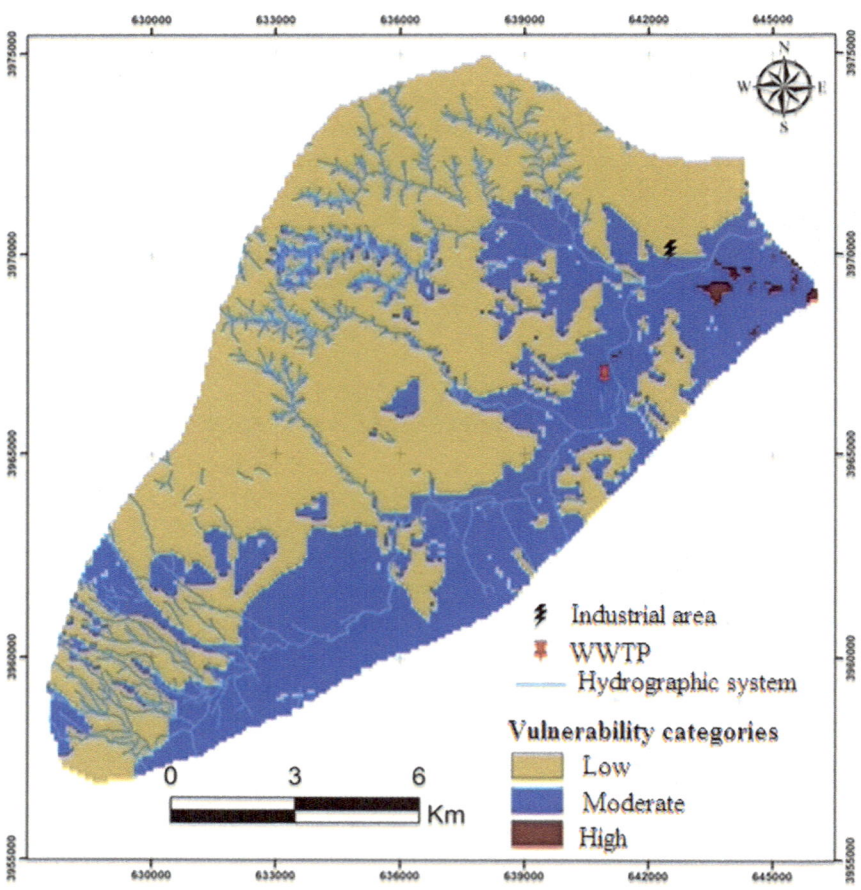

Fig. 8.5 Vulnerability mapping of Oued Laya aquifer (Pesticide DRASTIC method)

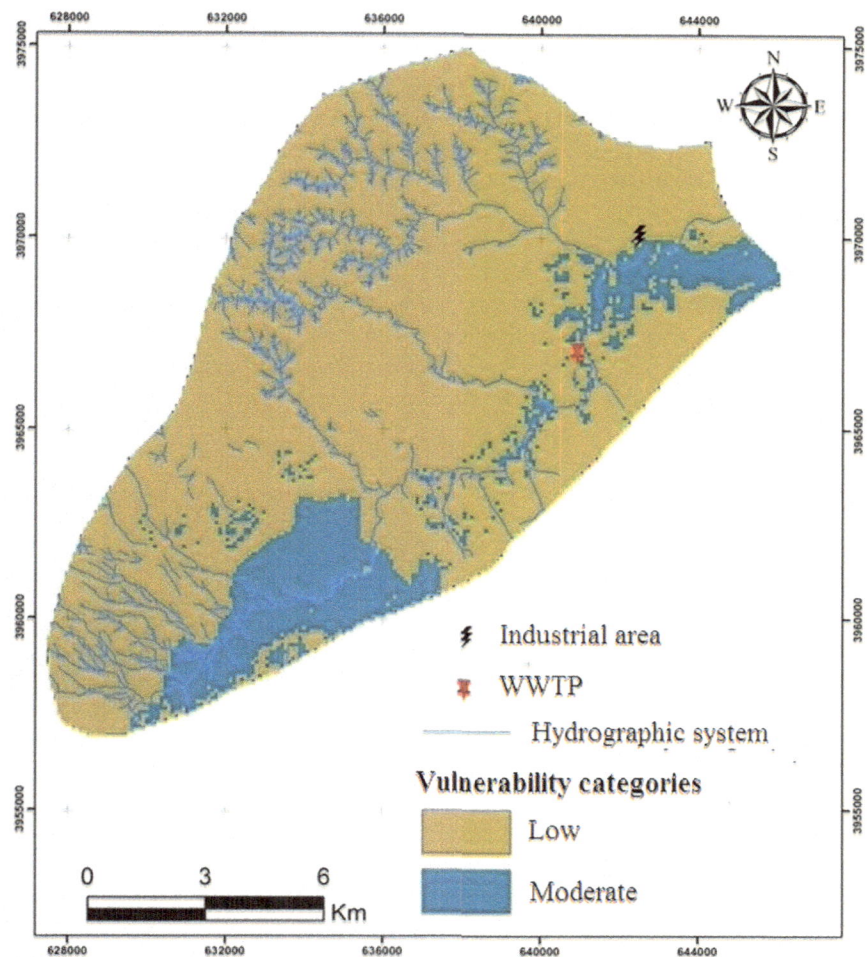

Fig. 8.6 Vulnerability mapping of Oued Laya aquifer (SI method)

index between 0 and 45, and between 46 and 64, respectively. The low vulnerability is still predominant, representing 85% of the total aquifer area. This low susceptibility is essentially explained by the effect of the small recharge and the moderate thickness of the vadose zone (between 21 and 100 m). Areas with moderate vulnerability cover about 15% of the total area divided in three parts: North-East, Central East and South-East. In the Northeastern part (Sousse city, Northeastern part of Kalaa Sghira, 200 m from the WWTP and Akouda, 200 m from the industrial discharge area), the moderate vulnerability is mainly due to a thin vadose zone (<8 m), a flat topography (slope <2%) and a high potential polluting agricultural and urban land use. The cereals and vegetable agricultural area constitute a potential diffuse source of pollution given the excessive fertilizers and pesticides input. In the urban area, the uncontrolled industrial landfill (Akouda region) and the WWTP low quality effluents (Kalaa Sghira region) disposed at 401 m from Oued Laya is a major risk that affects the groundwater quality. Indeed [37], state that water concentrations of Oued Laya at 1.5 km from the WWTP and 3 km from the landfill exceeded the normative NT 106.03. In the Central East part and the South-East at Sidi El Hani, Msaken and a part of Kalaa Sghira, the moderate vulnerability is explained by the moderate thickness of the vadose zone (4 to 15 m in the Central East and 10.4 to 26.9 m in the South-East) coupled with a low slope (<2%) and a low pollution attenuation of the saturated zone and the soil. The presence of pollution sources in an area qualified as risky highlights an alarming situation of Oued Laya aquifer. This situation will deteriorate in the absence of protection measures. Therefore, questioning the place of the uncontrolled landfills and improved effluent treatment performance of WWTP become an absolute necessity.

8.7   Standard DRASTIC, Pesticide DRASTIC and SI Vulnerability Assessment

The resulting vulnerability maps for three methods show a clear difference in the revealed categories and the extent of each category. Standard DRASTIC and SI subdivided the aquifer area into low and moderate vulnerabilities level neverthe-less Pesticide DRASTIC reveals a third higher level. The intrinsic vulnerability Standard DRASTIC characterizes the aquifer mostly low vulnerable. The specific vulnerability SI and Pesticide DRASTIC methods confer higher susceptibility to pollution compared to Standard DRASTIC. Vulnerability category-based compar-ison between Pesticide DRASTIC and SI attests a good agreement, 77% of the total aquifer area (Table 8.4) situated all over the study area, and covering diverse physical and hydrogeological characteristics (Fig. 8.7).

The good concordance between Pesticide DRASTIC and SI is explained by the fact that both models assess the specific vulnerability related to anthropogenic activity and especially agriculture. The agreement in vulnerability categories is observed for a specific site when a high rating of soil media matches a high rating of land use and vice versa. The main reason behind this is the high weight given to soil media in the Pesticide DRASTIC index equation and to land use in the SI index equation,

Table 8.4 Number of medium and high nitrate content located in the low vulnerability areas

	Area	%
−1 (Pestide DRASTIC < SI)	76	0.4
0 (Pestide DRASTIC = SI)	13562	77.4
1 (Pestide DRASTIC > SI)	3888	22.2
2 (Pestide DRASTIC > SI)	6	0.0

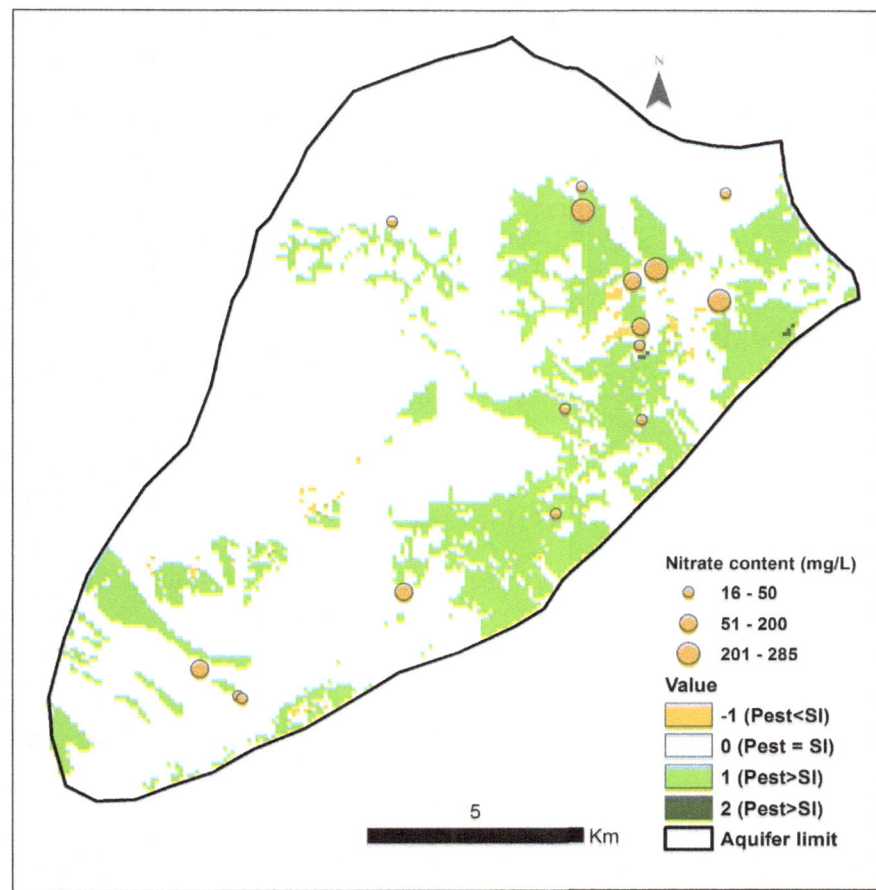

Fig. 8.7 Vulnerability category-based comparison between Pesticide DRASTIC and SI. −1: one vulnerability category lower for Pesticide DRASTIC compared to SI; 0: the same category assigned; 1: one vulnerability category higher for Pesticide DRASTIC compared to SI; 2: two vulnerability categories higher for Pesticide DRASTIC compared to SI

since land use is highly dependent on soil media characteristics. The discrepancy in the other hand is mainly due to the different weights attribution to the four common parameters. Pesticide DRASTIC enhances surface and vadose zone criteria (soil, table water depth, topography) and consequently it gives less importance to the impact of the saturated zone. However, in SI index, aquifer lithology, land occupation and net recharge have high weights meanwhile water depth and topography have low weights.

The sampled groundwater in fifteen wells show that the nitrate content in Oued Laya shallow aquifer varies from 16.9 to 414.6 mg/L. Eight samples have low nitrate content (below 50 mg/L), four samples have a medium concentration lying between

50 and 200 mg/L and three samples have high contents exceeding 200 mg/L. The first interval is located all over the aquifer extend, the second is observed along the Eastern part of the study area and the third is concentrated in the North-East part. The presence of medium and especially high nitrate content reveals that the aquifer is contaminated with nitrate, coming from agricultural fertilizers, Kalaa Sghira's wastewater treatment plant and the uncontrolled landfill located in the Oued Laya river bank close to Akouda. These high nitrate contents highlight the vulnerability of Oued Laya phreatic aquifer to anthropogenic activities and then compromise the quality of standard DRASTIC and SI maps in reflecting friendly the vulnerability of the aquifer. However pesticide DRASTIC seems that reflects better this vulnerability. Indeed, it assigns 8 ha as high vulnerable.

In addition, the number of medium and high nitrate content located in the low vulnerability category obtained from the maps of the three methods (Standard DRASTIC, Pesticide DRASTIC and SI) is higher for DRASTIC (four samples) followed by the SI (three samples) and the Pesticide DRASTIC with two samples. These findings enforce the fact that Pesticide DRASTIC better reflects the Oued Laya specific aquifer vulnerability than SI and standard DRASTIC. Standard DRASTIC, which is an intrinsic vulnerability method, reflect worst the specific vulnerability to anthropic activities. Indeed, despite its popularity, the DRASTIC method has some disadvantages and limitations as reported in previous studies such of [38–41] since it disregards the effect of regional characteristics.

8.8 Conclusion

This study maps the vulnerability to pollution of Oued Laya aquifer and the delimitation of risky areas through three parametric methods; Standard DRASTIC, Pesticide DRASTIC and Susceptibility Index coupled with GIS. The Standard DRASTIC and SI methods delineate two vulnerable categories, low and moderate while the Pesticide DRASTIC model presents low, moderate and high vulnerable categories. Moderate vulnerability occupies 3, 15 and 35.5% for the three methods, respectively. The high vulnerable area occupies only 0.5%. The vulnerability category-based comparison between Pesticide DRASTIC and Susceptibility Index reveal a good agreement, exceeding 77% of the total aquifer area. The repartition of groundwater nitrates contents in each vulnerability zone emphasises the goodness of both methods in assessing vulnerability of Oued Laya aquifer. The most vulnerable areas are concentrated mainly in the Northeastern part. These areas are already exposed to agricultural and urban pollution from the wastewater treatment plant of Kalaa Sghira and the uncontrolled landfill located in the Oued Laya river bank close to Akouda. Thus, protective measures against diffusion contamination are urgently needed. These measures consist on removing the uncontrolled landfill, increasing the performance of Kalaa Sghira's wastewater treatment plant, improving irrigation and agricultural production systems management and practices and best monitoring of ground water resources.

8.9   Recommendations

The findings of this study are a useful tool for groundwater pollution control and prevention. It could be used to assist the landuse planners in addressing the problems that groundwater of Oued laya aquifer might have as a result of the anthropogenic activities. The performance of the used models could be improved by using more samples of groundwater nitrate concentrations. Furthermore, nitrate concentrations and sensitivity analysis will allow modifying the recommended weighting values and propose adapted thresholds in the study region. This will improve the understanding of the aquifer reaction to the anthropogenic activities and predict the vulnerability and the groundwater pollution risk more accurately.

References

1. Babiker IS, Mohamed AA, Hiyama T, Kato K (2005) A GIS based DRASTIC model for assessing aquifer vulnerability in Kakamigahara Heights, Gifu Prefecture, central Japan. Sci Total Environ 345:127–140
2. Drias T, Toubal AC (2015) Mapping of the vulnerability to pollution of the Tebessa-Morsott alluvial aquifer (Oued Ksob watershed) Extreme is Algerian. Larhyss J 22:35–48
3. Civita M (1994) Aquifer vulnerability maps to pollution: Theory and practice Studies on aquifer vulnerability 7. Pitagora, Italy
4. Vrba J, Zaporozec A (1994) Guidebook on mapping groundwater vulnerability in International Contributions to Hydrogeology, 16, I. A. Hydrogeologists 131 p
5. Aller L, Lehr JH, Petty R, Bennett T (1987) DRASTIC: A standardized system to evaluate groundwater pollution using hydrogeologic settings. J Geol Soc India 29(1):23–37
6. Rodney CS (2006) Groundwater vulnerability to agrochemicals: a GIS-based DRASTIC model analysis of Caroll, Chariton, and Saline counties. University of Missouri, Colombia, MO, p 147
7. Foster S (1987) Fundamental concepts in aquifer vulnerability, pollution risk and protection strategy. In: Van Duijvenbooden W, Van Waegeningh HG (eds) Vulnerability of soil and groundwater to pollutants. Committee on Hydrological Research, The Hague, pp 69–86
8. Ribeiro L (2000) SI: a new index of aquifer susceptibility to agricultural pollution, Internal report, ER-SHA/CVRM Lisbon Portugal
9. Stempvoort V, Ewert L, Wassenaar L (1993) Aquifer vulnerability index: GIS compatible method for groundwater vulnerability mapping. Can Water Resour 18:25–37
10. Doerfliger N, Jeannin PY, Zwahlen F (1999) Water vulnerability assessment in karst environments: a new method of defining protection areas using a multi-attribute approach and GIS tools (EPIK method). Env Geol 39(2):165–176
11. Badaoui I (2014) Use of GIS to assess the environmental vulnerability of groundwater aquifers: Case of the Jurassic aquifer (Mougheul, Southeast Algeria). Inter J Environ Glob Clim Change 2(3):109–116
12. Brou D, Lazare KK, Innocent KK, Seraphin KK, Moussa S, Brice K-W, Dago G (2013) Assessment of the vulnerability to pollution of the aquifers of the alterite formations using the DRASTIC and SYNTACS methods: case of the town of M'bahiakro, Center of Côte d'Ivoire. Inter J Innovation Appl Stud 2(4):464–476
13. Sinan M, Maslouhi R, Razack M (2003) Use of GIS to characterize the vulnerability and sensitivity to pollution of groundwater. Application to the Haouz tablecloth in Marrakech, Morocco, Maroc. 2nd FIG Regional Conference Marrakech, Morocco

14. Rosen L (1994) A study of the DRASTIC methodology with emphasis on Swedish conditions. Groundwater 32(2):278–285
15. Navulur KCS, Cooper BS, Engel BA (1995) Groundwater vulnerability evaluation to nitrate and pesticide pollution on a regional scale using GIS. Conférence Internationale d'été ASAE à Chicago, IL, 20 p
16. Pukkoneni E, Teiter K (1995) Compilation of groundwater vulnerability map of Islcmd Saaremaa by GIS means. GIS-Baltic sea states 95, Exhibition, Conference-Tallinn, 1995
17. Anane M, Abidi B, Lachaal F, Limam A, Jellali S (2013) GIS-based DRASTIC, Pesticide DRASTIC and the Susceptibility Index (SI): comparative study for evaluation of pollution potential in the Nabeul-Hammamet shallow aquifer, Tunisia. Hydrogeology J 21(3):715–731
18. Hamza MH, Added A, Francès A, Rodrıguez R, Ajmi M, Abdeljaoued S (2008) Assessment of the vulnerability to potential pollution of the metline-Ras Jebel-raf-Raf alluvial coastal aquifer (North-East Tunisia) according to the parametric Drastic, Sintax and SI methods. Revue de Sciences de l'Eau 21(1):75–86
19. Hentati I (2004) Study and mapping of the environmental vulnerability of the Sfax-Agareb water table. Master thesis, Faculty of Sciences, 102 p
20. Smida H, Abdellaoui C, Zairi M, Ben Dhia H (2010) Mapping of areas vulnerable to agricultural pollution by the DRASTIC method coupled with a Geographic Information System (GIS): the case of the Chaffar water table (south of Sfax, Tunisia). Sécheresse 21(2):131–146
21. Banton O, Villeneuve JP (1989) Evaluation of groundwater vulnerability to pesticides: a comparison between the land-use index and the przm leaching quantities. J Contam Hydrol 4:285–296
22. Croskrey A, Groves CG (2008) Groundwater sensitivity mapping in Kentucky using GIS and digitally vectorized geologic quadrangles. Environ Geol 54:913–20
23. Schnebelen N, Platel JP, Nindre LE, Baudry D (2002) Groundwater management in Aquitaine Year 5. Sectoral operation. Protection of the Oligocene aquifer in the Bordeaux region, Report BRGM/RP-51178-FR
24. Ake GE, Kouassi D, Kouadio BH, Dibi B, Saley MB, Biemi J (2009) Contribution of DRASTIC and GOD intrinsic vulnerability methods to the study of nitrates pollution in the Bonoua region (South-East of Côte d'Ivoire). Eur J Sci Res 31:157–171
25. Kazakis N, Voudouris KS (2015) Groundwater vulnerability and pollution risk assessment of porous aquifers to nitrate: modifying the DRASTIC method using quantitative parameters. J Hydrol 525:13–25
26. Teixeira J, Chamine´ HI, Espinha Marques J, Carvalho JM, Pereira AJSC, Carvalho MR (2014) A comprehensive analysis of groundwater resources using GIS and multicriteria tools (Caldas da Cavaca, Central Portugal): environmental issues. Environ Earth Sci 73 (6):2699–2715
27. CNEA "Centre National des Etudes Agricoles" (2008) Report of the first phase of impact study of the work of CES in the Governorate of Sousse, Tunisia, 120 p
28. DGRE "Direction Générale des Ressources en Eau" (2004) Annuaire de la qualité des eaux souterraines en Tunisie, 385 p
29. Mehdouani M (2003) The hills around Sousse: study and mapping of forms linked to water erosion. DEA Fac SC Hum, University of Tunis, 101 p
30. Goodchild MF (1996) GIS and environmental modeling: progress and research issues. GIS World Books, Fort Collins, CO
31. De Marsily G (1981) Quantitative hydrogeology. Masson, Paris, p 216
32. Engel BA, Navulur KCS, Cooper BS, Hahn L (1996) Estimating groundwater vulnerability to non-point source pollution from nitrates and pesticides on a regional scale. Inter Assoc Hydrol Sci 235:521–526
33. Chow VT (1964) Handbook of applied hydrology. McGraw-Hill, New York
34. Rodriguez R, Reyes R, Rosales J, Berlin J, Mejia JA, Ramos A (2001) Estrucuration de mapas tematicos de indices de vulnerabilidad acuifera de la mancha urbana de salamanca guanajuato. Technical Report, Municipio de salamanco, CEAG, IGF-UNAM, 120 p
35. Ahmed A (2009) Using generic and pesticide DRASTIC GIS based models for vulnerability assessment of the Quaternary aquifer at Sohag, Egypt. Hydrogeol J. http://doi.org//10.1007/s10040-009-0433-3

36. Sener E, Sener S, Davraz A (2009) Assessment of aquifer vulnerability based on GIS and DRASTIC methods: a case study of the Senirkent-Uluborlu Basin (Isparta, Turkey). Hydrogeol J 17(8):2023–2035
37. Dridi L, Majdoub R, Ghorbel F, Ben Hlima M (2014) Characterization of water and sediment quality of Oued Laya (Sousse/Tunisia). J Mater Environ Sci 5(5):1500–1504
38. Almsari MN (2008) Assessment of intrinsic vulnerability to contamination for Gaza coastal aquifer, Palestine. J Environ Manag 88:577–593
39. Bai L, Wang Y, Meng F (2012) Application of DRASTIC and extension theory in the groundwater vulnerability evaluation. Water Environ J 26:381–391
40. Neshat A, Pradhan B, Pirasteh S, Shafri HZM (2013) Estimating groundwater vulnerability to pollution using a modified DRASTIC model in the Kerman agricultural area, Iran Environ Earth Sci. https://doi.org/10.1007/s12665-013-2690-7
41. Pórcel RAD, Schüth C, De León-Gómez HL, Hoppe A, Lehné R (2014) Land-use impact and nitrate analysis to validate DRASTIC vulnerability maps using a GIS platform of Pablillo River Basin, Linares, N.L., Mexico. Inter J Geosci 5:1468–1489

Part IV
RS and GIS for Natural Risks Applications

Chapter 9
Mapping Environmental Risk Degradation Under Climate Stress and Anthropogenic Pressure: Case Study of Abdeladim Watershed, Tunisia

Olfa Riahi

Abstract Nowadays, natural resources are exposed to an increased risk of degradation due to climatic constraints and human pressure. In this context, our research aims to study of the sensitivity of agricultural land in the Abdeladim watershed by mapping the evolution of the hydrographic network (between 1963 and 2016) and assess the susceptibility to soil quality and loss degradation. The objective of this typology was the selection of typical or most representative catchment areas of the Tunisian Dorsale. Among the watersheds studied, the latter is the least watered but the most eroded. It is typical of the aggressiveness of the semi-arid climate of central Tunisia and the sensitivity of the environment to water erosion. To explain this situation, our research was based on a diachronic study of the water system and on a multi-factor study (pedology, slope, land use…) all within the framework of a GIS. Thus, the study of the evolution of the river system (1963–2016) showed a tendency to increase the number of wadis, their length and their width. This development is mainly achieved by the processes and mechanisms of water erosion carried out by concentrated flow (linear, regressive and lateral erosion). This results from the action of two main factors, one is anthropogenic the other is natural. On the other hand, the study of the potential sensitivity of agricultural land in the Abdeladim watershed reveals a growing sensitivity to the degradation of its agricultural potential both by soil loss and by the reduction of quality of the latter. This is the result, on the one hand, of climatic conditions (aridity) and, on the other hand, of anthropogenic use not adapted to the potential of soils. Thus, several measures must be taken to combat the erosion of agricultural land in the Abdeladim watershed. The main aim is to rationalise the human use of soil and water resources and to treat the slopes of waterways by installing anti erosion scheme.

Keywords Climate stress · Sensitivity to degradation · Risk degradation · Mapping environmental · Abdeladim · Tunisia

O. Riahi (✉)
Lr CGMED (LR 99ES02), University of Tunis, 94 Boulevard 9 Avril 1938, 1007 Tunis, Tunisia
e-mail: olfariahi@gmail.com

High Institute of Human Sciences, University of Jendouba, Avenue de UMA, Jendouba North BP. 104, 8189 Jendouba, Tunisia

© Springer Nature Switzerland AG 2021
F. Khebour Allouche et al. (eds.), *Environmental Remote Sensing and GIS in Tunisia*, Springer Water, https://doi.org/10.1007/978-3-030-63668-5_9

9.1 Introduction

Erosion is controlled by two factors. The first is bio-physical. The second is anthropogenic [1]. This is a risk that threatens natural resources and development [2]. Over the years, water erosion has led to a reduction in the area of agricultural land and its fertility [3], as well as in the retention capacity of dams and hill lakes [4]. As a result, the threat also affects socio-economic development in rural areas [5]. Moreover, in the Tunisian arid domain, agricultural land is subject to often intense water erosion [6] due to the bio-physical characteristics of these environments as well as inappropriate human exploitation methods. Thus, torrential rains, low vegetation cover [7], heavy exploitation of natural resources and little conservative farming techniques increase the runoff coefficient and the risk of erosion. In Tunisia, for better exploitation of agricultural land and surface water, hill dams and lakes have been designed and built in this field. However, the soil resource in this area requires for its conservation and optimal use the protection against its potential degradation by minimizing this risk, which, over the years, leads to the reduction of the surface area of agricultural land and its fertility and thus to the reduction of its productivity [7]. In this context, we tried to identify agricultural land (in a watershed with a hillside lake) according to its sensitivity and the type of risk it presents. Sensitive areas are starting points for sediments, and therefore priority areas for the location of anti-erosion developments facilities. Hence the great interest in carrying out this type of detailed studies in cultivated watersheds. In this context, several studies have focused on the analysis and monitoring of water erosion, particularly in arid areas of Tunisia, such as the study of water erosion in the watersheds of the Tunisian Dorsale equipped with lakes and dams hills. Our study of the risk of agricultural land degradation is part of this area of research. It is based on the implementation of a GIS in which several types of multidate data are integrated and exploited. It aims to locate the most threatened agricultural land and identify the risk that threatens it. Finally, our research leads to the proposal of more than recommendations refined to limit the risk of degradation. Indeed, In order to limit the acceleration of the degradation of agricultural land in the Abdeladim watershed, it is necessary to popularize the information with the population, rationalize land use and implement adequate erosion control in risk areas.

9.2 Methodological Framework

9.2.1 GIS: The Study's Relevance to Agricultural Productivity

The objectives of geographic information systems are the preservation, verification and retrieval of all information that can be collected in the field or from maps, satellite images or aerial photographs. They also allow for digital data capture and storage, spatial and temporal data management, comparison and overlapping of different information layers with the ability to update them quickly.

The study of potential sensitivity to degradation consists, after creating a GIS database, in exploiting it by performing several operations of an automatic combination of the different sensitivity factors and classification of the resulting modalities. This methodology allows the conservation, verification and storage of all the information that can be collected in the field or from maps, satellite images or aerial photographs. It also allows the digital capture and storage of data, spatial and temporal data management, comparison and superposition of different layers of information with the possibility of quickly updating them.

9.2.2   The Steps of the Work

Our work consists, first, in analysing the evolution of the action of water erosion in a global way over the entire 53-year period (from 1963 to 2016). Secondly, the study focuses on the pace of change specific to each period (1963/1989, 1989/2000, 2000/2003 and 2003/2016). Thirdly has potential sensitivity to degradation depending on the biophysical and anthropogenic characteristics of the field of Search as well as the trend of evolution of the water system.

In other words, the methodology adopted to carry out this work allows the location of the areas that are the most vulnerable to potential degradation. So, this methodology results in the development of the potential sensitivity map to degradation. This is of obvious interest. It can be used as a reference for the proposal of a development plan based on a good knowledge of the dynamics of agricultural land. In order to achieve these goals, our work has been structured in three main steps:

- data collection (Fig. 9.1) consists of gathering all types of information necessary to monitor the evolution of agricultural land in the Abdeladim watershed from the bibliography, already existing map documents (topographical, geological, pedological…), aerial photographs (1963, 1989 and 2000) field data (2003) and satellite image (2016).
- The creation of the GIS database consists of processing the different types of information collected (homogenization, turnaround and geo-referencing of maps…) and their integration into the GIS (Fig. 9.1).
- and the exploitation of the database (see Fig. 9.1) consists of classifying agricultural land in the Abdeladim watershed according to their vulnerability to. Thus, the work leads to thematic maps that will serve as tools to help decision-making.

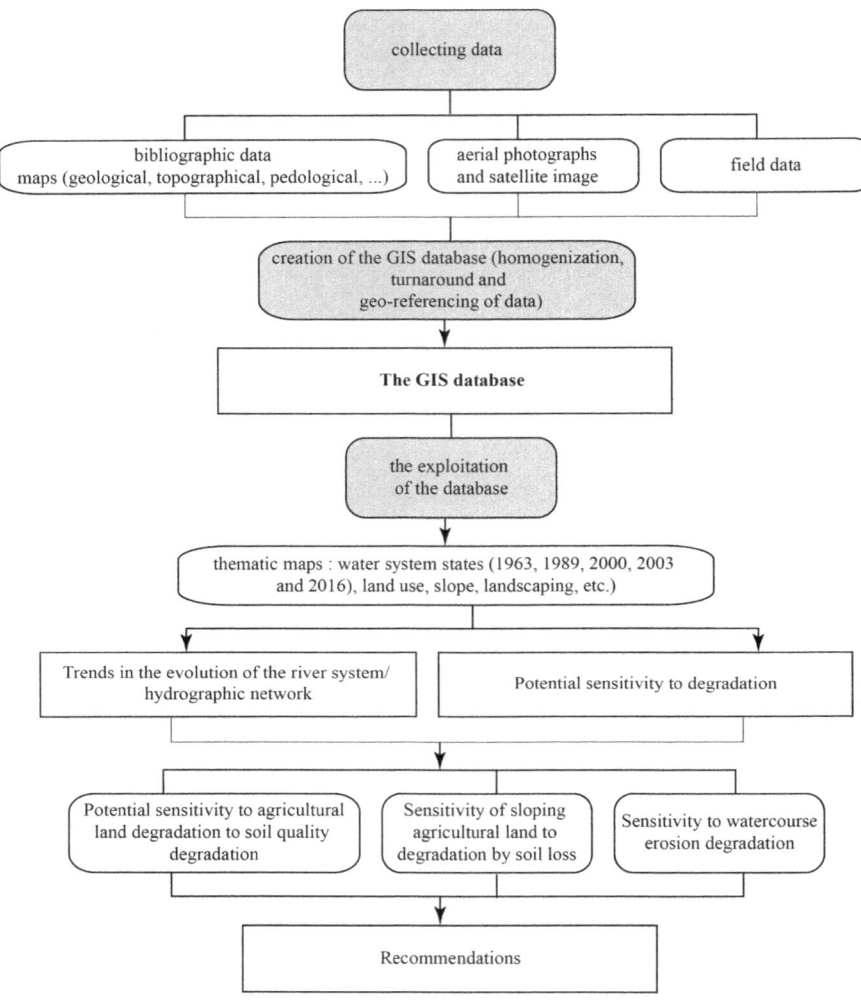

Fig. 9.1 The steps of the study of the Environmental Risk Degradation in Abdeladim watershed

9.3   Conceptual and Methodological Framework

9.3.1   Sensitivity and Potential Sensitivity to Degradation: The Study's Interest in Preserving Agricultural Productivity

9.3.1.1   Degradation and Potential Degradation

Degradation is the disruption of an already established balance in relation to a previous situation. It is, therefore, any form of degradation projected in the future on the assumption that current conditions will not change, in other words, it is the continuation of the current rate of degradation under the same land use and management conditions [7]. Thus, in this research, to understand the phenomenon of potential degradation of agricultural land, we distinguish three types of degradation. The first is the degradation of soil quality the second type is degradation through the loss of sloping agricultural land. The third type is degradation due to the various processes and mechanisms of water erosion implemented by the concentrated flow in wadis draining agricultural land, degradation through the loss of sloping agricultural land and by various processes and mechanisms of water erosion driven by concentrated flow in streams draining agricultural land.

9.3.1.2   Degradation of Soil Quality

Soils function above all as an open system whose fertility can be degraded either by water or by crops [8]. They are subject to more or less irreversible degradation risks [9]. The Food and Agriculture Organization of the United Nations defines land degradation as a process that leads to a decrease in the actual or potential quantitative and/or qualitative productive capacity of the soil [10].

Soil quality degradation is an insidious phenomenon that can be defined as a process that reduces the potential for soil production or the usefulness of natural resources [11 cited by 12], particularly as a result of the vegetation that absorbs the elements necessary for soil development. This is all the more intense in the case of cultivation that is inadequate to the soil's potential. Thus, agriculture is the main cause of soil depletion in chemical terms [8].

So, land use that is not appropriate to the soil's capabilities leads to soil quality degradation and thus to accelerated degradation of agricultural land. Indeed, the decrease in soil fertility leads to a reduction in soil productivity, resulting in the deterioration of fauna and flora [13].

To study the sensitivity of soil quality to potential degradation, we based ourselves on a comparison of the soil characteristics of agricultural land and its suitability for cultivation with its current uses.

9.3.1.3 Degradation of Sloping Agricultural Land Due to Soil Loss

Degradation by loss of sloping agricultural land can even lead to the total disappearance of the soil if the rate of degradation is faster than that of pedogenesis [13 and 14]. The mapping of the sensitivity of the sloping agricultural land consists of classifying the agricultural land of the Abdeladim catchment area according to its vulnerability to water erosion. To segment the watershed of Abdeladim, a set of intrinsic factors was used. Bio-physical factors are slope values, lithology, vegetation density, vegetation cover periodicity. Anthropogenic factors are, first, cultivation techniques and second, natural resource use practices and the state of erosion control development [7].

9.3.1.4 Water Erosion Degradation by Concentrated Flow

Erosion is a complex phenomenon that threatens water and soil resources in the first place, and therefore agricultural productivity [15]. The processes and mechanisms of water erosion by runoff water are triggered as soon as the soil and overlying substrate are no longer able to absorb precipitated water. This occurs when the intensity of the rain is greater than the infiltration or when the rain falls on saturated soil [16]. The degradation of agricultural land under the effect of water erosion is the result of a combination of a set of bio-physical (rainfall, temperature…) and anthropogenic (agrarian techniques, land use patterns, inappropriate development…) factors.

The study of water erosion consists in tracking the evolution of its action over 53 years (between 1963 and 2016). It is based on a comparison of five states in the river system. The first three are mapped with reference to the corresponding aerial photography missions in 1963, 1989 and 2000 respectively. The fourth is based on field measurements and surveys conducted in 2003. The last one is based on the satellite image extracted from Google Earth dated 20 March 2016.

First, it is necessary to analyze the evolution of the action of water erosion in a global way over the entire 53-year period (from 1963 to 2016). Secondly, the Search focuses on the rate of change specific to each period (1963/1989, 1989/2000, 2000/2003 and 2003/2016).

The analysis of the diachronic evolution of water erosion is based on the mapping of the evolution of the total length of the hydrographic network, the length of the incised wadi and the evolution of the surface area of the wadi beds. This analysis results in the classification of wadi sections according to their vulnerability to water erosion.

9.4 Abdeladim Watershed: A Fragile Natural Balanced Environment Undergoing Significant Anthropic Pressure

Located in a mountainous area (Fig. 9.1), the Abdeladim watershed (655 ha) shows a contrast between an intera-mountainous plain with low slope values framed by steep mountain slopes and a system of hills. It is dominated by limestone and lower Senonian marls of marine origin. The uncompacted outcrops (marl and quaternary alluvia) account for 71.5% of the total area of the Abdeladim watershed. In addition, the soils, on the other hand, are rendzina with limestone horizons or brown, black or grey rendzina and soils that have not evolved much. The former are characterized by a low organic matter content on the surface (2%) and zero at depth [17]. They are mainly exploited by cereal farming (wheat and barley), despite the quality and average suitability of these soils for dry cultivation. The second, on the other hand, has a relatively high organic matter content that is always above 5%. On the other hand, the soils that are not very developed are characterized by "The shallow depth and uneven topography of the places they occupy(and which) make them soils that are not very interesting for cultivation, so they are left to the forest domain" [17].

In addition, the Abdeladim watershed is characterized by a semi-arid climate with a cool winter at high altitude and a dry season that lasts five months. Rainfall is characterized by inter-annual variability and aggressiveness. This Torrentiality appears both on a monthly and daily basis. The interannual variability of their aggressiveness and the torrentiality of the rains directly influence the interannual and event-to-event water regime [7].

In addition, the Abdeladim catchment area is formerly occupied by men, as evidenced by the Roman ruins of Hanchir Abdeladim. This former occupation is most likely the cause of a long and intense exploitation of natural resources. Thus, almost all the inhabitants (96% in 2003) use the potential available in the catchment area for both domestic needs and grazing. During the summer season the herds graze the straw in the fields. Grazing in the fields takes place during the two months between harvest and ploughing (late July to late September). To this is added an agrarian exploitation takes place within the framework of a three-year or biennial cultural cycle. However, the frequency of fallow return varies not only according to the crop cycle but also according to farmers' financial means, income from the previous cropping season and the size of the plots. In addition, 30% of the farmers surveyed do not let the soil rest since they only own small plots of land that do not exceed 2.5 Hectare (ha) in total. 48% of them let the soil rest for 1 year out of 3. The remaining 22% let their land rest every other year. However, in the case of poor harvests following the succession of poor agricultural years, farmers are sometimes forced not to fallow in order to meet their needs [7].

Table 9.1 Trend in the evolution of the Abdeladim river system due to different types of water erosion

Date	Linear erosion				Regressive erosion		Lateral erosion
	Number		Length in km		Length in km	Density of drainage in km^{-1}	Area of notches in ha
	Incised	Not incised	Incised	Not incised			
1963	33	57	13.8	15.7	29.5	0.045	1.3
1989	50	77	14.7	22.2	36.9	0.056	3.1
2000	114	56	28.6	17.1	45.7	0.069	8.6
2003	119	56	29.1	17.1	46.2	0.070	9.1
2016	150	75	59.8	14.9	74.7	0.114	20

9.5 Trends in the Evolution of the River System/Hydrographic Network

9.5.1 Linear Erosion

The 53-year follow-up of the evolution of incised and non-incised wadis is an indicator that makes it possible to specify the action of linear erosion. Between 1963 and 1989, the number of people in this category increased (Table 9.1) by 51.5% and its length increased at a rate of about 0.03 km/year. Between 1989 and 2000, the number of incised wadis increased by 128%. Their length has increased at an average speed of about 1.3 km/year. Field measurements (2003) show a development of 4.4% of the incised wadi population at an average incision rate of 0.1 km/year (see Table 9.1), while the length of the incised wadi increased by 35% between 1963 and 1989 and 5.5 km in length at an average development rate of 0.3 km/year [7]. Finally, during the last period (2003–2016), the length of the incised wadis increased by 30.7 km (2.2 km/year) and 79.3% of its workforce.

9.5.2 Regressive Erosion

The length of the hydrographic network (see Fig. 9.2) draining the agricultural lands of the Abdeladim watershed underwent a major change between 1963 and 1989 (see Table 9.1). Indeed, the wadis have annexed 7.4 km in 26 years with an average speed equal to 0.3 km/year. Between 1989 and 2000, it increased by 8.8 km at an average speed of (0.8 km/year). In 2003, its length increased by 0.5 km at an average speed of 0.2 km/year. Finally, in 2016, the hydrographic network experienced the rate of regressive erosion (2 km/year), which allowed it to add 28.5 km to its length compared to 2003 (see Table 9.1).

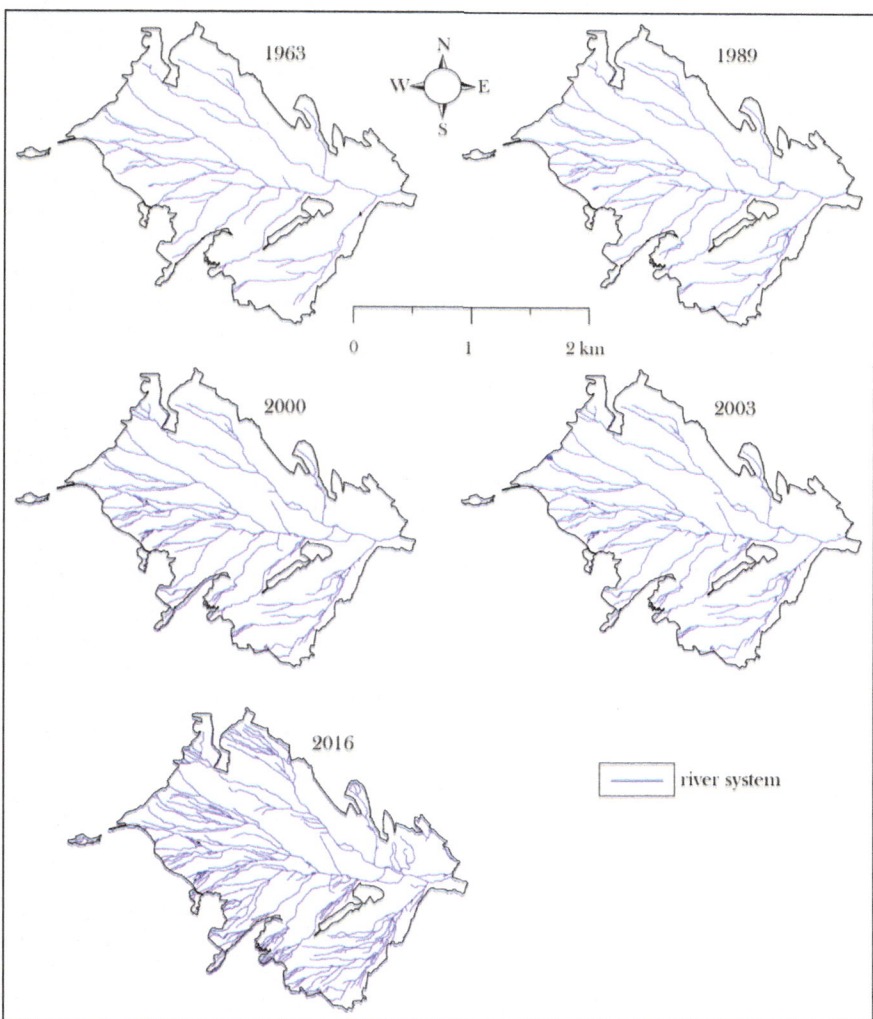

Fig. 9.2 Evolution of the river system of Agricultural land of Abdeladim watershed between 1963 and 2016. *Source* 1963, 1989: aerial photographs (OTC; Tunisia), modified; 2000 aerial photography (Ministry of Agriculture, Tunisia), modified; 2003 personal field surveys; 2016 Google earth modified

In addition, the total drainage density has increased differently from one period to the next. Indeed, it went from 0.084 km^{-1} in 1963, to 0.1 km^{-1} in 1989, 0.130 km^{-1} in 2000, 0.131 in 2003 [7] and 0.212 km^{-1} in 2016. Thus, the evolution of drainage density shows a clear upward trend (see Table 9.1).

9.5.3   Lateral Erosion

With regard to lateral erosion, in 1989, the area of notches increased the average annual area of the river system by 0.7 ha/year; an overall increase of 135.6% compared to the area of notches in 1963. During 1989 and 2000, the area occupied by notches has increased by 5.6 ha in 11 years, with an evolution of 181% at a rate of about 0.5 ha/year. Similarly, between 2000 and 2003, the notches continued to widen but at a slower rate. Indeed, the area of the large notches annexed 0.4 ha with an average speed of 0.1 ha/year. Finally, in 2016, 10.9 ha will be added to the surface area of the notches, giving lateral erosion a speed of 0.8 ha/year. Thus, the 53-year mapping of the development of the hydrographic network draining agricultural land in the Abdeladim watershed revealed a general trend in the evolution of its various characteristics. Indeed, between 1963 and 2016, the total length of the hydrographic network increased by 45.2 km at an average speed of 0.8 km/year. Similarly, the length of the incised wadis a in 2016 is 46 km longer than in 1963, i.e. an average speed equal to 0.9 km/year. In addition, during the same period, the area of large notches showed a similar trend with an increase of 18.7 ha, resulting in an average speed equal to 0.35 ha/year.

9.6   Potential Sensitivity to Degradation

9.6.1   Potential Sensitivity to Agricultural Land Degradation to Soil Quality Degradation

9.6.1.1   Agricultural Land Not Sensitive to Potential Soil Quality Degradation

These lands are slightly larger than those sensitive to potential soil quality degradation. They cover 201.76 ha or 57% of the total area of land under cultivation in the Abdeladim watershed. This sensitivity class mainly occupies the center of the watershed and a narrow corridor in Fidh ash Shraga (Fig. 9.3). It includes two types of agricultural land use. The first is characterized by an appropriate use of the soil's potential. This use exerts fewer constraints compared to other modes of use. This can ensure the sustainability of the resource. These are agricultural lands cultivated as field crops on soils of average quality for annual crops [17]. The second is characterized by inadequate land use and development capacity. These are lands used for arboriculture with annual intercropping crops. Nevertheless, soils that are moderately suitable for annual crops can withstand multiple tillages, which mitigate the impact of inappropriate use on the ability to develop the soil.

9.6.1.2 Agricultural Land Sensitive to Potential Soil Quality Degradation

Agricultural land sensitive to potential soil quality degradation represents 150.45 ha, that is 43% of the total agricultural land area in the Abdeladim watershed. They cover almost the entire lower reaches of the watershed (see Fig. 9.3). These lands are exclusively composed of soils for which the forest is the only alternative [17]. However, they are all cultivated despite their ability. Thus, the degree of sensitivity to the potential degradation of the high soil quality of these agricultural lands is due to human use that is inadequate to the resources and potentials offered by the soil. These are land used for annual crops and land that are used for arboriculture with annual crops in between.

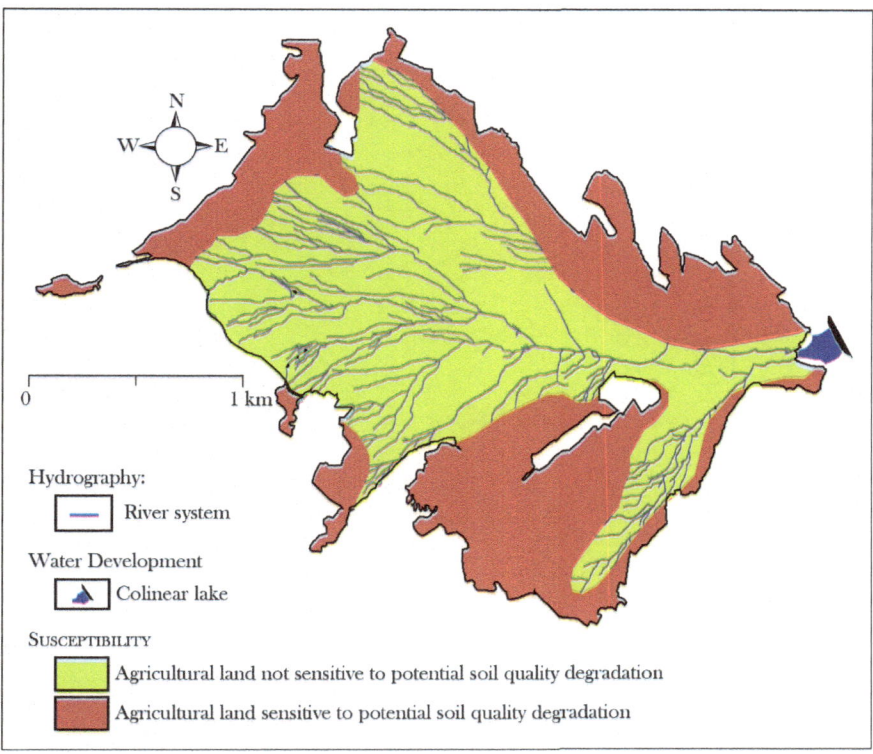

Fig. 9.3 Potential sensitivity to agricultural land degradation to soil quality degradation. *Source* Map of the ability to dry crops of Kasserine, 1970 (Service Soil, Tunisia), modified. Soil Map of the delegation of Kasserine, 1970 (Service Soil, Tunisia), modified

9.6.2 Sensitivity of Sloping Agricultural Land to Degradation by Soil Loss

9.6.2.1 Agricultural Land not Sensitive to Degradation by Soil Loss

The majority of agricultural land classified in this sensitivity category is located on the right bank side of the Abdeladim river (Fig. 9.4). The agricultural land not very sensitive to water erosion is mainly located in Fidh ash Shraga with some agricultural land on the left bank side of the main artery (north of the watershed) in addition to some other agricultural land at the entrance to the basin on the northeast side. They represent only 10% of the total watershed area. It consists of two classes of slopes: the first is less than 3% and the second is between 3 and 6%. On the other hand, agricultural land with slope values of less than 3% is mainly located upstream of

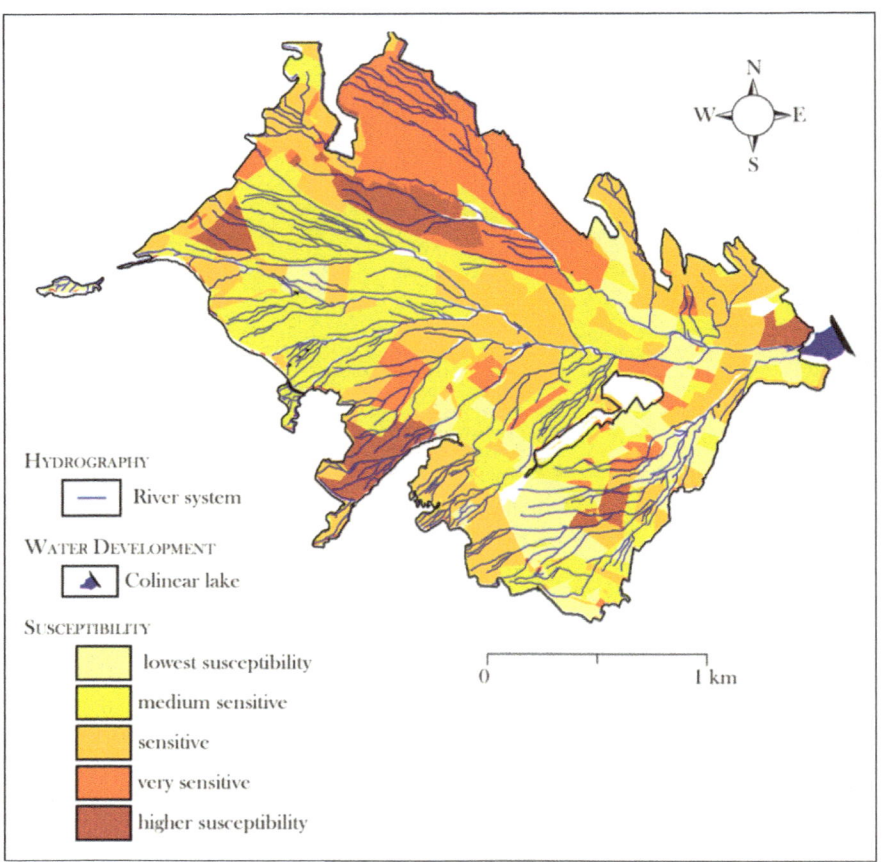

Fig. 9.4 Sensitivity of sloping agricultural land to degradation by soil loss

erosion control structures. They are characterized by a lack of action of erosion processes with a potential tendency to deposition. Their low degree of sensitivity is due to the low slope values that do not allow the flow to organize itself and acquire a threshold velocity necessary to perform an erosion action. As for the agricultural land with slope values of between 3 and 6%, is subdivided into two categories according to lithology: first, that characterized by hard rock outcrops, and second, that with a soft substrate. The first category is characterized by a low density of vegetation cover both on agricultural land with permanent vegetation and on land with seasonal vegetation. The low erosion potential of these agricultural lands is due, in addition to the low value of the slope, to the nature of the outcrop. As for the second category, its low sensitivity is the result of the combination of a high to medium density vegetation cover on the one hand, and the existence of well-maintained erosion control measures that fulfil their protective role on the other.

9.6.2.2 Agricultural Land that Is Moderately Sensitive to Degradation by Soil Loss

The agricultural land that makes up this part (Fig. 9.4) is distributed throughout the watershed. They occupy 24% of the total area of the watershed. They are divided into two classes of slopes: the first is that of slopes below 3%, the second class is that of slopes between 3 and 6%. As regards agricultural land with slope values of less than 3% is characterized by the dominance of the outcrop of soft substrates. They only benefit from a seasonal vegetation cover materialized by annual crops (mainly field crops). The vegetation cover has a low density on hard substrates and a variable density (low to high) on soft rocks. By exploiting these agricultural lands, the occupants of the Abdeladim watershed adopt environmentally protective farming techniques (such as ploughing perpendicular to the slope). Overall, these characteristics are similar to those of the previous class (not very sensitive to erosion).

However, the lack of maintenance of erosion control facilities increases the sensitivity of these agricultural lands to erosion. Damages related to control structures can trigger the action of erosion processes by creating slope failures. This explains the average sensitivity of these agricultural lands. As for the agricultural lands with a slope between 3 and 6% with an outcrop of hard substrate all have a seasonal vegetation cover (annual crops) that is generally of low density. The farming practices carried out on these agricultural lands are protective of the environment (ploughing perpendicular to the slope). However, anti-erosion facilities do not benefit from regular care on agricultural land with dense vegetation cover. Medium-density agricultural land, on the other hand, is not equipped with anti-erosion structures. The average sensitivity of this category of plot is justified by the damage to erosion control facilities on agricultural land with a high coverage rate. In addition, the medium high sensitivity of the rest of the agricultural land is due to low-density seasonal vegetation associated with the absence of erosion control structures. On the other hand, agricultural land with a slope aspect between 3 and 6% located on a soft substrate has a generally seasonal vegetation cover (annual crops). Vegetation density is variable.

The cultivation practices and occupation techniques are broadly protective of the environment with the exception of uprooting, cutting and grazing. Agricultural land with low vegetation density is equipped with anti-erosion facilities in good condition and fulfilling their protective role. This reduces the effect of the combination of sparse and seasonal vegetation with a soft substrate. In addition, the rest of the agricultural land lacks erosion control facilities and structures. This explains the average sensitivity of these agricultural lands despite the high density of vegetation.

9.6.2.3   Agricultural Land Is Highly Sensitive to Degradation by Soil Loss

These agricultural lands form a kind of halo separating the slopes of the jbals from the intra-mountainous plain (Fig. 9.4). A few other agricultural lands are located in the centre of the watershed. They represent 24.25% of the total area of the watershed. They include agricultural land with slope values ranging from 3 to 6% and 6 to 15%. As regards agricultural land with slope values between 3 and 6% is dominated by seasonal vegetation cover on a soft rock outcrop. Vegetation density is low to high. The farming practices carried out in these agricultural lands are protective of the soil (ploughing perpendicular to the slope). Agricultural land with dense vegetation cover is equipped with damaged erosion control facilities, which can trigger water erosion processes. This explains their vulnerability. Agricultural land with a low to high vegetation density is nowhere equipped with anti-erosion facilities. This leaves this category of agricultural land without protection against erosion, which well explains their sensitivity to erosion. As for the agricultural land with slope values of 6 to 15% and a hard substrate has a high to medium seasonal vegetation cover rate. The cultivation techniques carried out on these agricultural lands protect the soil against water erosion (with the exception of herbaceous plant uprooting, logging and grazing). Everywhere, these agricultural lands are marked by the absence of anti-erosion facilities. The high sensitivity of this category of agricultural land is the result of three factors: first, relatively high slope values, second, seasonal vegetation cover (despite sometimes high density), and third, the absence of erosion control measures. This last factor leaves these agricultural lands without any protection from erosion processes. Otherwise, agricultural land with slope values of 6 to 15% and an outcrop of soft rocks is characterized by seasonal vegetation cover (annual crops) and generally low density. Agricultural land in this category is equipped with well-maintained anti-erosion facilities that fulfil their protective role. This reduces the risk of combining relatively high slope values with soft substrates and sparse, seasonal plant cover.

9.6.2.4 Agricultural Land with a Very High Sensitivity to Degradation by Soil Loss

The agricultural lands that make up this part are located near the hillside lake at Fidh ash Shraga, north of the watershed and on the lower slopes (Fig. 9.4). They represent only 9.75% of the surface area of the watershed. They are divided into two classes of slopes: the first is between 3 and 6%, the second class varies between 6 and 15%. As regards agricultural land with slope values between 3 and 6% is all located on an outcrop of soft rock. The vegetation cover is variable and seasonal. Despite the relatively moderate slope values, these agricultural lands are very sensitive to degradation by soil loss. They owe their high sensitivity to human intervention. Indeed, the state of the anti-erosion installations (damages and lack of maintenance) amplifies their vulnerability to erosion. This situation is even more serious in the case of farming practices that damage the natural environment (ploughing in the same direction as the slope). On the other hand, agricultural land with slope values between 6 and 15% is located on a soft substratum. Their vegetation cover is of a seasonal type with a variable density. The farming techniques and the use of natural resources are generally of a protective nature. In addition, these agricultural lands are devoid of erosion control structures. Thus, their very high sensitivity is justified by the combination of relatively high slope values with a soft substrate and the seasonal aspect of the vegetation cover.

9.6.2.5 Agricultural Land Highly Sensitive to Degradation by Soil Loss

They represent 32% of the surface area of the watershed. The agricultural land that forms it covers the slopes of the jbals and Kodiaat in the Nhal. They belong to two classes of slopes: the first is characterized by values from 6 to 15%, the second is characterized by values higher than 15% and which can reach 45%. Most of the agricultural land with slope values between 6 and 15% is located on soft bedrock. The dominant vegetation cover is seasonal (annual crops). Vegetation density is generally low. This situation is aggravated by techniques of using natural resources that upset the balance of the environment (ploughing in the same direction as the slope, uprooting, cutting and grazing). There is a lack of anti-erosion facilities on these agricultural lands. Thus, the type of substrate, the high slope values, the techniques that do not respect the balance of the environment and the absence of erosion control structures are all factors that weaken this category of agricultural land and make it highly sensitive to degradation by soil loss. Unlike this, agricultural land with slope values between 15 and 45% has the highest slope values in the entire watershed. The latter are strictly above 15% and can reach 45%. Some agricultural land develops on a hard substrate and others on an outcrop of soft rocks. On the other hand, the vegetation cover is of a permanent type (Aleppo Pine forest and scrubland) but with a variable density. Human practices on these agricultural lands are all harmful to the natural environment (logging, herbaceous uprooting and grazing). In addition, anti-erosion facilities are totally absent. The high sensitivity of these agricultural lands to degradation by soil loss is mainly due to the slope factor. Indeed, with very high slope values, the risk of environmental degradation by water erosion becomes very high. These are all factors that weaken this category of agricultural land and make it highly susceptible to degradation by soil loss.

9.6.3   Sensitivity to Watercourse Erosion Degradation

9.6.3.1   The Downstream Course of the Abdeladim Watershed: The Least Sensitive Tributaries to Water Erosion

The lower river is characterized by cradle or asymmetrical banks. The banks, generally not very marked in the topography, with the exception of the Fidh ash Shraga wadi, are characterized mainly by rarely steep slopes. As a result, the banks in this area do not have large slope values that can influence the erosive dynamics. The downstream course of the Abdeladim Watershedis distinguished by banks with a cradle or asymmetrical profile. The banks, generally not very marked in topography, with the exception of the wadi Fidh ash Shraga, are characterized mainly by rarely high slopes. As a result, the banks at the level of this zone do not have large slope values that can influence the erosive dynamics (Fig. 9.5). In addition, the maximum depth of the wadis at the level of the downstream river only exceeds 2 m locally. Thus, the height of the waterfall from agricultural land is relatively moderate. In addition, this part of the watershed is marked by the outcropping of a soft substratum (usually alluvium). Human intervention is limited to the installation of a few anti-erosion structures. These are all handmade by the farmers. Despite the presence of a soft substrate, all other physical or anthropogenic factors do not favour the action of erosion. This explains, indeed, the low degree of sensitivity to erosion of the downstream river. Furthermore, the work of regressive erosion, linear erosion and lateral erosion is the most limited compared to the other two parts of the Abdeladim watershed. Lateral erosion mainly reaches the banks of the Fidh ash Shraga wadi. The other two types of erosion occur on the right bank side of the Abdeladim wadi.

9.6.3.2   The Upstream Course of the Abdeladim Watershed: Moderate Arteries Sensitive to Water Erosion

In the upstream section, only the banks of the main artery and the ravine that extends it upstream are of the steep type. The depth of the wadis in this part of the catchment area is relatively high that it can reach 4.5 m in the main artery (Fig. 9.5). In addition, anti-erosion measures in this part of the Abdeladim catchment area are all carried out by farmers using traditional methods. The development of these structures is generally lacking. Thus, the combination of the outcrop of a soft substrate, steep slopes, relatively high head heights and unmaintained erosion control facilities increase the sensitivity of the upstream waterways. In this area under consideration, regressive

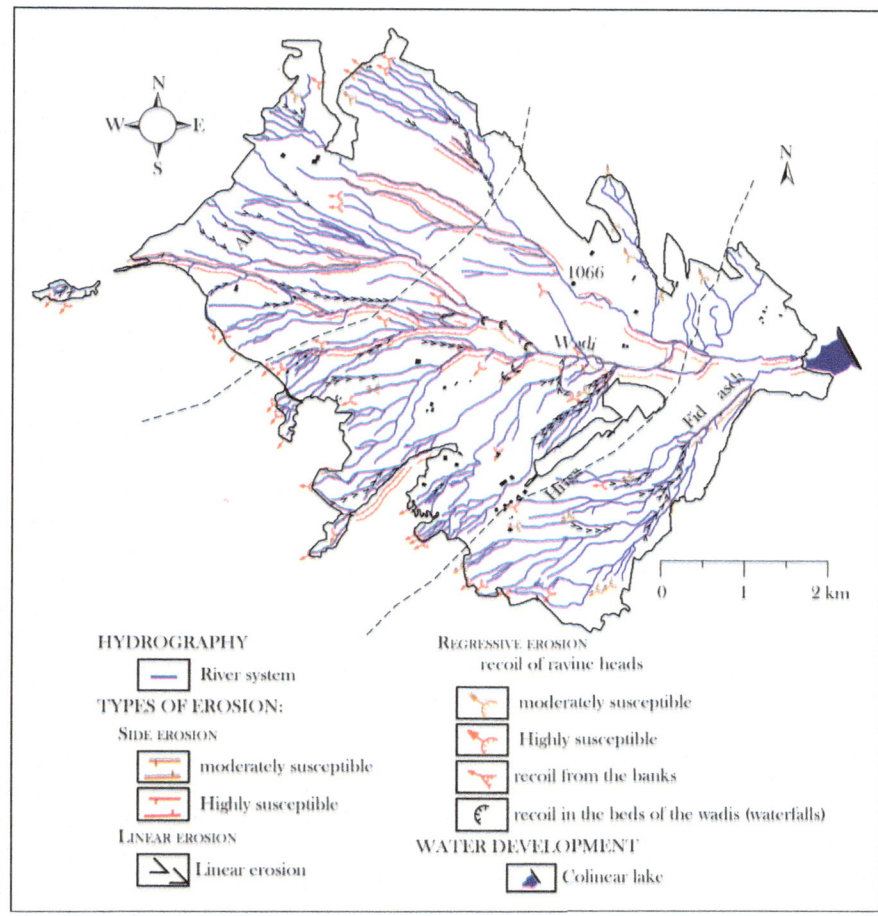

Fig. 9.5 Sensitivity to watercourse erosion degradation of Agricultural land of Abdeladim watershed

and lateral erosion are the two most active types of erosion. On the other hand, the action of the first is mainly carried out by the retreat of the ravine heads. Lateral erosion, on the other hand, is mainly carried out in the main artery and a few other wadis on the left bank side of the Abdeladim wadi.

9.6.3.3   The Average Course of Abdeladim Watershed: The Most Sensitive Arteries

In the middle course, the banks of the Abdeladim wadi and its main tributaries are cut vertically. Slope breaks between the walls of the banks and the surfaces of agricultural land favour the emergence of ravines from the banks as well as the gradual retreat of their heads and their branching into the agricultural land. In addition, the breaks in the slope of the river bed materialized by the waterfalls (which generally appear following the outcropping of the limestone banks), favour the retreat of the heads carried out in the wadis. In the middle course, the banks of the Abdeladim wadi and its main tributaries are cut vertically. Slope breaks between the walls of the banks and the surfaces of agricultural land favour the emergence of ravines from the banks as well as the gradual retreat of their heads and their branching into the agricultural land. Moreover, the breaks in the slope of the river bed materialized by the waterfalls (which generally appear following the outcropping of the limestone banks), favour the retreat of the heads carried out in the wadis (Fig. 9.5). The average course appears to be the least developed area in the watershed. It is equipped with a few artisanal thresholds and a bench built through a tributary. The installation of this bench, which is a structure originally designed to control areal flow, testifies to a poor design of erosion control measures. In addition, In this part of the landscape, several trails are provided to allow passage through the wadis. They channel runoff water and encourage the opening and deepening of new ravines, as is the case with the runway in the centre of the basin. This led to the rapid opening of two ravines on either side of the main river. In this account, the average price is the only part of the watershed where the banks are extracted by the population for use as terracotta. The extraction of marl from the banks leads to the formation of notches. These can contribute to the acceleration of lateral erosion by promoting the action of vertical axis vortex movements of the flow. Thus, the factors in the average Courseincrease the sensitivity of the waterways that drain this part of the watershed. Indeed, the combined actions of lithology, the slope of the banks and the action of man in the environment make the waterways of the middle course the most sensitive to the erosion of the hydrographic network of the Abdeladim watershed. In this area of the watershed, linear and lateral erosion are the two most active types of erosion. Lateral erosion mainly affects the banks of the Abdeladim wadi and its main tributaries. Similarly, linear erosion work is carried out in the beds of the wadis that lead to the right bank. The action of regressive erosion varies according to the type of retreat. Indeed, the retreat of the heads characterizes the ravines located mainly on the right bank side of the main artery. In the middle course of the Abdeladim wadi, regressive erosion concerns the ravines that are created and developed from the banks as well as the waterfalls materialized by a rupture of the slope linked to the limestone outcrop.

9.7 Measures to Combat the Degradation of Agricultural Lands

In order to cope with the high sensitivity of the Abdeladim watershed, it is necessary to implement developments that are appropriate to the conditions of the natural environment while at the same time taking a series of accompanying measures (easing anthropic pressure on natural resources and raising public awareness).

9.7.1 The Extension of Information

The dissemination of information to residents is an important measure before the implementation of any anti-erosion measures. This encourages farmers to maintain it and preserve control structures. In addition, the participation of the inhabitants of the catchment area in the construction, monitoring and maintenance of anti-erosion facilities is necessary to ensure their proper functioning and efficiency. In addition, the assistance, training and advice that CRDA's technical services can provide to farmers in the field of erosion control are incentives for the population to build and maintain facilities adapted to their natural environment.

9.7.2 Control of Potential Soil Quality Degradation

In order to combat soil leaching, agricultural land whose suitability does not allow farming should be protected. This will contribute to the improvement of their fertility and decrease their potential sensitivity to degradation.

9.7.3 Control of Potential Degradation by Loss of Soil

In the Abdeladim watershed, total retention benches are the most common type of erosion control structure on sloping land. They are all made on soft substrates (marl and alluvial). As a result, they can run the risk of slipping due to the plasticity of the marl, especially in rainy years. This can trigger erosion processes (as a result of slipping or breaking of the buckets). Replacing these erosion control structures with partial retention benches reinforced with rows of stones can reduce the risk of water erosion on sloping land. However, the use of this type of structure should not be generalized to the entire watershed. It is better suited to the lithological and orographic conditions of very limited areas such as Fidh ash Shraga, where slope values are less than 3% and alluvial outcrops are present.

Similarly, the installation of dry stone strips is recommended to reduce the risk of degradation by water erosion. This type of anti-erosion development corresponds to a type of dry-stone obstacle placed on sloping land (6–8%) following contour lines in order to slow down the rate of runoff [18].

9.7.4   Control of Potential Degradation by Stormwater

9.7.4.1   Regressive Erosion

It is based on the biological treatment of ravine heads and the implementation of weirs.

In the Abdeladim catchment area, the fixation of the ravine heads can be achieved by planting vegetation adapted to the soil and climatic conditions of the Abdeladim catchment area. Among the species that can be used to fix the heads of ravines are Prosopis and Diss (Bupleurumspinosum). It is noted biological treatment is only proposed for ravine heads located under the Aleppo Pine forest and in rangelands. The other ravine heads are located in agricultural plots. Defending some of these (usually small in size) may yield the opposite results. Indeed, the population, which considers their financial income too low [7] will refuse any reduction. The mechanical control against the retreat of ravine heads which is carried out in the lilies of watercourses and from the banks is based on the realization of weir threshold. The location of these structures is based on two priority classes. The first priority class concerns the areas that have experienced the greatest decline. The second priority class corresponds to areas that have experienced a significant rate of retreat but remains lower than the first one.

9.7.4.2   Control of Linear and Lateral Erosion

The implementation of this type of structure is designed to fight linear erosion in the first place. This filtering structure, which allows the passage of water while capturing sediments, seems to be adapted to the conditions of the wadis of the upstream course and of some arteries in the middle course of the Abdeladim watershed. The implementation of this type of development is a priority in the wadi sections that have experienced the most active erosive dynamics.

9.7.4.3   Control of Regressive Erosion

Gabion sills are, like dry stone sills, designed for fixing the river bed in the first place and for fixing the banks in the second place. These are structures located transversely in the bed of the watercourse. They are built of dry stones strung with iron. This type of threshold can be installed in the wadis of the middle course of the Abdeladim watershed, which are marked by banks carved from a soft outcrop and up to 7 m deep. However, in this part of the landscape, two priorities are established based on the rate of erosion. The first order concerns the wadi sections most affected by regressive erosion. The second corresponds to those affected by this natural risk but with a lower degree than the first.

9.8 Discussions

It is certain that understanding the functioning of erosion processes and mechanisms and knowledge of the factors that guide them can help to combat this phenomenon and to stop or at least minimize the degradation of natural resources and damage to urban infrastructure (roads, habitats,…) and water(dams, hill dams, hill lakes). In addition, in the case of lake and dam watersheds, the majority of water erosion studies are conducted at watershed outlets or on experimental plots. Studies at the outlet level give us an idea of the total contribution of sediments carried by dripping water [4, 19–21]. The second approach, carried out on larger or smaller plots, raises the problem of extrapolating results [22–24]. However, these two approaches cannot locate areas of risk within watersheds. In other words, it is not possible to identify exactly the areas of sediment departure, priority areas for the location of anti-erosive developments, hence the great interest of conducting detailed studies in watersheds. This type of work can identify erosion triggers, estimate the rate of erosion, and finally locate areas at risk.

9.9 Conclusion

Thus, the watershed of Abdeladim is a natural environment under climatic stress and anthropogenic pressure. This makes it an environment marked by a fragile balance. Indeed, its bio-physical characteristics make it a vulnerable environment subject to degradation. The watershed is dominated, on the one hand, by steeply sloping land and, on the other hand, by soft outcrops. These are surmounted by thin and poorly structured soils. This situation is aggravated by a torrential rainfall and hydrological regime. Also, the vegetation cover is threatened by degradation due to several natural and anthropogenic factors. In addition, anthropogenic pressure on the environment is caused by a combination of farming techniques and the use of natural resources, which, on the whole, do not preserve the balance of the natural environment. The synchronic study showed that the shaping of the landscape of the Abdeladim watershed is currently carried out under the influence of the mechanical and chemical processes of rock material preparation and especially by the action of water erosion. Furthermore, the diachronic of the hydrographic network from 1963 to 2016 of the watershed studied revealed a general trend towards the amplification of the action of water erosion through the various mechanisms of linear, regressive and lateral erosion. This general trend should be seen in relation to anthropogenic intervention, materialized by an extension of crops land by clearing. Nevertheless, a detailed study of this general trend reveals a variation in erosion rates over four periods (1963–1989; 1989–2000; 2000–2003 and 2003–2016). Indeed, the last period shows an acceleration of water erosion that is linear, lateral and regressive. The increase in the action of erosion is explained by the increase in precipitation during the fourth period compared to the other two periods. In the same vein, a diagnostic of the potential sensitivity of agricultural land to soil quality degradation shows that 43% of the total area of land under cultivation is sensitive to this risk.

On the other hand, the Abdeladim watershed agricultural land study highlighted the sensitivity of sloping agricultural land to soil loss degradation. Indeed, 66% of its agricultural surface area is occupied by land with high, very high and very high sensitivity. In addition, the wadis of the middle course of the Abdeladim catchment area are the most sensitive to water erosion. It is in this area that the most important manifestations of the action of the mechanisms of linear and lateral erosion are recorded in relation to the upstream and downstream rivers. Thus, the diagnostic of the evolution of water erosion over time and its spatial distribution has highlighted the high sensitivity of the watershed from Abdeladim to water erosion. In order to cope with such a situation, it is necessary to develop the watershed by carrying out a series of works and applying certain measures. The rationalization of human exploitation of the environment can reduce human pressure on its natural environment. The improvement of rangelands involves defending the most degraded parts of the forest and reforesting them.

9.10 Recommendations

- To implement adaptation to the conditions of the natural environment while taking a series of accompanying measures.
- To popularize the information with the population, rationalize land use and implement adequate erosion control in risk areas.

Acknowledgements This research was carried out as part of a research agreement between several institutions for which I thank on behalf of their directors. This is, on the one hand, the Faculty of Human and Social Sciences in Tunis (FSHST); the Laboratory For Geomorphological Mapping Of Environments, Environments and Dynamics (CGMED) and, on the other hand, the Institute for Research for Development (IRD ex-ORSTOM), Laboratory for the Study of Interactions between Sol-Agrosystem-Hydrosystem (UMR-LISAH).

References

1. Mtimet A, Attia R, Derouiche Ch, Pontanier R, Agrbeoui S (1996) Assessment of the physico-hydric parameters of soil sensitivity to water erosion (referential on the erodibility of land in Tunisia). Tunisia, Ministry of Agriculture, Directorate of Soils
2. Dron D, Guérin A-J (2018) Soils: preserve this base of life for the next centuries. In: Responsibility & Environment: Soils at risk: reducing artificialisation. Annals of Mines. N° 91. France
3. http://www.fao.org/publications/highlights-detail/fr/c/1194411/

4. Boughattas NH, SfarFalfoul M, Boussema MR, Snane M-H (2003) Relationship between the efficiency of the hill lakes and the sensitivity to gullying of their watershed en Central Tunisia. Conference Communications Collection: Water Risk Management in Semi-Arid Countries; Tunis 20–22 Mai 2003. Tunisia
5. Valentin C (2018) The stakes and the roles of soils in the living economy 06 Soils at the heart of the critical zone of the Earth. Responsibility & Environment: Soils at risk: reducing artificialisation. Annals of Mines. # 91
6. Boufaroua M, Ghedoui S, Albergel J, (2003) Quantification of erosion on the small Watershed of the hill lakes of the Tunisian Dorsale. Conference Communications Collection: Water risk management in semi-arid countries. Tunis 20–22 Mai 2003. Tunisia
7. Riahi O (2005) Study of erosion and landscape evolution through the development of a geographical information system case of Abdeladim watershed (Kasserine, Tunisia). Master in Geography of Faculty of Humanities and Social Sciences of Tunis. Tunisia
8. Tessier D, Bruand A, Le Bissonnais Y, Dambrine E (1996) Chemical and physical soil quality: Spatial variability and evolution. In: Study and Management of soils. Special number
9. Lemercier B, Berthier L-L, & Walter C (2011) The stakes of soil protection. In: Water and soils. France
10. Gaddas R, le Floc'h JL (1966) Pedological map of the delegation of Kasserine FAO. Pedological Service Printing. Leaf background IGN N° 91
11. Barrow C-J (1991) Land degradation: development and breakdown of terrestrial environments. Cambridge University Press, UK
12. Sylla D (2012) Spatio-temporal modeling of the vulnerability of the environment to the degradation of semi-arid solids from radar data. University from Sherbrooke
13. Riahi O (2017) The dynamics of geomorphological landscapes and recent evolution of an arid natural environment: the region of Hadej-BouHedma (south Tunisia). Ph.D. thesis from the Faculty of Human and Social Sciences in Tunis (FSHST) University I. Tunisia
14. Cointepas J-P (1971) Explanatory note about the legend of Tunisia's pedological map at 1/1000000. In: Soils of Tunisia Bulletin of the Soil Division N° 3
15. Khlifi S, Godart MF, Bahri H, Ben Haha M-N (2003) Quantification of water inputs and siltation rates of TathizalRasRmal and Snedhill lakes, Jendouba. In: Water risk management in semi-arid countries. Tunis, Tunisia
16. Ouvry J-F (2012) Soil degradation by water erosion: what remedies in the field crop. In: Water and agriculture: what challenges today and tomorrow
17. Le Floc'h J (1967) Soil report, focal area of Kasserine—Feriana. The Directorate of Soils. Tunisia
18. Water and Soil Conservation Directorate: "Water and Soil Conservation Guide" 1995
19. Boufaroua M, Ghedoui S, Dbebri A, Ben Youssef M, Albergel J, Guiguen N, Pepin Y, Ben YounesLouati M, Selmi S, Jenhaoui Z, (1999) Hydrological directory of hill lakes 1997/1998. Ministry of Agriculture, Directorate of Water and Soil Conservation and IRD Tunis. Tunisia
20. Boufaroua M, Ghedoui S, Dbebri A, Ben Youssef M, Albergel J, Pepin Y, Ben YounesLouati M Jenhaoui Z (2000) Hydrological directory of hill lakes 1998–1999. Ministry of Agriculture, Directorate of Water and Soil Conservation and IRD Tunis. Tunisia
21. Boufaroua M, Smaoui A, Ghedhoui S, El Batti F, Dbebri A, Albergel J, Guiguen N, Pepin Y, Ben YounesLouati M, Jenhaoui Z, Selmi S, Rahingomanana N (1997) Hydrological directory of hill lakes 1995–1996. Ministry of Agriculture, Directorate of Water and Soil Conservation and IRD Tunis. Tunisia
22. Casenave A, Valentin C (1986) The surface states of the Sahelianzone: influence on infiltration. Ed. ORSTOM
23. Collinet J, Tastouri Jeber IS (2000) Experimental study of runoff and erosion on Siliana farmland (Tunisia). INRGREF, Soil Directorate and IRD Tunis
24. Collinet J, Zante P, Balieu O, Ghesmi M (2001) Mapping the erosive risks in the Zanfour Hill Dam watershed (North Tunisian Dorsal). CRDA of Kef and IRD Tunis, Tunisia

Chapter 10
Application of Remote Sensing and GIS for Risk Assessment in Monastir, Tunisia

Felicitas Bellert, Konstanze Fila, Reinhard Thoms, Michael Hagenlocher, Mostapha Harb, Davide Cotti, Hayet Baccouche, Sonia Ayed, and Matthias Garschagen

Abstract Climate change and urbanization have increased disaster risk in cities and urged the need for effective disaster risk management and risk-informed urban planning. However, up-to-date data that can support risk assessments is often lacking. The ever increasing spatial and temporal resolution of remote sensing sensors offers tremendous opportunities to support risk assessments in cities. In a pilot project for the coastal city of Monastir, Tunisia, multi-temporal optical remote sensing and spatial analysis have been used to support the assessment of current and future exposure, vulnerability, and risk associated with flash floods and coastal erosion. The results were made available in a web-based information system that enables stakeholders to develop response mechanisms and to integrate risk information into urban

F. Bellert (✉) · K. Fila · R. Thoms
IABG, Einsteinstraße 20, 85521 Ottobrunn, Germany
e-mail: bellert@iabg.de

K. Fila
e-mail: fila@iabg.de

R. Thoms
e-mail: thoms@iabg.de

M. Hagenlocher · M. Harb · D. Cotti
Institute for Environment and Human Security, United Nations University, Platz der Vereinten Nationen 1, 53113 Bonn, Germany
e-mail: hagenlocher@ehs.unu.edu

M. Harb
e-mail: Harb@ehs.unu.edu

D. Cotti
e-mail: Cotti@ehs.unu.edu

H. Baccouche
16, El Amal City, 5000 Monastir, Tunisia

S. Ayed
Municipalité de Monastir, Avenue Habib Bourguiba, 5019 Monastir, Tunisia

M. Garschagen
Department of Geography, LMU Munich, Luisenstraße 37, 80333 Munich, Germany
e-mail: m.garschagen@lmu.de

© Springer Nature Switzerland AG 2021   191
F. Khebour Allouche et al. (eds.), *Environmental Remote Sensing and GIS in Tunisia*, Springer Water, https://doi.org/10.1007/978-3-030-63668-5_10

planning in order to meet the challenges associated with urban disaster risk. The chapter focusses on the role of remote sensing and GIS for urban risk assessments, drawing on lessons from Monastir, and discusses the potential transferability to other urban settings.

Keywords Coastal erosion · Flash flood · Disaster risk · Urban growth · Future scenarios · Exposure · Vulnerability · Risk · GIS · Modelling · Water depth · Velocity

10.1 Introduction

Today, the majority of the global population lives in cities, which are hubs of the global economy [1]. Estimates of the United Nations project that the trend towards urbanization is likely to continue in the future—with high urbanization rates in Africa and Asia, notably in small and mid-sized cities [2]. While cities provide access to jobs, relevant infrastructure and services, and hence continue to attract people, urbanization and climate change are exacerbating disaster risk in cities [3, 4, 5, 6]. Urban risk is expected to rise in the future as urban areas continue to expand into hazard-prone areas, while climate-related hazards such as floods, storms, heatwaves or sea level rise increase in frequency and intensity in many parts of the world [5, 7].

Tunisia faces major risks associated with natural hazards and climate change. Rising temperatures and changing precipitation patterns combined with the increasing frequency and intensity of floods and droughts are threatening the agricultural sector, human health and the local economy [8]. As a result of rapid population growth from 7 million people in 1984 to 11.5 million (estimate) in 2017[1] and the fact that 80% of the country's economic activities and two thirds of its population are concentrated along its coastline, exposure to coastal hazards such as sea level rise and coastal erosion has also increased significantly. It is estimated that without targeted adaptation measures, and under a high emissions scenario, more than 78,000 Tunisians could be affected by sea level rise and flooding annually in the last quarter of the 21st century [8]. At the same time, the country is highly susceptible to climate-related hazards, given its high reliance on the agricultural sector and tourism, as well as widespread poverty—notably in rural areas in the southern and western parts of the country.

In response, the Tunisian Ministry of Environment has adopted various plans to address climate change impacts and has motivated government action towards a national risk map project. The Tunisian Nationally Determined Contributions (NDCs) which were submitted to the United Nations Framework Convention on Climate Change (UNFCCC) in 2015 mentioned mitigation, vulnerability, and adaptation challenges. Moreover, the NDCs declared disaster risk management (DRM) as a part of the national objectives defined in the future Master Plan. In this context, risk-informed sustainable urban planning is considered vital to make cities more

---

[1] https://data.worldbank.org/country/tunisia.

resilient. Accordingly, the Tunisian "Ministère de l'Equipement de l'Habitat et de l'Aménagement du Territoire" plans to devise a national urban Risk Information and Analyzing System (RIAS) to assess the impact of natural hazards on people and the economy, and to reduce their effects.

Risk assessments are widely acknowledged to provide relevant baselines for DRM and adaptation planning [9, 10, 11]. However, the lack of up-to-date, relevant geospatial data, as well as an insufficient understanding of the underlying drivers of disaster risk, pose significant challenges to the assessment of disaster risk in cities, as well as to risk-informed urban planning.

Earth observation in combination with geographic information systems (GIS) offer a wide range of options to support comprehensive risk analysis and prevention planning. With its ability to capture and link different levels of information, GIS is regarded as the ideal tool to provide all the necessary information for the three components relevant to risk assessment: hazard, exposure, and vulnerability [12, 13, 14]. Taubenböck et al. [15] reviewed the use of remote sensing and GIS to support flood risk assessments in urban areas in different regions of the world. They concluded that multi-source spaceborne imagery can greatly strengthen urban risk and vulnerability analysis. At the same time, they warn of the potential mismatch between the spatial accuracy of remote sensing and the expectations of stakeholders, which in the case of urban planners is often based on cadastral records. In the context of data scarce urban environments in Africa, Michellier et al. [16] stress the potential of integrated remote sensing–based spatial analysis and field surveys in enhancing stakeholders' engagement and participation. However, relatively few studies so far have applied a combined approach in the North African urban context with the exception of Snoussi et al. [17] who performed a remote sensing/GIS-only coastal vulnerability assessment of several mid-size Moroccan cities, and Eckert et al. [18] who conducted a remote-sensing-based tsunami risk assessment in Alexandria, Egypt, though the originally planned social vulnerability investigation through field survey integration was unsuccessful due to political instability. Applications of integrated remote sensing/GIS risk assessments in North Africa—and in Tunisia in particular—remain scarce.

The chapter presents a methodology for urban risk assessments with a particular focus on the role of remote sensing and GIS, drawing on lessons from an application in Monastir, Tunisia. A standardized procedure has been developed to provide reliable data and information on urban growth and disaster risk trends in this pilot area. Relevant steps for the multi-risk assessment include the integration of geodata from high-resolution satellite images with socio-economic data and information elicited from expert interviews. In addition, the chapter discusses the potential and gives recommendations for a nation-wide upscaling to support the current endeavors of the Tunisian government to make cities more resilient.

10.2   Case Study: Monastir, Tunisia

The city of Monastir is located in eastern Tunisia about 160 km south of the capital Tunis. It covers an area of about 1020 km^2, which corresponds to 0.6% of the country's surface, and has a population of approximately 93,306 (Census 2014). The coast of Monastir expands over 15 km and is limited by the Hamdoun wadi in the North and the Khniss wadi in the South [19]. The city is located in a semi-arid area that exhibits mild, rainy winters and hot, dry summers. Short and intense precipitation events occur at irregular intervals on a regional scale with high inter-seasonal and inter-annual variability [1]. Monastir is prone to multiple natural hazards, namely flash floods, coastal erosion, and earthquakes. Flash floods are commonly recurring events, which require frequent interventions from the local civil protection, with up to 297 interventions recorded at the governorate level in 2016, indicating a significant disruption of livelihoods. Since Monastir is a touristic destination with numerous coastal hotels, the process of coastal erosion impacts a particularly valuable economic sector of the city. The main factors influencing this process are the terrain properties (e.g. erodibility), waves, wind and intense precipitation. Both the consequences of rising sea levels and heavy rainfall events threaten important infrastructure in Monastir. For this reason, the municipality and the Agency for Coastal Protection and Management (APAL) have already undertaken structural safety measures along some sections of the coast. The frequency of earthquakes is relatively low, with major events occurring in 1988 and 2013, and mainly affecting the population in terms of structural damage to buildings and relocation necessities. Figure 10.1 shows an overview of the study area and its main land cover classes.

10.3   Methodology

10.3.1   Conceptual Risk Framework

Over the past decades, a range of conceptual frameworks have been developed and applied to assess disaster risk in cities. Here, the risk definition and framework proposed by the Intergovernmental Panel on Climate Change (IPCC) in its Fifth Assessment Report (AR5; [7]) is applied (see Fig. 10.2).

According to the IPCC, risk (of climate-related impacts) is defined as the likelihood over a specified period of severe alterations in the normal functioning of a community or a society [7] and results from the interaction of (climate-related) hazards, exposure, and vulnerability. The conceptual risk framework (Fig. 10.2) shows that vulnerability and exposure are, to a great extent, consequences of socioeconomic pathways and societal conditions [7].

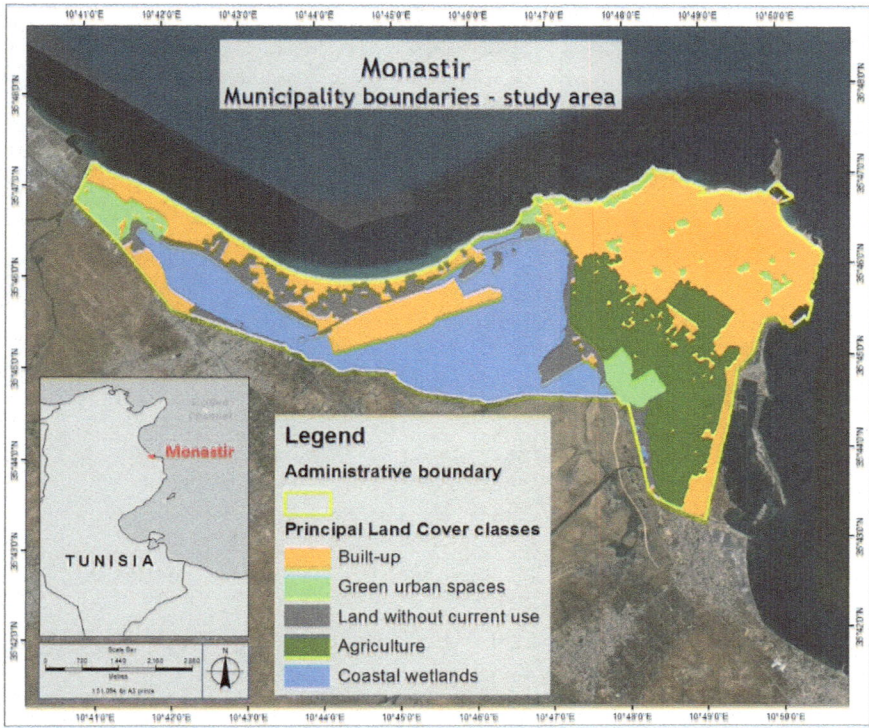

Fig. 10.1 Location of the study area in Tunisia with the main land cover classes. *Source* IABG/UNU—EU Project UDRASP 2017–2018

10.3.2   *Workflow*

Figure 10.3 shows the overall workflow for risk assessment. Multi-temporal remote sensing data from WorldView3 and Landsat was collected and analyzed to produce a series of land use/land cover (LULC) maps and a digital terrain model (DTM), which served as inputs for an urban growth model and the modelling of flash floods respectively. Additional input for the latter included precipitation data, whereas the urban growth model also integrated the outcomes of two scenario workshops conducted with relevant stakeholders. These two models informed the analysis of the hazard and exposure components of the risk assessment. The combination of stakeholder consultations with an extensive household survey and census data provided the basis for the assessment of social vulnerability to flash floods.

   The final geodatabase includes the results of the risk assessment, a digital version of the urban development plan (Plan d'Aménagement; PAU), a satellite-derived mosaic product, and additional data sources to support the visualization of the results

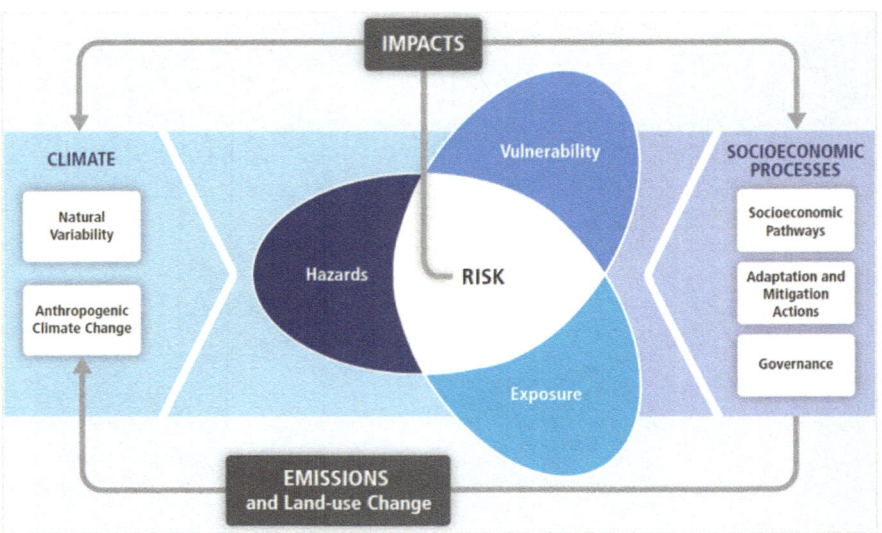

Fig. 10.2 Conceptual framework for the risk analysis [7]

in the web-based information system. The following subsections describe the main analysis steps in detail.

10.3.3   Data Acquisition and Development of a Geodatabase

An essential task for every spatial risk assessment is the acquisition, pre-processing and analysis of relevant data and information. Especially in data scarce environments like Monastir, remote sensing and GIS technology can be used to collect and process data and to provide the relevant information for subsequent risk assessments. In order to tackle the complexity of disaster risk in Monastir, high- and medium-resolution satellite images, socio-economic data, and information obtained in expert interviews were integrated (Table 10.1). Further, the urban development plan (PAU), which is updated every 10 years and builds the basis for urban development in Tunisia, served as a principal input for the analysis. Additional primary data was collected during stakeholder consultations (workshops and focus group) and a household survey. The information obtained from these sources was quantified and analyzed statistically, then integrated into the GIS environment.

The diversity and inhomogeneity of the input data made it necessary to establish appropriate and accessible tools for the exchange of data and relevant information, and as a basis for decision-making. Thus, a centralized platform was developed to facilitate the collection, analysis, and sharing of data and information with relevant stakeholders.

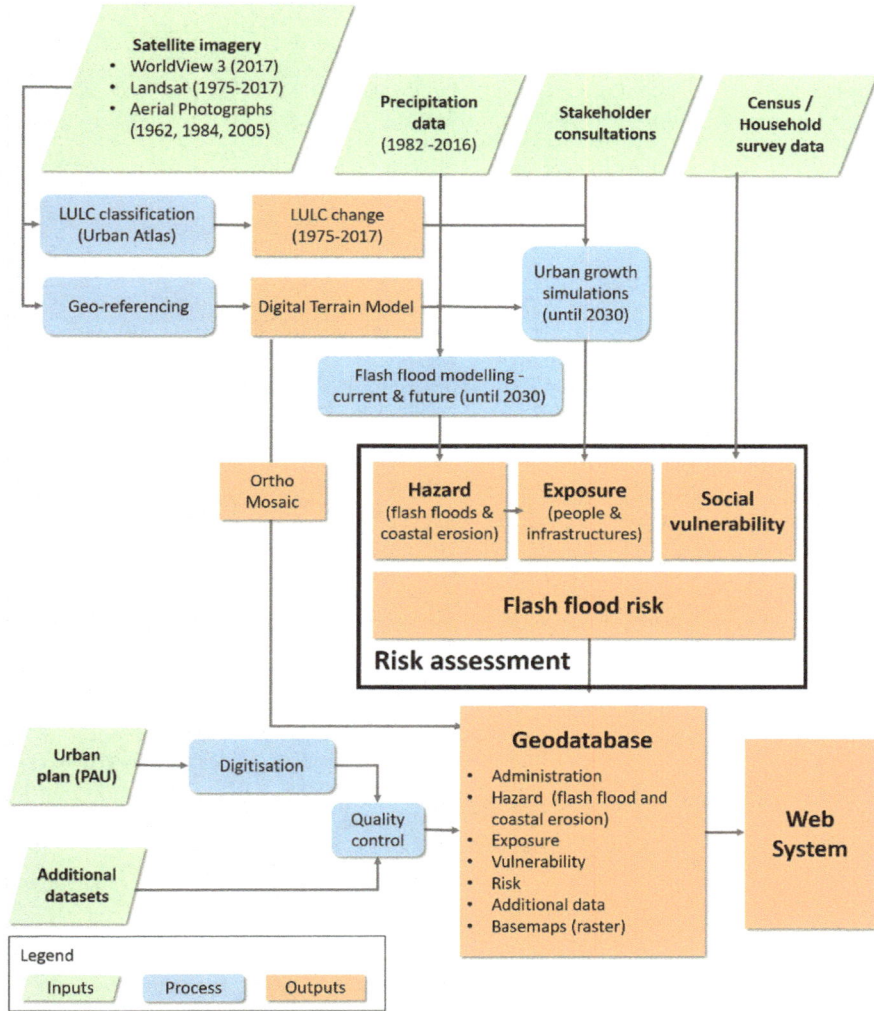

Fig. 10.3 Workflow for remote sensing/GIS-based urban risk assessment. *Source* IABG/UNU—EU Project UDRASP 2017–2018

Quality controls have been carried out to comply with the data quality guidelines as an essential part of a harmonised data set. Technical quality control is mostly designed to review the captured features by automated checks and routine. In order to validate all captured features regarding their positional and thematic accuracy, logical consistency, and completeness, automated data reviews with simple-to-use checks were applied to detect, correct and verify geometric or attributive errors.

Table 10.1 Input data

Type	Characteristics	Application	Source
Expert input	Three focus group discussions during stakeholder workshops and expert interviews	Information on relevant drivers of risk (as a basis for the risk assessment) and socio-economic scenario (as a basis for urban growth modelling)	Primary data
World View 3 (WV3)	Stereoscopic imagery, "ortho ready stereo" type, acquisition date in 2017, spatial resolution: 0.3 m	The baseline for mapping; derivation of digital terrain model (DTM) by dense image matching to create a digital height model (DHM) and for ortho-rectification with a suitable DTM	European Space Imaging (EUSI)[a]
Landsat	Period 1975–2017, Landsat 4-5-7-8: with 30 m multispectral spatial resolution	Derivation of past urban development by multi-temporal analysis and mapping	US Geological Survey/Earth explorer[b]
Aerial photographs	Periods 1962, 1984 and 2005; Resolution: 10–50 cm	For cartographic purpose, suitable in an urban context to show the historical situation and support the past urban development	Municipality of Monastir and «Centre national de cartographie et télédétection (CNCT)»
Urban development plan (Plan D'Aménagement Urbain—PAU)	Computer-Aided Design (CAD) format	Converted to GIS as a baseline for urban planning and the official dataset of a Tunisian municipality	Municipality of Monastir
Other relevant official geodata	Administrative units of the city and region, Thematic data on hydrology, geology, and climate	Input for the risk assessment	Municipality of Monastir
Census data	Socioeconomic data at sub-municipal level	Input for the vulnerability and risk assessment	Institut National de la Statistique (INS)
Household survey	Socioeconomic data at household level	Input for the vulnerability and risk assessment	Primary data

[a]www.euspaceimaging.com
[b]https://earthexplorer.usgs.gov/

Following the quality checks, all spatial datasets were integrated and managed in a GIS database according to a customized data model, which ensures a harmonized, standardized data structure and fulfils the requirements of the guidelines of the International Organization for Standardization (ISO)[2] and the standards of the Open Geospatial Consortium (OGC).[3] Table 10.1 shows the input data used as relevant information and data source for the study.

10.3.4 Land Use/Land Cover Analysis Based on the Urban Atlas Standard

Changes in LULC, infrastructure and urban areas were identified as necessary missing information to fulfil the following modelling requirements. In this context, GIS and remote sensing were used to:

- Compare the city's hazard situation in the past and present by analyzing satellite images over the last 40 years
- Update and extract missing or incomplete data in the main database by interpretation or classification of current satellite imagery
- Generate a DTM by processing high-resolution stereo satellite imagery from WorldView3
- Generate image mosaics as a base layer for visualization purposes
- Provide classification results and terrain elevation information for flash flood and urban growth modelling.

In order to document the temporal development of rural and urban areas in the city of Monastir and to detect changes, image data from the Landsat satellites were acquired and semi-automatically classified. Classification was performed according the nomenclature of Urban Atlas [20] which was chosen to serve as a baseline for city land cover data. To verify the results, visual checks were performed using historical aerial photographs from the last four decades.

10.3.5 Hazard Analysis

10.3.5.1 Flash Flood Modeling

In the course of the flash flood hazard modeling, water depth and flow velocity in case of a 100-year precipitation event were estimated, and it was defined that a flow velocity of 2 m/s and/or a water depth of 1.5 m would endanger human life [21].

---

[2]http://www.iso.org (ISO 19113:2002; ISO 19110:2002; ISO 19126:2009; ISO 19115:2003).
[3]http://www.opengeospatial.org/docs/is.

Table 10.2 Flash flood
hazard classification

Hazard class	Range	Description
Moderate	0.75–1.5	Dangerous for some (i.e. children)
Significant	1.6–2.5	Dangerous for most people
Extreme	>2.5	Dangerous for all

Surface runoff and flow paths were modeled using the physical process-based modeling suite MIKE FLOOD.[4] Based on high-resolution optical satellite imagery (World View 3), a DTM with 1-meter spatial resolution was derived and rescaled to 5-m resolution for modeling purposes. Height information about the buildings and roads was added to generate a base for runoff modelling. Different roughness values were assigned to main elements like road networks or natural areas depending on their drainage characteristics to account for differences in surface runoff velocities [22]. Additionally, parameters like infiltration values of the soil, porosity, groundwater floor distance and saturation were considered in the model. Infiltration values were derived from soil maps for the unsealed areas in the study area. Furthermore, parameters such as porosity, groundwater floor distance, discharge into lower layers and (pre)-saturation were defined. In the next step, precipitation data from 1982 to 2016 was integrated into the model. Hazard-prone areas of past and present extreme events were modeled using data from high-intensity rainfall events in 1982 and 2016. Furthermore, records of past extreme events and their precipitation values were sorted and used to extrapolate and create a 100-year flooding hazard event. To simulate the HQ100 scenario (100-year return period) in Monastir, a 12-h rainfall was applied over the study area. The modelling was conducted according to an EULER II distribution, which is based on the observation that the highest rainfall intensity occurs at the end of the third part of a rainfall event. Afterwards, a flood hazard rating was calculated by using an equation which combines water depth, flow velocity and a debris factor [23]:

$$HR = d * (v + 0.5) + DF$$

where HR is the flash flood hazard severity, d the depth of flooding (in meters), v the velocity of floodwaters (m/s), and DF the debris factor. The debris factor takes into account the effect of debris, which is carried by a flash flood.

The generated hazard layer was classified into three categories that correspond to increasing hazard severity, as depicted in Table 10.2.

Referring to a study by DEFRA [23], these ranges describe different degrees of danger for exposed people. By applying the classification for several flash flood hazard scenarios, the impacts on affected environment and people are more visible and better comparable.

---

[4]MIKE FLOOD—Modelling Software *MIKE21* (source: DHI—Wasy).

10.3.5.2 Coastal Erosion

In a first step, the coastal area of Monastir was divided into sectors according to the different investigation phases that were part of the national coastal protection program [24–25, 26]. Coastal erosion was analyzed based on high-resolution World View 3 satellite imagery and the DTM. The extension of the hazard area was defined depending on slope and maximum flow velocities that were modelled in the course of the flash flood analysis. The land use within the hazard area was classified into five main categories of land use (built-up areas, natural zone, coastal structures, transport, and water) to provide information concerning the exposure of the area. Furthermore the extension of the coastal protection was expressed as a development status as derived from local reports and interviews, which describes whether all coastal protection activities are completed, under construction or have not yet been examined. The exposure of the coastal area was examined and described considering the terrain conditions, land use cover and development status. The results of the GIS analyses are subsequently combined with results of reports by local authorities. Findings from a field survey were interpreted to gain a more detailed understanding of the coastal erosion processes in the study area. Further influencing factors like climate, geology, and hydrology were not considered in the analysis, but should be included in a more comprehensive coastal erosion analysis in the future.

10.3.6 Exposure Analysis

A fundamental concern of risk analysis is to estimate how many people and which assets (e.g., critical infrastructure, buildings, etc.) are located in areas that are prone to natural hazards. Therefore, buildings in residential urban zones and traffic routes that are located in areas prone to flash floods were identified and classified according to their degree of exposure. Since high-resolution gridded population data was not available for Monastir, data from the population census of 2014 was downscaled to the building block level in order to also assess how many people are potentially exposed to flash floods. The population of each building block was estimated statistically by combining the size of the building block and its density class as identified in the Urban Atlas classification. In order to assess exposure of people and the road network to flash floods, the results of the 100-year event flash flood modeling were spatially combined with the residential buildings and the road network. In addition to calculating the entire exposed population and the length of the road network (in km) potentially exposed to flash floods, the relative exposure (in %) was derived for each building block and sub-unit in the city. Finally, the spatial overlap between the exposed assets and the created hazard map were used to assess the present flash flood exposure.

10.3.7   Vulnerability and Integrated Risk Assessment

In addition to assessing hazard and exposure, an indicator-based assessment of social vulnerability to flash flooding was conducted at the sub-city scale. Relevant drivers of vulnerability and risk were identified based on a literature review [27], expert interviews and during a focus group discussion with relevant local stakeholders. Since census data was not available at the sub-city scale for all indicators, data for the vulnerability indicators was collected through a comprehensive household survey which was conducted in May 2018 (n = 696 households). In total 15 out of 78 sub-units were selected for the household survey, which were considered to be representative for the other sub-units of the city by the local municipality. Following, a number of statistical steps for index construction (e.g. missing data analysis, outlier treatment, normalization, weighted aggregation) the indicators were aggregated into an index of social vulnerability. Thereby, weights for the individual indicators were obtained from an online survey with relevant local experts ($n = 7$), among which were urban planners, statisticians, and representatives of various national and regional agencies.

The outcomes of the exposure analysis and the vulnerability analysis were then combined into an integrated risk assessment focusing on flash flood risk at the sub-city scale. Thereby, a multiplicative approach was used where risk is a function of exposure to hazards and vulnerability:

$$RISK = exposure\ to\ hazards * vulnerability$$

In addition to assessing current flash flood risk, future scenarios of flash flood risk until 2030 were also developed by integrating narrative storylines derived from expert workshops with statistical data analysis. However, this goes beyond the scope of this chapter, which focusses on the role of GIS and remote sensing.

10.4   Results

This section summarizes the major findings of the risk assessment. Additionally, a brief overview of the characteristics of the web-based information system and its potential is offered.

10.4.1   Flash Floods and Coastal Erosion in Monastir

The flash flood model showed a particular concentration of water depth in specific areas of the city for the 100-year return period event. Figure 10.4 shows the outcomes of the flash flood model for the city center and its immediate surroundings. The

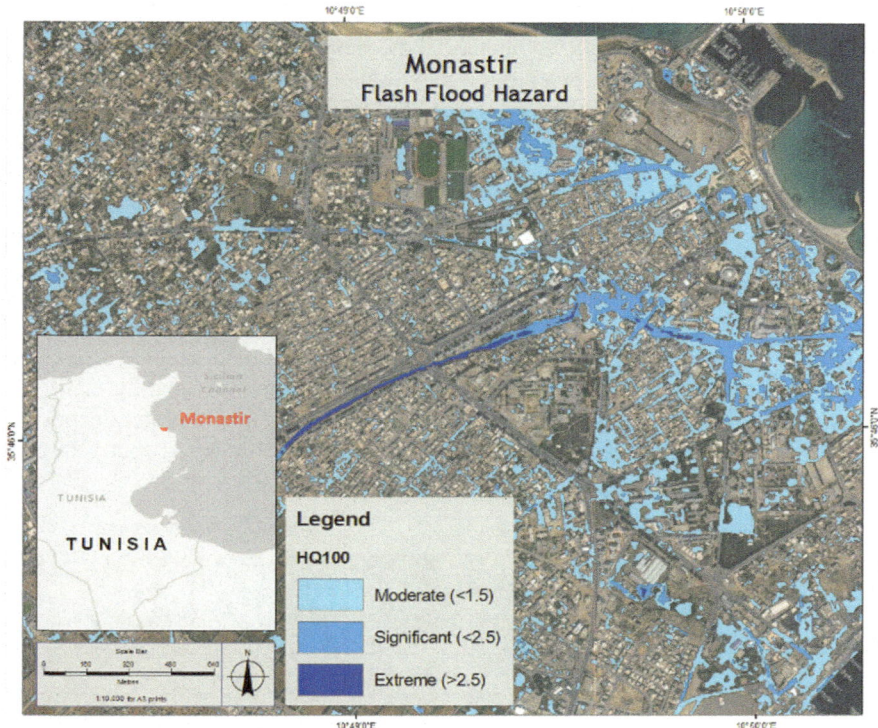

Fig. 10.4 Spatial extent and severity of a 100-year return period flash flood event in Monastir. *Source* IABG—EU Project UDRASP 2017–2018

residential areas in the eastern part of the city are affected particularly by the 100-year return period flash flood event. Moreover, the extreme hazard class (>2.5), accumulating along the railway tracks and the main train station, is identifiable. Both people and infrastructure in those areas are particularly prone to flash floods, as also revealed by the exposure analysis.

The results of the coastal erosion analyses provide an overview of the varying degrees of hazard along the coast of Monastir. The cliffs are currently not developed and are subject to various degradation processes. The morphology of the coastal stretch is dominated by cliffs that range from 5 m up to 20 m. The steep slopes are unstable and have collapsed in several areas. Landslides have also occurred at a number of locations. Deep gullies run through the slopes, which are occasionally channeled by drainage structures. Small stretches of sandy beaches are located on the shore, which have been subject to accelerated erosion.

In general, it can be observed that the coastal regions of the outer parts of the municipality are often more severely damaged, as fewer protective measures are installed and the regions are therefore more exposed to erosion processes. Especially the outer region in the southwest, which is characterized by hotel complexes and enclosed beaches, exhibits higher exposure to erosion. As several buildings in the touristic area are built within a few meters of the shore that violates the regulations, the effects can already be observed through partial or total structural collapse.

The beach areas closer to the city center have been protected against erosion by national structural measures [24–25]. As far as these are intact, a clear reduction of coastal erosion can be recognized.

10.4.2 Exposure to Flash Floods

Population exposure showed significant clusters in the old medina and the eastern part of the city center. Both areas are densely populated and affected by the flood extent, with one census sub-unit registering up to 57% of its population as exposed. Figure 10.5 visualizes the exposure of the population and transport routes in one

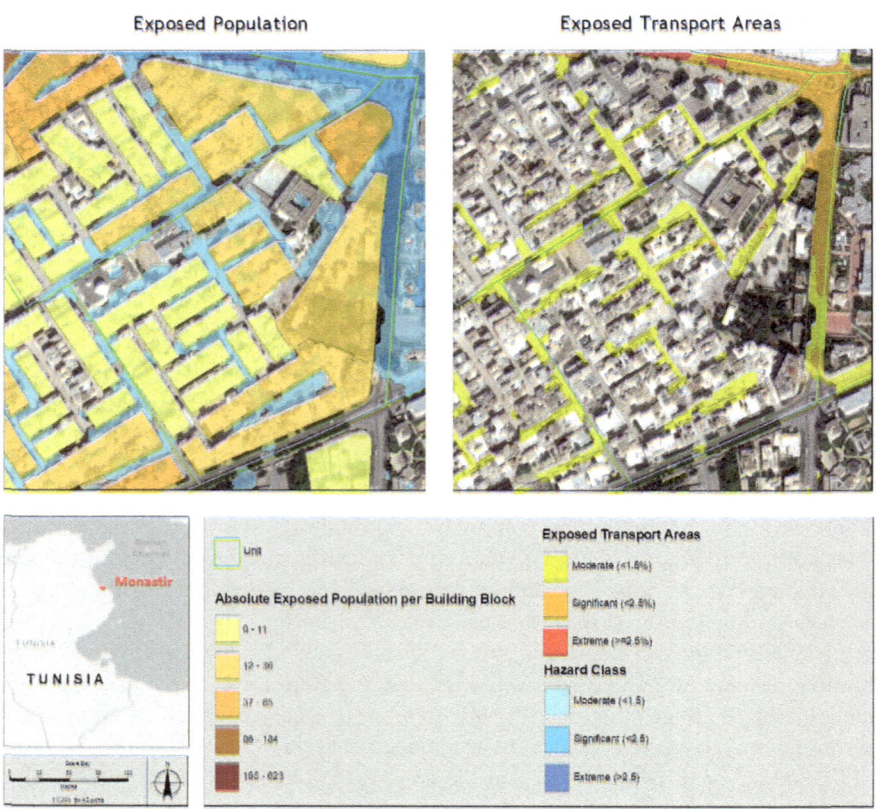

Fig. 10.5 Outcomes of the exposure analysis. *Source* IABG—EU Project UDRASP 2017–2018

census sub-unit through a series of thematic maps.

10.4.3   Social Vulnerability and Flash Flood Risk

Social vulnerability is medium to high throughout all 15 sub-units with very little variability (minimum score: 38; maximum score: 53). Figure 10.6 shows the spatial variability of flash flood risk across the 15 sub-units. Index scores of zero indicate no risk, while scores of 100 indicate extremely high risk Flash flood risk is low

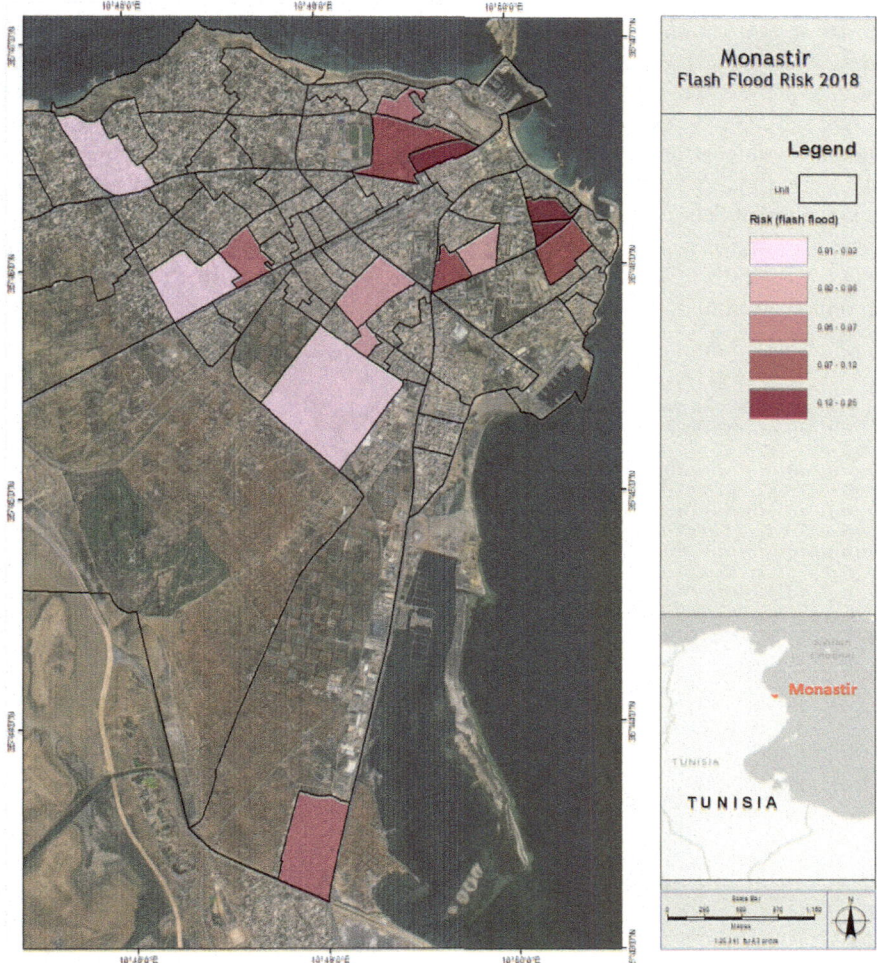

Fig. 10.6 Flash flood risk in Monastir. *Source* IABG/UNU—EU Project UDRASP 2017–2018

to medium depending on the specific neighborhood (minimum score: 0; maximum score: 24).

However, given that social vulnerability to flash floods is medium to high throughout the surveyed sub-units, risk could drastically increase if flash floods become more frequent in the future as a result of climate change and further surface sealing.

10.4.4 Web-Based Information System

In order to meet the stakeholders' request for a user-friendly and easily accessible tool for spatial information, the results of the integrated risk assessment for flash floods and the hazard mapping for coastal erosion were included in a web GIS platform (inclusive of a database, a web service and a web application, developed on PostgreSQL, GeoServer and NodeJS respectively). Data from all analyses is made available in the platform as separate layers, allowing a high level of customization of risk information by the end user. Additional spatial information is provided through the inclusion of data from the Open Street Map project and the Open Layers library. Figure 10.7 shows an exemplary view of the web-based information system.

The web platform, hosted by the Municipality of Monastir, will be made available for consultation of local decision makers and urban planners. Table 10.3 lists all available layers.

Fig. 10.7 Graphical user interface of the web-based information system. *Source* IABG—EU Project UDRASP 2017–2018

Table 10.3 Web platform—available layers, depicting present situation (2018) and future scenarios (2030)

Typology	Available layers
Administration	• Census sub-units • Administrative boundaries (city sectors, municipality, delegation, governorate)
Hazard and exposure	• Current hazard prone areas (coastal erosion, flash flood) and 100-year return period for flash floods • Current and future exposure • Course of the most critical drainage paths after heavy rainfall
Vulnerability and risk	• Current and future social vulnerability to flash floods (for 15 selected units) • Current and future flash flood risk (for 15 selected units)
Additional data	• Unrevised Tunisian datasets: climate zones for governorate of Monastir, land cover • Intermediate data layers: generated in the project and provided for further use in the risk assessment • Urban development layer: 1975, 1981, 1984, 1986, 1990, 1992, 1999, 2002, 2005, 2008, 2011, 2014, 2017 • Urban growth layers (according urban development layer and future scenarios (2030) • Critical infrastructures (e.g., roads, etc.) according to local specifications
Basemaps	• Satellite imagery mosaic (World View 3) • Open Street Map • Open Layers

10.5   Discussion and Lessons Learned

The use of Earth observation and GIS in the context of DRM is not new, and both the potentials and limitations have been highlighted by various authors [28], and more recently in a special issue entitled "Earth Observation to Support Disaster Preparedness and Disaster Risk Management" published in Remote Sensing.[5] However, their application in Tunisia—particularly in the context of urban disaster risk and sustainable urban planning—remains limited. Our analysis has shown that remote sensing-derived information and GIS analysis can provide extremely valuable inputs for urban risk assessments, even in data scarce environments such as Monastir. The analysis of multi-temporal imagery of high and medium resolution formed an essential part of the modelling of flash floods, the retrospective analysis of urban growth, as well as for the simulation of future urban growth in the city. The collection and inclusion of all available spatial information (including the digitization of the city PAU) into a standardized, quality-checked geodatabase significantly improved all phases of the indicator-based spatial risk assessment.

---

[5]https://www.mdpi.com/journal/remotesensing/special_issues/EO_disaster.

The use of the Urban Atlas Standard for the LULC characterization is considered useful, since it enables the comparability of results on a European and partly on a global level. This approach supports both a potential transfer to other urban contexts within or outside the region as well as the upscaling of the analysis to the national level. While remote sensing provided relevant data and inputs for the assessment of hazard and exposure, its potential in the field of social vulnerability assessment remains limited. The vast majority of socio-economic data for the vulnerability assessment remained dependent on secondary sources (e.g. census data), which were limited in spatial resolution and accessibility, or on primary data collection (household survey), which is demanding in terms of time, resources and processing. Further investigations into the definition of socio-economic proxies obtainable through earth observation analysis hence present a highly relevant field of future research, especially in data scarce environments.

The active participation of local partners proved to be extremely relevant for data collection and indispensable for future updates and usage of the produced database. Moreover, the latter generated the necessity of appropriate capacity building and knowledge transfer activities, which helped to form local expertise in the use of geographic information systems for urban planning.

Finally, the study showed the necessity of clearly defining in advance local stakeholders' needs and expectations in order to ensure a proper co-design and co-development of the risk assessment as well as of potential disaster risk management solutions. The inclusion of local stakeholders further allowed the development of solutions—such as the web platform for data visualization and dissemination—tailored to local needs as they were expressed during several consultations. Moreover, these consultations were vital in providing additional data that enriched the performed spatial analysis. With the finalized geodatabase handed over to the municipality of Monastir—the main beneficiary of the project—for continuous update and use, local decision-makers in Monastir can now rely on a working tool for accurate inclusion of risk assessment into future urban planning.

10.6 Conclusions

The experience in Monastir has shown that remote sensing and GIS have the potential to generate relevant information to support urban disaster risk assessments in data scarce environments, and ultimately risk-informed urban planning. These results are of particular significance in the North African urban context, where there is a general scarcity of reliable data while being simultaneously recognized as a global hotspot for environmental disaster risk. In addition, the experience in Monastir could prove to be an essential model for other coastal cities in Tunisia and North Africa, as the methods developed are fully transferable to other regional urban areas, all the while considering the differences in the urban environment, critical hazards and vulnerability conditions. Finally, an upscaling of the process from the city of Monastir to the national level is recommended as it would enforce the coordination among different administration levels and offer a more comprehensive risk-informed strategic planning not only in Monastir but also for all Tunisian cities confronted with natural hazards and climate change.

Acknowledgements The research is part of the project UD-RASP (grant no. ECHO/SUB/2016/740186/PREV19) funded by European Commission Directorate-General for European Civil Protection and Humanitarian Aid Operations (DG ECHO).

References

1. United Nations Human Settlements Programme, publisher (2016) Urbanization and development. Emerging futures: world cities report 2016. UN-Habitat, Nairobi, Kenya
2. Birkmann J, Welle T, Solecki W, Lwasa S, Garschagen M (2016) Boost resilience of small and mid-sized cities. Nature 537(7622):605–608. https://doi.org/10.1038/537605a
3. Adelekan I, Johnson C, Manda M, Matyas D, Mberu BU, Parnell S et al (2015) Disaster risk and its reduction: an agenda for urban Africa. Int Dev Plan Rev 37(1):33–43. https://doi.org/10.3828/idpr.2015.4
4. Dodman D, Leck H, Rusca M, Colenbrander S (2017) African urbanisation and urbanism: implications for risk accumulation and reduction. Int J Disaster Risk Reduction 26:7–15. https://doi.org/10.1016/j.ijdrr.2017.06.029
5. Garschagen M, Romero-Lankao P (2015) Exploring the relation-ships between urbanization trends and climate change vulnerability. Clim Change 133(1):37–52. https://doi.org/10.1007/s10584-013-0812-6
6. Ruocco A, Gasparini P, Weets G (2015) Urbanisation and climate change in Africa: setting the scene. In Urban vulnerability and climate change in Africa. Springer International Publishing, Cham (4)
7. IPCC (2012) Managing the risks of extreme events and disasters to advance climate change adaptation. A Special Report of Working Groups I and II of the Intergovernmental Panel on Climate Change [Field CB, Barros V, Stocker TF, Qin D, Dokken DJ, Ebi KL, Mastrandrea MD, Mach KJ, Plattner G-K, Allen SK, Tignor M, Midgley PM (eds)]. Cambridge University Press, Cam-bridge, UK, and New York, NY, USA, 582 p
8. USAID (2018) Climate risk profile Tunisia. Available from https://www.climatelinks.org/sites/default/files/asset/document/Tunisia_CRP.pdf
9. Hagenlocher M, Schneiderbauer S, Sebesvari Z, Bertram M, Renner K, Renaud FG, Wiley H, Zebisch M (2018a) Climate risk assessment for eco-system-based adaptation: a guidebook for planners and practitioners. Deutsche Gesellschaft für Internationale Zusammenarbeit (GIZ) GmbH, Bonn. Available from https://www.adaptationcommunity.net/wp-content/uploads/2018/06/giz-eurac-unu-2018-en-guidebook-climate-risk-asessment-eba.pdf
10. Klijn F, Kreibich H, de Moel H, Penning-Rowsell E (2015) Adaptive flood risk management planning based on a comprehensive flood risk conceptualisation. Mitig Adapt Strat Glob Change 20(6):845–864. https://doi.org/10.1007/s11027-015-9638-z
11. Serrao-Neumann S, Crick F, Harman B, Schuch G, Choy DL (2015) Maximizing synergies between disaster risk reduction and climate change adaptation: potential enablers for improved planning outcomes. Environ Sci Policy 50:46–61. https://doi.org/10.1016/j.envsci.2015.01.017
12. Armenakis C, Du EX, Natesan S, Persad RA, Zhang Y (2017) Flood risk assessment in urban areas based on spatial analytics and social factors. Geosci (Switzerland) 7(4):1–15. https://doi.org/10.3390/geosciences7040123
13. Hagenlocher M, Renaud FG, Haas S, Sebesvari Z (2018b) Vulnerability and risk of deltaic social-ecological systems exposed to multiple hazards. Sci Total Environ 631–632:71–80

14. Lianxiao, Morimoto T (2019) Spatial analysis of social vulnerability to floods based on the MOVE framework and information entropy method: case study of Katsushika Ward, Tokyo. Sustain (Switzerland) 11(2). https://doi.org/10.3390/su11020529
15. Taubenböck H, Wurm M, Netzband M, Zwenzner H, Roth A, Rahman A, Dech S (2011) Flood risks in urbanized areas—multi-sensoral approaches using remotely sensed data for risk assessment. Nat Hazards Earth Syst Sci 11(2):431–444. https://doi.org/10.5194/nhess-11-431-2011
16. Michellier C, Pigeon P, Kervyn F, Wolff E (2016) Contextualizing vulnerability assessment: a support to geo-risk management in central Africa. Nat Hazards 82(S1):27–42. https://doi.org/10.1007/s11069-016-2295-z
17. Snoussi M, Niazi S, Khouakhi A, Raji O (2010) Climate change and sea-level rise: Agis-based vulnerability and impact assessment, the case of the Moroccan coast. In *Geomatic solutions for coastal environments*. Nova Science Publishers, New York
18. Eckert S, Jelinek R, Zeug G, Krausmann E (2012) Remote sensing-based assessment of tsunami vulnerability and risk in Alexandria, Egypt. Appl Geogr 32(2):714–723. https://doi.org/10.1016/j.apgeog.2011.08.003
19. APAL Republique Tunisienne, Ministère de l'Environnement et du Developpement Durable, Agence de Protection et d'Aménagement du Littoral (2010) Expertise et elaboration des termes de references relative à l'erosion côtière du littoral de Monastir
20. European Commission (2016) Mapping guide for a European urban atlas v4.7, Product "Urban Atlas" from Copernicus "Urban Atlas Project" for the 2006 reference year and the "Urban Atlas" update and extension for the 2012 reference year
21. Kanonier A (ed) (2012) Raumplanung und Naturgefahrenmanagement, vol 19. LIT Verlag Münster
22. Engman ET (1986) Roughness coefficients for routing surface runoff. J Irrig Drainage Eng 112(1):39–53
23. DEFRA (2006) Flood risks to people. Phase 2: Defra Environment Agency
24. APAL Republique Tunisienne, Ministère de l'Environnement et du Developpement Durable, Agence de Protection et d'Aménagement du Littoral (2009) Etude de protection de la 2ème tranche de la falaise de Monastir
25. APAL Republique Tunisienne, Ministère de l'Environnement et du Developpement Durable, Agence de Protection et d'Aménagement du Littoral (2011) Etude de la frange littorale de Monastir Strategie de Rehabilitation
26. URAM (2003) Plan d'occupation de la plage de Monastir. L'etude du pop de Monastir
27. European Commission; Organisation for Economic Co-operation and Development (2008) Handbook on constructing composite indicators. Methodology and user guide. OECD, Paris
28. Ghaffarian S, Kerle N, Filatova T (2018) Remote sensing-based proxies for urban disaster risk management and resilience: a review. Remote Sens 10(11):1760

Part V
Soil Degradation and Drought

Chapter 11
Monitoring of Land Use-Land Cover Changes and Assessment of Soil Degradation Using Landsat TM and OLI Data in Zarzis Arid Region

Katar Achraf, Hammouda Aichi, and Bouajila Essifi

Abstract Land Use/Land Cover Change (LUCC) is recognized as a crucial driver of environmental change on all spatio-temporal scales. Zarzis region is reputed for its olive groves, vital for its socioeconomic development and because of severe climatic factors, it has a vulnerable ecosystem. Our study focuses on the monitoring of land use dynamics underpinning climate change and on the spatiotemporal assessment of the vigor of olive groves. Analysis of Landsat 5 TM image acquired in 2007 and Landsat 8 OLI image acquired in 2014 was performed under IDRISI software, by a remote sensing-based Land Change Modeler (LCM) method. Our results have shown a clear improvement in the vigor of the olive trees, mainly attributed to an increase in rainfall during the years 2010 and 2012. We have identified areas where land degradation has been attenuated both with Tabias, as a soil and water harvesting technique and with alley cropping between olive trees, as a promising recent agronomic practice. This analytical study is relevant for sustainable development.

Keywords Remote sensing · Climate change · Land Change Modeler · Olive groves · IDRISI · Landsat 8 OLI

11.1 Introduction

Climate change is a very serious phenomenon and has become a major global issue in recent years, especially in developing countries strongly affected by its impacts [1]. Climate change (CC) is a complex phenomenon that is reshaping the way we think about relationships between socio-political and biophysical systems [2]. Indeed, the climate varies naturally under the influence of various factors, including geographical

K. Achraf · H. Aichi
Agricultural Production Systems and Sustainable Development Laboratory, Higher School of Agriculture of Mograne, University of Carthage, 1121 Mograne, Zaghouan, Tunisia

B. Essifi (✉)
Laboratory of Eremology and Combating Desertification, Institut des Regions Arides (IRA), El Fje, 4119, Medenine, Tunisia
e-mail: Bouajila.Essifi@ira.agrinet.tn

(relief, ocean) and weather. This natural variation is accentuated by anthropogenic actions [3] and degradation (LD) is among the major environmental issues driven by a change in land use-land cover (LULC) and climate change worldwide especially in semi-arid regions [4]. "Land use term usually defined more strictly and referred to how, and the purposes for which, humans employ the land and its resources" [5]. Land cover refers to the habitat or vegetation type presents, such as forest and agriculture area. Land Use and Land Cover change (LULCC), also known as land change, is a term for the human modification of Earth's terrestrial surface [6]. It is widely admitted that LULCC has an important effect on both the functioning of the Earth's systems as a whole [7] and the majority of ecosystems [8, 9]. This change is based on the purposes of need, which is not necessarily only making the change in land cover but also change in intensity and management [10].

The climate is changing and tends to evolve in the Mediterranean basin in the 2030 and 2050 horizons towards an increase of the temperature and a fall of the precipitations on the background of the already great variability of the regional climate. During the 2011–2070 climatic period, this variability will increase on average by 5 to 10% compared to the situation of the past century [11].

The problem of the degradation of natural resources is highly felt in Southern Mediterranean countries with semi-arid and arid climates, where water and soil resources are scarce and vulnerable [12]. In Zarzis arid region, located in south-eastern Tunisia, rainfall intensity is subject to strong variations in time and space. Indeed, in autumn, storms can reach an intensity of more than 100 mm/h for laps of 5 min, which leads to the development of destructive runoff blades [13]. Also, the area is subject to strong anthropogenic pressure that accelerates land degradation. For instance, poor LULC practices identified in Zarzis such as deforestation, livestock overstocking, overgrazing, water stress, forest fires, retreat of grass cover, bush encroachment and arable land use intensification disturb the soil structure leaving the land vulnerable to water erosion and reactivation of sand dunes and subsequently the disruption of its olive agroecosystem.

The integration of remote sensing and GIS technologies is an effective approach for analyzing land use and cover changes [14]. Land change models can help scientists and users to understand change processes and design policies to reduce the negative impact of human activities on the earth system at scales ranging from global to local [15].

Contrariwise, Monitoring LULC changes, is still missing in many semi-arid regions worldwide [16] and particularly in Zarzis. However, Monitoring LULC change in this context is very relevant to help policymakers in mediating the negative consequences of LULC while sustaining the olive agroecosystem at the local scale. Based on LULC types, soils and/bare lands conditions within underpinning context of climate change, this paper aims to study spatiotemporal changes in the vigor of the olive trees in Zarzis region and consequently, attempt to assess land degradation vulnerability, using remote sensing data.

11.2   Materials and Methods

Two Landsat satellite images, covering Zarzis region, were used to perform the analysis and acquired from Thematic Mapper and Operational Land Imager sensors in 2007 and 2014, respectively. Both of them within the middle spatial resolution (30 m). As shown in the following flowchart (Fig. 11.1), we have used the ISOCLUST for classification method Land Change Modeler (LCM) for analyzing land cover changes under the IDRISI® software.

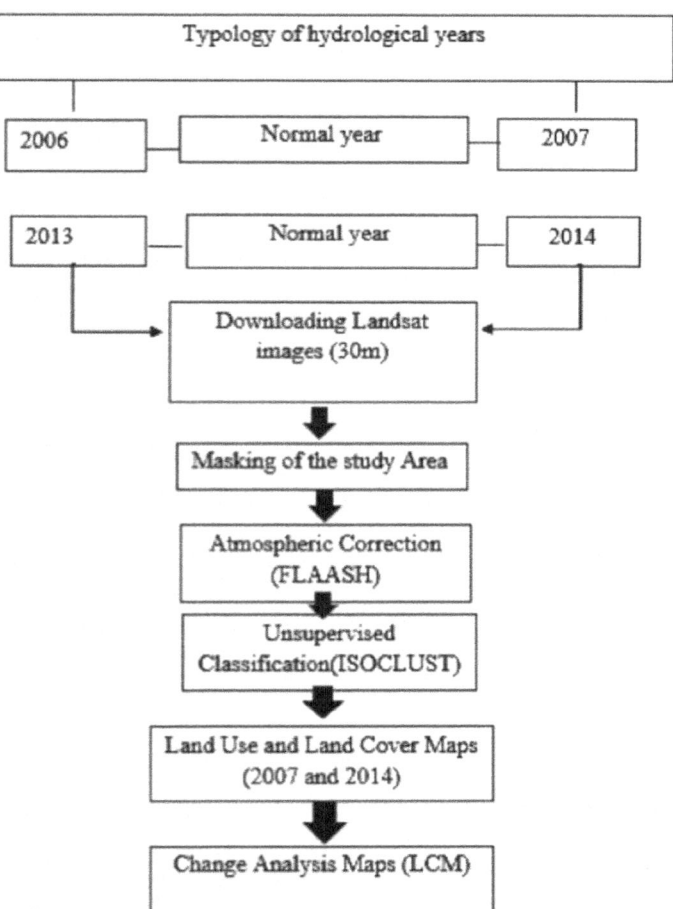

Fig. 11.1 Flowchart of the methodology

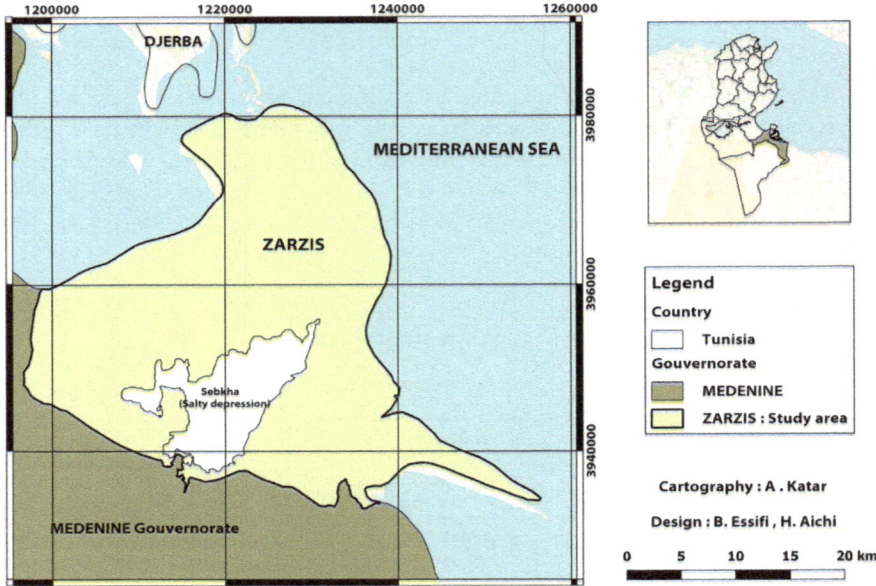

Fig. 11.2 Geographical location of the study area Basemap: OSM under QGIS OpenLayers Plugin [19]

11.2.1 Study Area

The region of Zarzis is located southeast of Tunisia it lies between 33°30′14″N latitudes and 11°06′43″E longitudes and its total area is about 340 sq. km (Fig. 11.2). It belongs to the arid bioclimate with temperate winter. The average temperature varies from one season to another. It ranges from 7.5 to 18.5 °C in winter and between 35 and 45 °C in summer. The average of the annual rainfall varies between 100 and 200 mm [17]. The area is known for its olive groves, which cover 63,420 hectares [18].

11.2.2 Land Degradation Constraints

The Zarzis region is characterized by a lower arid bioclimate (Fig. 11.3). The average annual temperature in the region of Zarzis is 22.5 °C. The potential evapotranspiration (PET) exceeds 1500 mm/year. The water balance (P-PET) is still in deficit [20]. Winds blow in average 100 days/year, mostly in the spring and in the autumn and can, therefore, cause dreadful sandstorms. In summer, a very hot and very dry wind blowing from the south, known as "Chehili," brings dry air, causing the dehydration of the vegetation [21]. These winds blow with a frequency of 20–30 days/year.

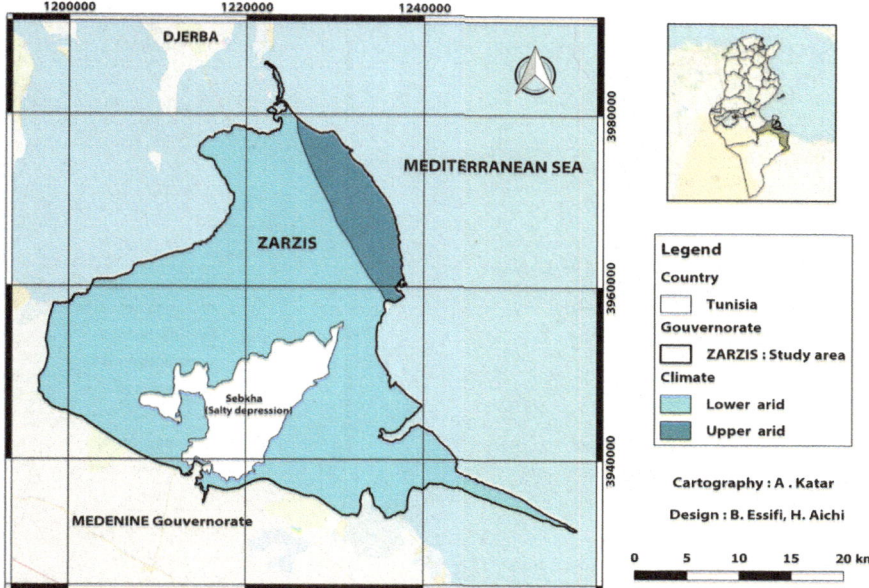

Fig. 11.3 Climatic map of Zarzis region Basemap: OSM under QGIS OpenLayers Plugin [19]

Local winds are responsible for local aeolian transportation, dunes formations, and translocations [22].

Despite their low organic matter content, the soils of Zarzis are relatively well-differentiated and diversified but also sensitive to wind erosion [23]. Figure 11.4 shows the soil classes of Zarzis region.

The current spontaneous vegetation is exposed to two major constraints, on the one hand, the anthropic exploitation consisting in the abusive cut of the wood, the pasture and the clearing for episodic cultures; and on the other the scarcity and the irregularity of the precipitations. However, thee constraints, the vegetation has a remarkable edapho-climatic adaptation and resilience characteristics.

11.2.2.1   Satellite Images Acquisition

The selection of two spaced acquisition dates for the satellite images to be used in our study was based on the calculation of the rainfall normality percentage of the hydrological years. This percentage of rainfall normality is calculated according to the following formula [24]:

$$PNi(mm) = (Pi/Pm) * 100 \tag{11.1}$$

where

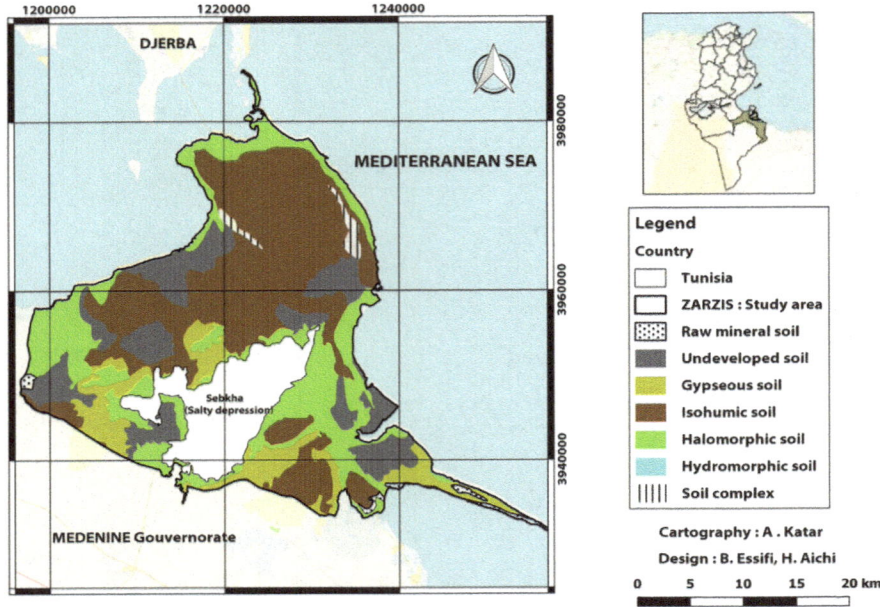

Fig. 11.4 Classes of soils in the Zarzis region Basemap: OSM under QGIS OpenLayers Plugin [19]

PNi (mm)   Percentage of normality of the ith year
Pi (mm)   Annual rainfall of the ith year
Pm (mm)   Average of rainfall of the series of considered years, i.e., from 1994 to 2015

A given year was classified as shown in Table 11.1.

In order to highlight the changes in land use and ecosystem services, we particularly selected two images acquired during the spring season to better appreciate the vegetation cover and its growth phases. For the selected dates, Landsat images were downloaded from US Geological Survey Global Visualization website. Landsat images are among the widely used satellite remote sensing satellite data and their spectral, spatial, and temporal resolutions made them useful input for mapping and planning projects [25]. Two Landsat TM images were acquired (Landsat 5 TM on

Table 11.1 Classification of hydrological years

Hydrological years type	PN_i
Wet	[110, +∞]
Normal	[80, 110]
Moderately dry	[55, 80]
Very dry	[0, 55]

July 4, 2007 and Landsat 8 OLI on July 7, 2014) to produce land use/cover categories respectively.

11.2.2.2 "Carte Agricole" GIS Data

The "Carte Agricole," Agricultural Map, is the official GIS document, published by the Ministry of Agriculture for the entire territory of Tunisia in 2004, which includes all data layers of the physical environment, soil and water resources, topography and bioclimate and records data on irrigated areas, urban areas, and infrastructure. This project was developed to procure, after combining the layers of available data, at the level of each elementary polygon, information on the parameters of the agronomic vocation and associated socioeconomic data. It is a decision support tool for local governance. The "Carte Agricole" GIS document of Medenine that include Zarzis was exploited to collect data that characterizes our study area.

11.2.3 Landsat Image Processing

11.2.3.1 Atmospheric Correction

Each image was separately processed. The two diachronic images were corrected atmospherically under ENVI® 5.3 using the FLAASH model. The corrected images were saved in GeoTiff format in order to be recognized under the IDRISI® software 17.0. Then, the two images were spatially resized to the administrative area of Zarzis region [19].

11.2.3.2 Principal Components Analysis

The two GeoTiff format images were imported under IDRISI 17, filtered by a 3 × 3 Median spatial filter to remove noise [26, 27]. The two original images have been transformed into principal components. Because multispectral data bands are often highly correlated, the principal components (PC) transformation is used to produce uncorrelated output bands. The first PC band contains the largest percentage of data variance and the second PC band contains the second largest data variance, and so on the last PC bands appear noisy because they contain very little variance, much of which is due to noise in the original spectral data. Principal components bands produce more contrasted and colorful color composite images than spectral color composite images because the data is uncorrelated [22, 28].

11.2.3.3 Unsupervised Classification

We opted for unsupervised classification. Unsupervised classification technique does not require the user to specify any information about the features contained in the images. We have used the ISOCLUST module under the IDRISI software environment. The ISOCLUST module is an Iterative Self-Organizing unsupervised classifier [29, 30]. Based on both of field visits and available maps of land use, we also explored the region of Zarzis and its olive groves under various environments and used different tools and software's, including QGIS, ArcGIS Earth and Google Earth. During field visits, we identified the number of classes in the study area along with a careful landscape description, and geolocalized interesting hotspots for the classification scheme.

11.2.3.4 Landsat Image Post-processing

"Land use dynamics play a vital role in the ecological sustainability of any region" [31]. Modelling land use change helps us unravel the dynamics of land use systems. The sensitivity of land-use patterns to changes in key variables can be tested. Sensitivity analysis can help to identify the most important mechanisms of change in a certain area that could not be identified from field observation [32]. To evaluate LULC change between the two dates, we have grouped the classes using the Assign module under IDRISI, which is a ranking tool that allows creating new image or vector files by linking the geography of elements defined in the input with the attributes defined in the output file. As we are only interested in soil surface condition within the olives groves, the output files had only three classes as follow: (i) Class 1: Olive trees on non-disturbed soil; (ii) Class 2: Olive trees on sandy soil; (iii) Class 3: Others LULC.

The analysis of the change in land use-land cover was done using Land Change Modeler. The Land Change Modeler for ecological sustainability is integrated software developed by IDRISI Selva. It helps one to analyze, measure, and project the impacts of land use-land cover changes on habitat and biodiversity. Use of such "a model also gives a better understanding of the functions of the land use systems and the support needed for planning and policy making. Such models can also predict the possible future change and use of the land cover under different scenario" [33–36]. LCM is used to compare pairs of classified images, which have identical legends, where the classification assigns the same unique and distinct identifier to each class on both dates.

The analysis of land cover change was performed between the pairs of Landsat images. The aim was to examine whether the areas fall into the same class on the two dates or a change to a new class has occurred. "The change analysis panel provides a rapid assessment of quantitative change by graphing gains and losses by land cover categories. A second option, net change, shows the result of taking the earlier land cover areas, adding the gains and then subtracting the losses. The third option is to examine the contribution to changes experienced by the single land cover" [26, 36]. Accordingly, the transitions and exchanges that took place between the various

LULC categories during the years were obtained in the map as well as graphical format.

11.3   Results and Discussion

As presented in Table 11.2, the hydrological classification has yielded six years identified as wet, five years as normal, four years as moderately dry and six years as very dry. We have chosen two normal years because we need to compare similar and comparable climatic conditions to avoid any potential error conclusions. Also, we cannot make a basic assumption of extreme events such as very dry or humid conditions. The first one is a Landsat 5 Thematic Mapper (TM), acquired in 2007 and the second one is a Landsat 8 Operational Land Imager (OLI), acquired in spring 2014 because perennial vegetation has its maximum photosynthetic activity to allow the extraction of reliable spectral indices: vegetation, soil, and others Land use.

Table 11.2   Typology of hydrological years from 1994/1995 to 2014/2015 (after, Medenine Weather Station)

Years	Percentage of normality	Typology
1994/1995	85.16	Normal
1995/1996	147.73	Wet
1996/1997	36.43	Very dry
1997/1998	15.18	Very dry
1998/1999	28.70	Very dry
1999/2000	100.66	Normal
2000/2001	47.37	Very dry
2001/2002	204.791	Humid
2002/2003	264.90	Humid
2003/2004	181.85	Humid
2004/2005	36.29	Very dry
2005/2006	84.25	Normal
2006/2007	108.99	Normal
2007/2008	76.65	Moderately dry
2008/2009	66.79	Moderately dry
2009/2010	55.26	Moderately dry
2010/2011	154.53	Humid
2011/2012	195.68	Humid
2012/2013	22.61	Very dry
2013/2014	107.30	Normal
2014/2015	78.78	Moderately dry

Table 11.3 Clusters identification between 2007 and 2014 based on field visits		2007	2014
	Olive Trees on Non-Disturbed Soil	Cluster 3-8	Cluster 3-5-7
	Olive Trees on Disturbed Soil	Cluster 2-6-7	Cluster 2
	Other LULC	Cluster 4-5	Cluster 4-6-8

11.3.1 ISOCLUST Classification

Using the ISOCLUST module in IDRISI, eight clusters were derived. However, three land cover types were retained in the final legend by aggregation of the original eight categories. These groups were identified as Olive trees on non-disturbed soil, Olive trees on sandy soil and Others LULC, in which we have grouped the classes: Sebkha (Salt depression), Urban area and coastal area (Table 11.3).

During field visits, we have noticed an abundance of olive groves in the different sectors of the region. Plantations areas, where olive trees are poorly vigor, were encountered on sandy soil, particularly near Ghrabet, located at 35 km from Zarzis city center and in the southern part of the Zarzis region (Fig. 11.5a). Plantations areas, with good vigor olive trees, were found on non-disturbed soil, particularly in Chamakh region and the northern part of Zarzis region (Fig. 11.5b).

11.3.2 Land Use and Land Cover Maps of 2007

In 2007, the LULC was characterized by the abundance of olive growing on disturbed sandy soil, particularly in the southern part of Zarzis region. This region is commonly characterized by sandy texture soils, poor in organic matter, and nutrients concentrated in the superficial horizon, so they are easily eroded by the wind (Fig. 11.6).

11.3.3 Land Use and Land Cover of 2014

The LULC map highlights an abundance of olive plantations with good vigor, especially in the northern part of Zarzis region (Fig. 11.7).This region is characterized by a temperate climate because of their position near the Mediterranean Sea, the abundance of olive groves with good vigor is due to the two wet years (2010–2011) and (2011–2012).

a-1 and **a-2:** Olive groves on sandy disturbed soil in the Ghrabet area

b-1 and **b-2:** Olive groves of good vigor on non-disturbed soil, in the area of Chamakh

Fig. 11.5 Photos of olive groves in the study area. **a-1** and **a-2** Olive groves on sandy disturbed soil in the Ghrabet area. **b-1** and **b-2** Olive groves of good vigor on non-disturbed soil, in the area of Chamakh

11.3.4 Change Analysis Using the Land Change Modeler Tool

"The change analysis panel provides a rapid assessment of quantitative change by graphing gains and losses by land cover categories. A second option, net change, shows the result of taking the earlier land cover areas, adding the gains and then subtracting the losses. The third option is to examine the contribution to changes experienced by the single land cover" [34, 36].

11.3.4.1 Change Analysis

The analysis of the changes gives estimations of profits and losses as well as the net change between the different classes between the two dates 2007 and 2014.

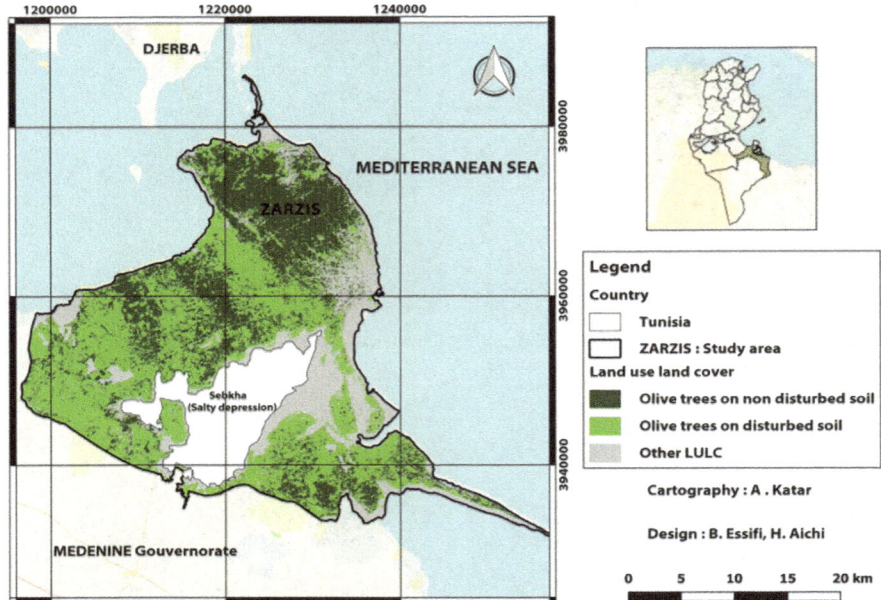

Fig. 11.6 Land-use and land-cover map of 2007, Zarzis region Basemap: OSM under QGIS 2.18.25 OpenLayers Plugin

– *Gain and loss analysis:* The analysis of profit and loss results shows that a net profit in terms of olive trees (good vigor) with a large amplitude of positive change (+47.54% of change) between the year 2007 and 2014 while for olive groves on silted soil decreased, the highest percentage of negative change (−36.7% of change) (Table 11.4).
– *Net change by category:* The analysis of the results of the net change by categories shows that the area for olive trees (Good vigor) increased by 11,386 hectares with a percentage of positive change (+32.86%) while the olive groves on disturbed soil decreased by 7535 hectares with a percentage of negative change (−27.38%) (Table 11.5).

11.3.4.2   Change Maps

Change Maps of LULC
Figure 11.8 illustrates the changes in land use between the different categories between 2007 and 2014 in Zarzis region.
Change Maps by Category

– **Olive groves on non-disturbed soil**

Figure 11.9 illustrates the positive and negative changes for olive groves on the non-disturbed soil.

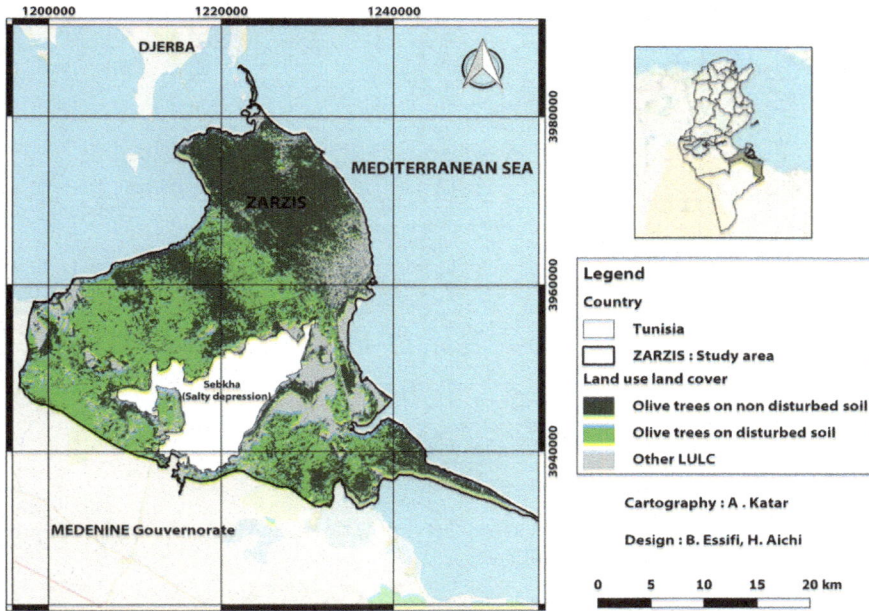

Fig. 11.7 Land-use land-cover map of 2014, Zarzis region Basemap: OSM under QGIS 2.18.25 OpenLayers Plugin

Table 11.4 Analysis of gain and loss between 2007 and 2014

	Gains by categories			Losses by categories		
	Hectare	% Change	% Area	Hectare	% Change	% Area
Olive groves on non-disturbed soil	16,471	47.54	7.23	−5085	−21.86	−2.23
Olive groves on disturbed soil	5331	19.37	2.34	−12,866	−36.7	−5.65
Other LULC	5083	19.44	2.23	−8934	−29.78	−3.92

Table 11.5 Net change by categories between 2007 and 2014

	Net change by categories		
	Hectare	% Change	% Area
Olive groves on non-disturbed Soil	11,386	32.86	5
Olive groves on disturbed Soil	−7535	−27.38	−3.31
Other LULC	−3851	−14.73	−1.69

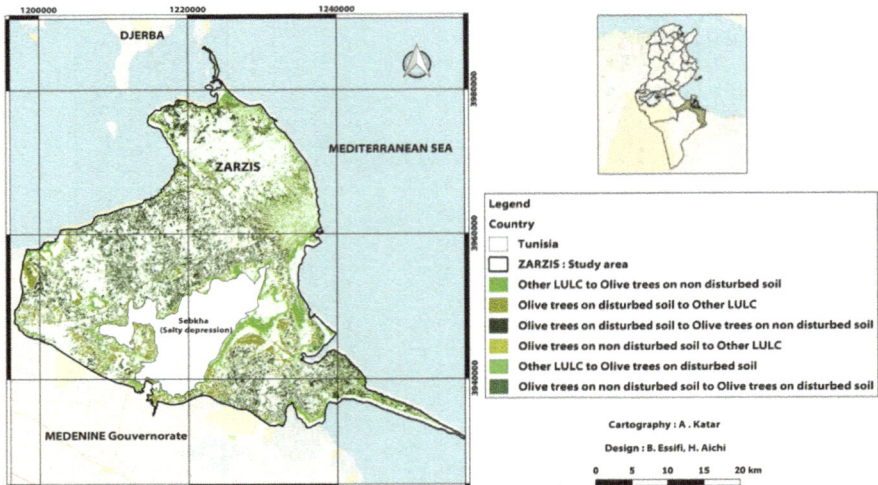

Fig. 11.8 Change map of land use-land cover between 2007 and 2014, Zarzis region Basemap: OSM under QGIS 2.18.25 OpenLayers Plugin

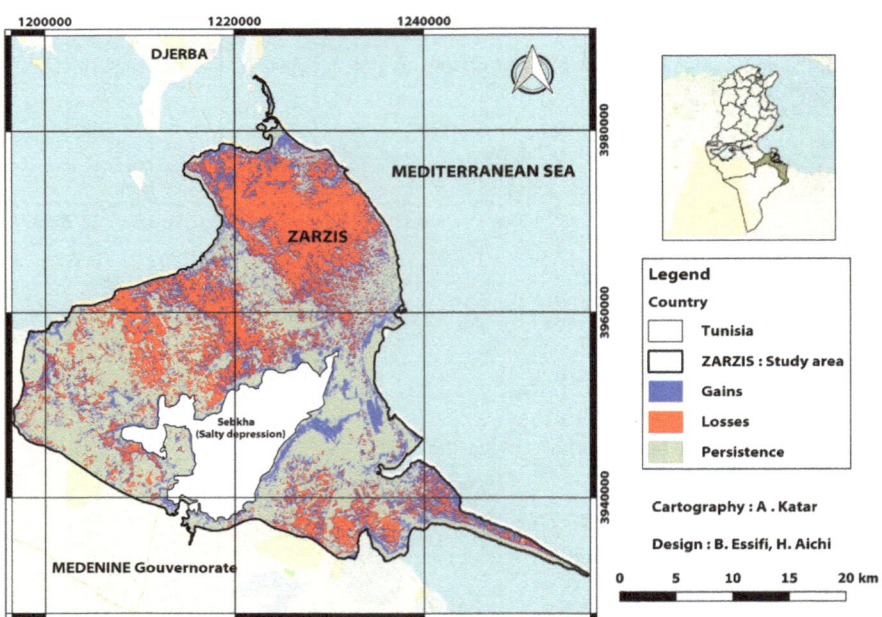

Fig. 11.9 Change map of olive groves on non-disturbed soil, Zarzis region Basemap: OSM under QGIS 2.18.25 OpenLayers Plugin

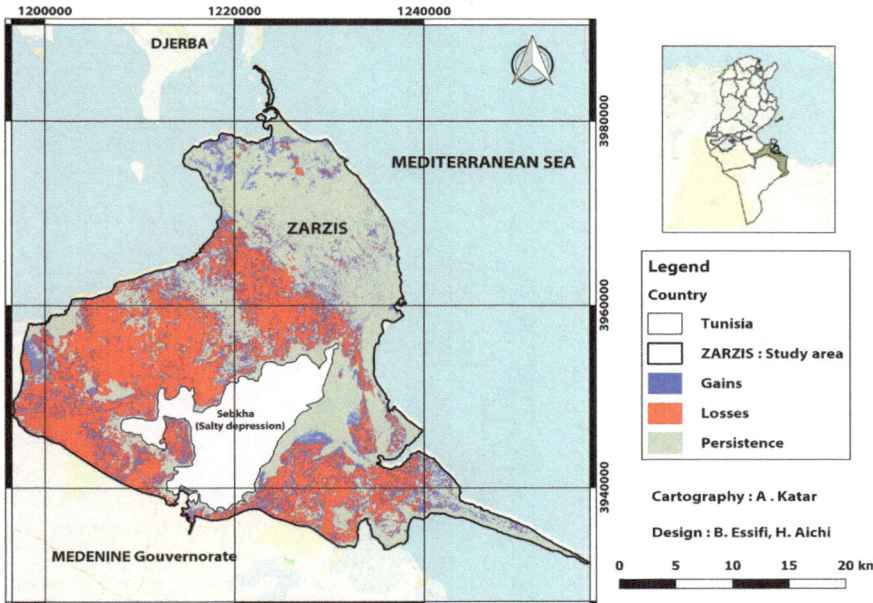

Fig. 11.10 Change map of olive groves on disturbed soil, Zarzis region Basemap: OSM under QGIS 2.18.25 OpenLayers Plugin

In the southeast of Zarzis region, positive change is visible in olive groves on non-disturbed soil. However, there is a slight negative change in the disturbed soil near the sebkha (Salt depression) (Fig. 11.9).

– **Olive groves on disturbed soil:** The positive and negative changes for olive groves on disturbed soil shows that the negative change is very clear and widespread over the entire area of the Zarzis region between 2007 and 2014 (Fig. 11.10).

Transition Maps

- From olive groves on the non-disturbed soil to olive groves on disturbed soil: The transition from olive groves on non-disturbed soil to olive groves on disturbed soil shows that transitional regions are limited to highly sandy regions riparian to the sebkha and areas.
- Those area suffer from a very dry wind blowing from the south, known as "Chehili", brings dry air, causing the dehydration of olive leaves (Fig. 11.11).
- From Olive groves on the disturbed soil to olive groves on non-disturbed soil: The transition of olive groves, from disturbed soil to non-disturbed soil, is valuable in the studied region since the olive trees have been cultivated for centuries for its socioeconomic benefits for farmers and consumers. In fact, Zarzis is ranked among the first productive areas of olive trees in Tunisia (Fig. 11.12).

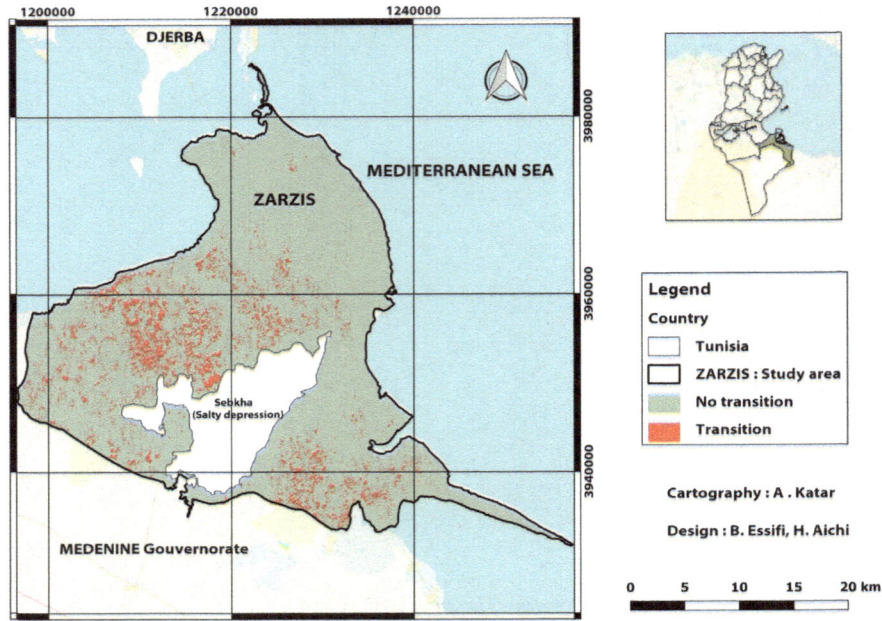

Fig. 11.11 Transition map from olive groves on non-disturbed soil to olive groves on disturbed soil

11.4   Conclusions

Assessing LULC variation, over both time and space, is vital for better planning, decision-making, and sustainable land management from local to a global scale. Earth observation is a valuable tool for evaluating the dynamics of agroecosystems, LULC changes, and land degradation status based on multi-temporal studies. The study of the vigor dynamics of olive groves in the Zarzis region was carried out in order to detect the impact of climate change on the environment and to map the land use using Landsat TM and OLI data between the years 2007 and 2014. The results indicate an improvement in the vigor of olive groves in the region.

This improvement is attributable to an increase in precipitation during the years 2010 and 2012, which are considered as wet years. Also, this region has benefited from the implementation of the national strategy for soil and water conservation, the first phase of which was carried out between 1990 and 2001 and the second phase between 2002 and 2011. Particularly through creating irrigated areas, managing land with Tabias, and promoting agroforestry consisting of integration of market gardening crops as intercropping between the lines of the olive trees. The present study demonstrated the efficiency of remote sensing data in the study of land use and land cover changes. It gives a good understanding of land use/land cover changes for

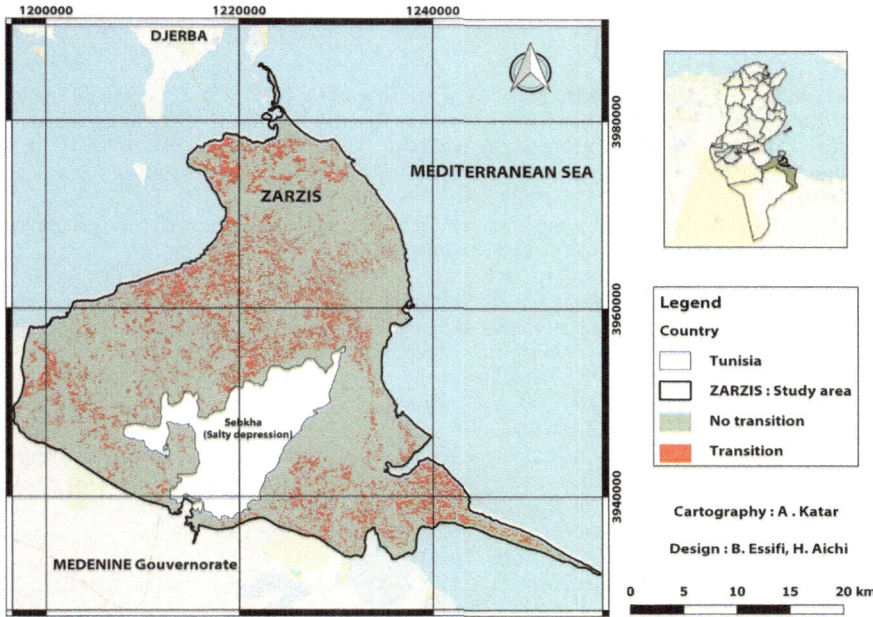

Fig. 11.12 Transition map from olive groves on the disturbed soil to olive groves on non-disturbed soil

a period between the years 2007 and 2014. Modelling and predicting LULC provides valuable information regarding natural resources degradation and utilization.

11.5   Recommendations

Modelling and predicting LULC provides valuable information with regard to natural resources degradation in southern Tunisia regions. The resulting maps from this study demonstrated the efficiency of remote sensing data, based on Land Change Modeler (LCM), give a good understanding of land use/land cover changes over a decade, which in turn will be very helpful for local agricultural development agencies for decision makings on combating desertification in the area.

Acknowledgements   This research was undertaken thanks to the valuable support of the Laboratory of Eremology and Combating Desertification at the Institut des Regions Arides (IRA), Ministry of Agriculture, Water Resources and Fisheries, Tunisia.

References

1. Adham A, Wesseling JG, Riksen M, Ouessar M, Ritsema CJ (2019) Assessing the impact of climate change on rainwater harvesting in the Oum Zessar watershed in Southeastern Tunisia. https://doi.org/10.1016/j.agwat.2019.05.006 (2019)
2. Hulme, M (2009) Why we disagree about climate change: understanding controversy, inaction, and opportunity. Cambridge University Press, Cambridge, 432 pp
3. Khaldi A (2005) Impacts of drought on the regime of underground flows in the limestone massifs of western Algeria "Monts de Tlemcen – Saida". University of Oran
4. Mashame G, Akinyem F (2016) Towards a remote sensing-based assessment of land susceptibility to degradation: examining seasonal variation in land use-land cover for modeling land degradation in a semi-arid context. ISPRS Annals of the Photogrammetry, Remote Sensing and Spatial Information Sciences, Volume III-8, 2016. XXIII ISPRS Congress, 12–19 July 2016, Prague, Czech Republic
5. William N (2000) Agricultural and small watershed hydrology: watershed characteristic. College of Engineering, Michigan State University
6. Mishra VN, Rai PK, Mohan K (2014) Prediction of land use changes based on land change modeler (LCM) using remote sensing: a case study of Muzaffarpur (Bihar) India. J Geogr Inst Cvijic 64(1): 111–127. https://doi.org/10.2298/ijgi1401111m
7. Lambin EFG, Helmut J (2006) Land-use and land-cover change: local processes and global impacts. Springer, Berlin and New York. (2006)
8. Millennium Ecosystem Assessment (MEA) (2005) Ecosystems and human well-beings: biodiversity synthesis. World Resources Institute, Washington, DC
9. Fischlin A, Midgley GF, Price JT, Leemans R, Gopal B, Turley C, Rounsevell MDA, Dube OP, Tarazona J, Velichko AA (2007). Ecosystems, their properties, goods and services. In: Parry ML, Canziani OF, Palutikof JP, Vander Linden PJ, Hanson CE (eds) Climate change 2007 impacts, adaptation and vulnerability. Cambridge University Press, Cambridge, pp 211–272
10. Verburg PH, Chen Y, Soepboer W, Veldkamp A (2000) GIS-based modeling of human environment interactions for natural resource management. Applications in Asia In: Proceeding of 4th International Conference on Integrating GIS Environmental Modeling: Problems, Prospects and Research Needs, Banff, Alberta, Canada, September 2–8, 2000, 1–18.5
11. King LH, Almohamad ZN (2006) Climate projections in Tunisia for 2030 and 2050. Tunisian Climate Change Adaptation Strategy Working Document. Tunis
12. Trabelsi K (2011) Mapping of the land cover of the Oued Chiba watershed in 1987 and 2010 from Landsat 5 TM images. End of Studies Project, the Higher School of Agriculture of Mograne
13. Gasmi I (2013) Study of the Jessours in the Monts de Matmata (El-Jouabit-Toujène microwatershed). National Agronomic Institute of Tunisia
14. Chaudhry A, Sharma S (2015) Remote sensing and GIS based approaches for LULC change detection-a review. Int J Curr Eng Technol
15. Camacho Olmedo MT, Paegelow M, Mas JF, Escobar F (2017) Geomatic approaches for modeling land change scenarios. An introduction. Lecture Notes in Geoinformation and Cartography, 1–8. https://doi.org/10.1007/978-3-319-60801-3_1
16. Mashame G, Akinyemi F (2016) Towards a remote sensing-based assessment of land susceptibility to degradation: examining seasonal variation in land use-land cover for modelling land degradation in a Semi-Arid context
17. Béchir R, Ounalli N, Sghaier M (2011) Territorial inequality in the Governorate of Medenine (south-eastern Tunisia): analysis of regional indicators for improving living conditions
18. ODS (Development Office of the South) (2014) Medenine Governorate in figures
19. Agricultural map of the governorate of Medenine (2004) Regional Agency of Agricultural Development (CRDA) of Medenine. Ministry of Agriculture, Water Resources and Fisheries, Tunisia

20. Escadafal R (1989) Characterization of the surface of arid soils by field observations and remote sensing. Applications: Example from the Tataouine region (Tunisia). Doctoral thesis in pedology, University of Paris 6
21. Gamoun M (2012) Impact of resting on plant cover dynamics: application to the sustainable management of Saharan pastoral spaces in southern Tunisia. Doctoral thesis, Faculty of Mathematical, Physical and Natural Sciences of Tunis
22. Essifi B (2008) Comparative study of the risks of erosion around water points in the Dahar and El Ouara rangelands in Tataouine region: application of Remote Sensing and GIS. Specialized master in geomatics. Faculty of Letters, Arts and Humanities of Manouba/Institut des Regions Arides (IRA) of Medenine
23. Floret C, Pontanier R (1982) The aridity in pre-Saharan Tunisia: climate, soil, vegetation and development. University of Montpellier
24. Rognon P (1996) Drought and aridity: their impact on desertification in the Maghreb, Volume 7–4, December 1996. Lien: http://www.jle.com/fr/revues/sec/e-docs/secheresse_et_aridite_leur_impact_sur_la_desertification_au_maghreb_270740/article.phtml?cle_doc=00042194&type=text.html
25. Sadidy J, Firouzabadi PZ, Entezari A (2009) The use of radarsat and landsat image fusion algorithms and different supervised classification methods to use map accuracy-case study: Sari Plain-Iran
26. Eastman JR (2009) Idrisi release 2: tutorial, manual version 32, Clark Labs, Clark University, USA, p 342
27. Patidar P, Gupta K, Srivastava S, Nagawat AK (2010) Image de-noising by various filters for different noise. Int J Comput Appl 9:(4) (November)
28. Richards JA (1999) Remote sensing digital image analysis: an introduction. Springer-Verlag, Berlin, Germany
29. Ball GH, Hall DJ (1965) ISODATA, a novel method of data analysis and pattern classification. Technical Report, Stanford Research Institute, Menlo Park, CA, USA
30. Mourtala B (2013) Dynamics of ecosystem services in the Oum Zessar watershed (South East Tunisia). Master thesis, National Agronomic Institute of Tunisia
31. Kumar S, Bhaskar PU, Padmakumari K (2015) Application of land change modeler for prediction of future land use land cover. A case study of Vijayawada city. In 2nd International Conference on Science, Technology and Management. University of Delhi, New Delhi, India
32. Verburg PH, Kok K, Pontius JRG, Veldkamp A (2006) Modeling land-use and land-cover change. In: Lambin EF, Geist HJ (eds) Land-use and land-cover change: local processes and global impacts. Springer, Berlin
33. Costanza R, Ruth M (1988) Using dynamic modeling to scope environmental problems and build consensus. Environ Manag 22:183–195
34. Clark Labs (2009) The land change modeler for ecological sustainability. IDRISI Focus Paper, Worcester, MA: Clark University. www.clarklabs.org/applications/upload/Land-Change-Modeler-IDRISI-Focus-Paper-pdf
35. Ahmed B, Ahmed R (2012) Modeling urban land cover growth dynamics using multi-temporal satellite images: a case study of Dhaka, Bangladesh. ISPRS Int J Geoinformation 1:3–31. https://doi.org/10.3390/ijgi1010003
36. Rai P, Narayan Mishra V, Mohan K (2014) Prediction of land use changes based on land change modeler (LCM) using remote sensing: A case study of Muzaffarpur (Bihar), India. http://www.doiserbia.nb.rs/img/doi/0350-7599/2014/0350-75991401111M.pdf

Chapter 12
Drought Assessment in Tunisia by Time-Series Satellite Images: An Ecohydrologic Approach

Hedia Chakroun

Abstract Global and regional monitoring of drought are becoming an active research subject during the last decades. In the Middle East and North Africa (MENA) region drought episodes highly control water availability and the functioning of both forested and cultivated ecosystems. The ecohydrologic approach represents a relatively new trend in the holistic assessment of these limited water resources ecosystems as it explains the equilibrium between the components of the soil-vegetation-climate complex. On the other hand, during the last two decades, the models used in the assessment of drought causes and manifestations combine more and more indicators from multi-sensors satellite images. In the present study, the ecohydrology concepts and their methodological basis are presented, and an overview of biophysical and energetic variables derived from remote sensing data at the regional scale are exposed. A general review of the use of remote sensing in ecohydrology during the last two decades is also addressed as well as various methods using satellite images in the ecohydrologic modelling. These methods are divided into two major groups: the direct use of remote sensing in drought and humidity assessment, and the integration of satellite images with other data in water balance models for hydric stress assessment. The ecohydrological modelling integrating remote sensing data makes use of different types of models such as statistical, empirical or physical based models. The availability of free time series satellite images such as MODIS sensors since the year 2000 had allowed the exploration of various models for drought assessment where multispectral images are combined to derive drought indicators of the vegetation in a Tunisian Mediterranean ecosystem. Finally, data quality of time-series images and their calibration and correction are discussed to highlight the required processing for convenient use of these data in drought monitoring at the regional scale.

Keywords Ecohydrology · MODIS images · Northern Tunisia · Drought

H. Chakroun (✉)
National Engineering School of Tunis (ENIT), University of Tunis El Manar,
BP 37, 1002 Tunis, Tunisia
e-mail: hedia.chakroun@enit.utm.tn

© Springer Nature Switzerland AG 2021
F. Khebour Allouche et al. (eds.), *Environmental Remote Sensing and GIS in Tunisia*, Springer Water, https://doi.org/10.1007/978-3-030-63668-5_12

12.1   Introduction

Drought types are classified into meteorological drought due to a lack of precipitations over a period, hydrological drought with inadequate water resources required
for established water uses, agriculture drought due to vegetation water stress, and
socioeconomic drought with the failure of water resources systems to meet water
demands [1]. In limited water resources regions, all these drought manifestations are
interconnected resulting in a high control of drought episodes on water availability
and functioning of forested and cultivated ecosystems. In Tunisia, as well as in the
South Mediterranean region countries, the manifestations of drought affect both agricultural and forestry sectors. In this context, various approaches had been developed
for drought assessment, drought monitoring and mitigation. Many actions have been
initiated to make regular drought monitoring such as a good network of rainfall, flow
measurements stations, and statistics bulletins over the country. Also, actions related
to drought management are undertaken by identifying principal drought indices such
as SPI (Standardized Precipitation Index), and by adopting a strategy of intervention
in case of drought. These existing methods and strategies can be supported by a better
knowledge of the effect of drought on vegetation at a more regional scale such as the
ecosystem limits.

   During the last two decades, the models used in the assessment of drought causes
and manifestations had combined more and more indicators from multi-sensors satellite images. On the other hand, a relatively new trend in the holistic assessment of
ecosystems known as ecohydrologic approaches, have shown real potential in the
assessment of the limited water ecosystems, especially as it explains equilibrium
between the components of the soil-vegetation-climate complex. In this chapter, a
description and a discussion of the research methods based on remote sensing integrated to ecohydrologic approaches are presented. The ecohydrology concepts and
their methodological basis are first explained, and then, a general review of biophysical and energetic variables derived from remote sensing data at the regional scale
are exposed. Various models using time-series satellite images are presented, and
different remote sensing-based drought indices are proposed at the regional scale.
Finally, the data quality issues and their effects on drought assessment are discussed.

12.2   Principles of the Ecohydrology

The ecohydrology explains the mutual influence of hydrological and ecological
process and seeks to elucidate how hydrological processes influence the distribution, the structure, the functioning, and the dynamics of biological communities, and
how feedbacks from biological communities affect the water cycle [2, 3]. Studies
that have focused on ecology and hydrology interactions are not new if one refers
to Van den Honert's publications in 1948 on soil-plant interactions [4] that later
inspired other scientists such as [5] who proposed the hypothesis of ecohydrological

equilibrium, and [6] who studied the influence of climate on soil moisture. Nevertheless, in the nineties of the twentieth century, ecohydrology was introduced as a new discipline when the term "ecohydrology" was first adopted in the fifth UNESCO program on Hydrology [7]. This discipline, considered by some authors as new and by others as an interface between ecology and hydrology, explains the feedbacks between ecology and the physical processes related to the distribution of precipitation between transpiration, evaporation, infiltration and runoff [2, 8, 9]. The main developments in ecohydrology can be summarized in two approaches:

- The approach based on the hypothesis of the ecohydrological equilibrium explaining the interactions of the vegetation-soil complex as a response to a stochastic climate in semi-arid to arid zones. This trend is essentially presented by [5, 10] who proposed the hypothesis of ecohydrological equilibrium (or optimality) in areas with limited water resources.
- The approach based on ecohydrology as an adequate tool for integrated water management that has the potential to explicitly integrate hydrological processes with the biotic dynamics of ecosystems [7].

In limited water resources regions, Eagleson in [5, 10] formalized the interaction of the soil-vegetation-climate complex according to a dynamic approach that reproduces the hydrological behaviour based on the hypothesis of an equilibrium between biological and hydrological processes at the ecosystem level. This model known by "The Equilibrium Water Balance" roughly consists of an equation modelling the different mean annual flows of the soil-vegetation-climate complex at a one-dimensional representation of soil moisture dynamics (Eq. 12.1). This approach represents the dynamic of water balance conditioned by a stochastic climate; it is scaleless and can be applied at a plot or a regional level.

$$E[P] = E[ET(s_0, Climate, Soil, Vegetation)] + E[R_g(s_0, Climate, Soil)]$$
$$+ E[R_s(s_0, Climate, Soil)] \qquad (12.1)$$

E: Expected value of the variable
P: Annual precipitation
ET: Annual bare soil evaporation and transpiration
R_s: Annual surface runoff
R_g: Annual groundwater runoff
s_0: Equilibrium soil moisture concentration

The hypothesis of Eagleson in a natural, undisturbed ecosystem in equilibrium as stated in [11] are:

1. Within a few generations of vegetation development, soil moisture reaches a state of equilibrium between water balance components (precipitation, evaporation, transpiration, infiltration and runoff).
2. At a medium time-scale, the vegetation evolves towards a state of equilibrium such that the persistent plant species are those whose potential transpiration

efficiency results in the maximum equilibrium soil humidity; this explains the influence of soil on vegetation.

3. Over a longer timescale, the influence of the vegetation on the properties of the soil induces the evolution of the vegetation towards maximization of the canopy; this explains the influence of vegetation on the soil.

12.3   Remote Sensing Integration in Ecohydrological Approaches

12.3.1   Overview of Remote Sensing Principles

Remote sensing is the scientific discipline that brings together all the knowledge and techniques used to observe, interpret and manage geographical space from measurements and images obtained by sensors use. The basic data is the physical energy reflected or emitted in specific spectral ranges covering an interval of wavelengths (λ) where the measurements are made. Extraction of biophysical and thematic information from this data requires digital image processing and corrections known as geometric, radiometric and atmospheric corrections. Generally, calibration of images based on field observations is required especially when the images are used to extract biophysical variables. The biophysical status variables of the soil-vegetation complex (e.g. the Leaf Area Index (LAI), the humidity, the surface temperature, the net radiation) are used in ecohydrological models at different scales (plot, catchment, ecoregion). The dynamics of the soil-vegetation complex is principally measured by the surface reflectance in the visible spectral range (VIS) (where λ vary from 0.4 to 0.7 μm) and the Near Infra-Red range (NIR) (where λ vary from 0.75 to 0.8 μm). For the vegetation water state, the reflectance in the Short Wave Infra-Red (SWIR) is used (where λ vary between 1.6 and 2.2 μm). For much longer wavelengths known by the Thermal Infra-Red (TIR) (where λ vary between 8 to 14 μm), the surface temperature of soil-vegetation can be determined through the well-known Stefan-Boltzman equation. In the hyperspectral waves known by the radar domain where λ are centimetric, remotely sensed data allow the monitoring of soil humidity.

12.3.2   Concepts of Remote Sensing Uses in Ecohydrology

The spatio-temporal variability characterising the components of water balance has made the subject of various studies carried out both in ecology and in hydrology as distinct domains. In these fields, remote sensing had been widely used as it offers products that could be integrated into analytical tools to represent spatio-temporal aspects at different scales. Therefore, the multiple approaches of ecohydrology discipline could also take advantage of remote sensing. Moreover, data collected over the years using sensors constitute long series observations spreading over two or three

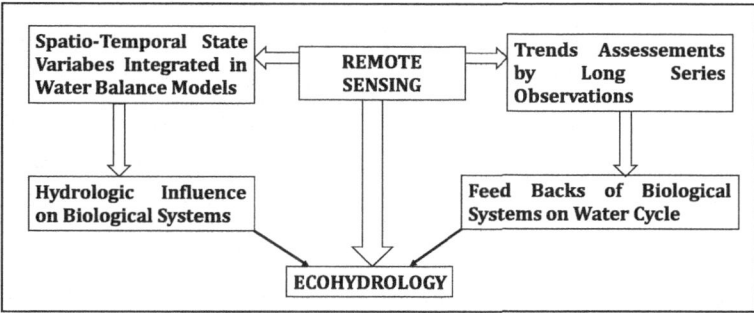

Fig. 12.1 Remote sensing potential in ecohydrology

decades, thus bringing more knowledge on the ecohydrological dynamic in space and time.

As stated earlier, the ecohydrology explains the mutual influence of hydrological and ecological process. Two main aspects could take advantage of the use of remote sensing products as illustrated in Fig. 12.1:

– A better understanding of the hydrological influence on an ecosystem: remote sensing is used as a source of information on the spatio-temporal dynamics of state variables related to soil and vegetation and their integration in water balance models.
– A better explanation of biological systems feedbacks on the water cycle: long time-series observations of the vegetation dynamics integrated into the water balances offer potential in analysing trends applied at the regional ecosystem levels.

More specifically, remotely sensed data are generally combined with ancillary data like digital elevation models, soil, land use maps, and field observations to produce dynamic state variable of the soil-vegetation complex such as the vegetation indices, the water status of soil and vegetation and the climate-driving energetic parameters as temperature and water evaporative factors. The state variables are integrated into functional models that could be empirical, physical, or statistical, to simulate the ecohydrologic functioning in an ecosystem. For example, the spatio-temporal dynamics of the vegetation is usually integrated into surface energy balance models through times-series images to evaluate the vegetation water consumption. This spatio-temporal information is also useful in monitoring the response of soil-vegetation to climate conditions through drought indicators follow-up that is of great interests in integrated water management. These processes are summarized in Fig. 12.2.

The next sections present case studies of ecohydrology dynamic modelling by remote sensing within the Khroumirie ecoregion covering a forested band along the Mediterranean coast with a total area of 2553 km^2. In Sect. 12.4, a study case of the canopy hydric status modelling by a statistical approach is presented. In Sect. 12.5, the integration of remote sensing and ancillary data in a water balance model is

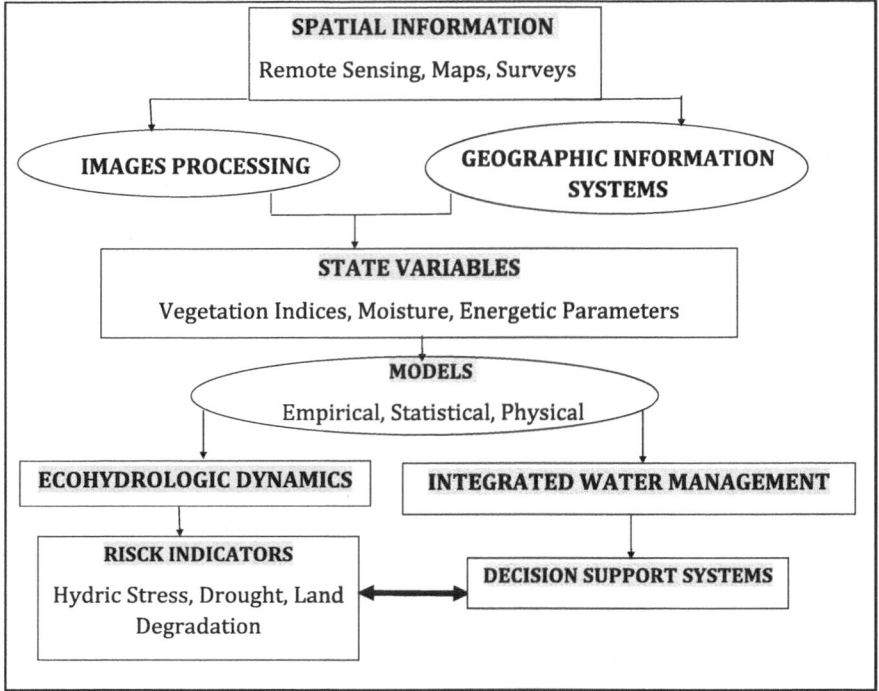

Fig. 12.2 Integration of remote sensing and ancillary data in ecohydrologic modelling

described as well as the ecohydrology equilibrium assessment at the ecoregion scale. A case study on the contribution of ecohydrology in integrated water management was published previously in a chapter book [12].

12.4 Humidity Canopy Assessment at the Ecoregion Level

The vegetation spectral response is correlated with its water content, mainly in the spectral bands of the SWIR. A multiplicity of indices developed and used in different ecosystems have been linked to soil and vegetation water content using in situ measurements coinciding with the satellite's passage whose sensors are measuring the spectral responses (for complete synthesis of these indices, refer to [13]).

A jointly based approach using leaf humidity measurements and weekly MODIS images MOD09A1 and MOD15A2 were developed to propose a model for hydric state evaluation at the canopy level in a forested region in northern Tunisia. The Equivalent Water Thickness at the canopy level (EWT_{CAN}) is an indicator of the leaf amount of water based on the fuel moisture content (FMC) which is the mass of water contained within vegetation in relation to the dry mass [14]. The adaptation of the formulation of EWT_{CAN} to the types of forested ecoregion in Northern Africa is expressed in (Eq. 12.2) which takes into account the specificity of the open and heterogeneous forest ecosystem in the Khroumirie ecoregion which is typical of North Africa forested areas (for more explanation on the foundation of this equation, refer to [15]). The EWT_{CAN} had been computed in different plots over the study region varying by vegetation density.

$$EWT_{CAN} = \frac{10^{-4}}{\rho_W} \left[\begin{array}{l} (COV)(LAI_{min}) \left(\dfrac{\sum_{i=1}^{N}(FMC_i)(LMA)_i}{N} \right) \\ + (1 - (COV))((LAI) - (LAI_{min})) \\ (FMC_{LITTER})(LMA_{LITTER}) \end{array} \right] \tag{12.2}$$

EWT_{CAN}: Equivalent water thickness of the canopy in each plot ($gH_2O \bullet cm^{-2}$)
LAI: Leaf area index ($m^2 \bullet m^{-2}$)
FMC: Fuel moisture content ($\%gH_2O \bullet gC^{-1}$)
COV: % vegetation cover of each plot
N: Total number of species in each plot
ρ_w: Density of pure water

The objective of this study was to find out a relation between the EWT_{CAN} and the vegetation indices derived from MODIS sensors. It is possible to distinguish two classes of vegetation indices: the Biomass Vegetation Indices (BVI) computed from the (VIS) and the (NIR) spectral MODIS bands (B1, B2, B3, B4), and the Moisture Vegetation Indices (MVI) computed from (SWIR) MODIS bands (B2, B5, B6, B7) that are particularly sensitive to the water content of vegetation [16]. The computed (BVI) were: the NDVI (Normalized Difference Vegetation Index) [17], the EVI (Enhanced Vegetation Index) which does not saturate at high biomass compared to NDVI [18]. Other vegetation indices combining vegetation and soil spectral response were also determined: the SAVI (Soil Adjusted Vegetation Index) [19], the MSAVI (Modified Soil Adjusted Vegetation Index) [20], and the ANDVI (Adjusted Normalized Difference Vegetation Index) [21]. The most used moisture vegetation indices (MVI) reported in the literature are the NDWI (Normalized Difference Water Indice) [22], the NDII (Normalized Difference Infrared Indice) [23] and the GVMI (Global Moisture Vegetation Indice) [24].

Statistical analyses were conducted to explore the relations between EWT_{CAN} from in situ measurements and biomass and moisture vegetation indices determined from remotely sensed data. These analyses show that the relationship between EWT_{CAN} and the vegetation indices (BVI) and (MVI) have significant linear relationships (R^2 between 0.6 and 0.7; $p < 0.01$) (Details of these analyses are in [15]). Statistical analyses also showed that for low vegetation density areas (LAI < 2), the adequate inversion model should integrate the LAI values. Therefore, two inversion models were proposed to make the EWT_{CAN} regionalisation from remotely-sensed vegetation indices:

- MOD1 model (Eq. 12.3) based on the only (BVI) or (MVI) within a relatively dense vegetation cover (validity of the model for LAI > 2)
- MOD2 model (Eq. 12.4) integrating the effect of plant cover density through the use of LAI.

$$EWT_{CAN\_MOD1} = \frac{1}{A}(VI) - \frac{B}{A} \qquad (12.3)$$

$$EWT_{CAN\_MOD2} = \frac{(VI) - B}{\alpha(LAI) + \beta} \qquad (12.4)$$

A, B: Slope and intercept of the linear relation between (EWT$_{CAN}$, VI)

α, β: Slope and intercept of the linear relation between [LAI, SLOPE (EWT$_{CAN}$, VI)]

VI: The BVI or the MVI vegetation Indice

The regionalisation of the EWT$_{CAN}$ models derived from temporal series of MODIS images at a regional scale enabled dynamic monitoring of the canopy water content thus contributing to the ecohydrological characterization of this ecosystem. Figure 12.3 represents the difference Δ (EWT$_{CAN}$) computed between two successive weeks for key periods of the dry season (from May to October 2010) in the Mediterranean climate (late spring, mid-summer, late summer and early autumn). This approach based exclusively on the time-series of images is attractive for the operational monitoring of the dynamics of the canopy water state and to detect hydric stress of the vegetation at a regional scale which is of great interest in the management of these areas and the prevention of drought effects such that forest fires. Modelling the water content of the canopy can also constitute a forcing data in the water balance models such as the one proposed by [25].

Fig. 12.3 Time course of regional maps of Δ (EWT$_{CAN}$; g•cm^{-2}) between May and October 2010 [15] (Reproduced by kind permission of MDPI Publishing). Before this figure it would be nice to add a general map of Tunisia to understand where this specific place is located

12.5   Ecohydrology Equilibrium Assessment by Remote Sensing

The dynamics of an ecosystem vegetation influence the water balance. This can be evaluated by the LAI which is an index allowing the measurement of the proportion of solar energy intercepted by the foliage. In multi-layer vegetation conditions, the interception of radiation in the canopy follows the Beer-Lambert extinction law $(1-\exp^{-k.LAI})$ where (k) is an extinction coefficient that can be calibrated according to the ecosystem species [26]. At the regional level, LAI satellite products have been widely used in water balance models in different ecosystems with natural and cultivated vegetation.

In the present study, our objective was to evaluate the validity of the hypothesis of ecohydrological equilibrium, initially proposed at the plot level, within the Khroumirie ecoregion. Our reasoning was based on the fact that the LAI is an integrated measure of the density of a canopy that varies with the availability of water according to the theory of the ecohydrological equilibrium [27]. Therefore, it is possible to integrate LAI times-series products into a water budget model to explore the interaction between water availability and the soil-vegetation dynamic.

12.5.1   Water Balance Modelling with Times-Series LAI-MODIS

In the Khroumirie ecoregion, LAI-MODIS time-series images were processed and integrated into a water balance model for the simulation of water stress spatio-temporal dynamic. Before their use, images were calibrated to avoid overestimation of LAI values (report to Sect. 12.6.2 for details on calibration process). We have integrated the calibrated (LAI-MODIS-CALIB) serial images into a water balance model called SIERRA (SImulator for MediterrERRanean landscApes), which is a physical-based grid model simulating water and carbon balances in Mediterranean ecosystems with natural vegetation [28]. In this model, water flows are explicitly described for vegetation and soil. The potential evapotranspiration (PET, Eq. 1 in Fig. 12.4) was calculated by the Priestley-Taylor formula [29] considered adequate for regional simulations in the Mediterranean regions [30]. The actual transpiration of the vegetation depends on the PET, but it is also strongly conditioned by the LAI [31] (Eq. 3 in Fig. 12.4), and by the stress inherent to the water potential of the soil [32] that was evaluated by the Saxton and Rowls model [33] using soil texture and organic matter contents (Eqs. 2 and 4 in Fig. 12.4).

Hydric stress simulations over the period 2003–2008 at the regional scale were achieved with a time step of 8 days and a spatial resolution of 1 km. A daily water stress index (WSI) defined as the ratio between actual and maximum transpiration was determined for each day of the simulation (Eq. 5 in Fig. 12.4). Computation of the

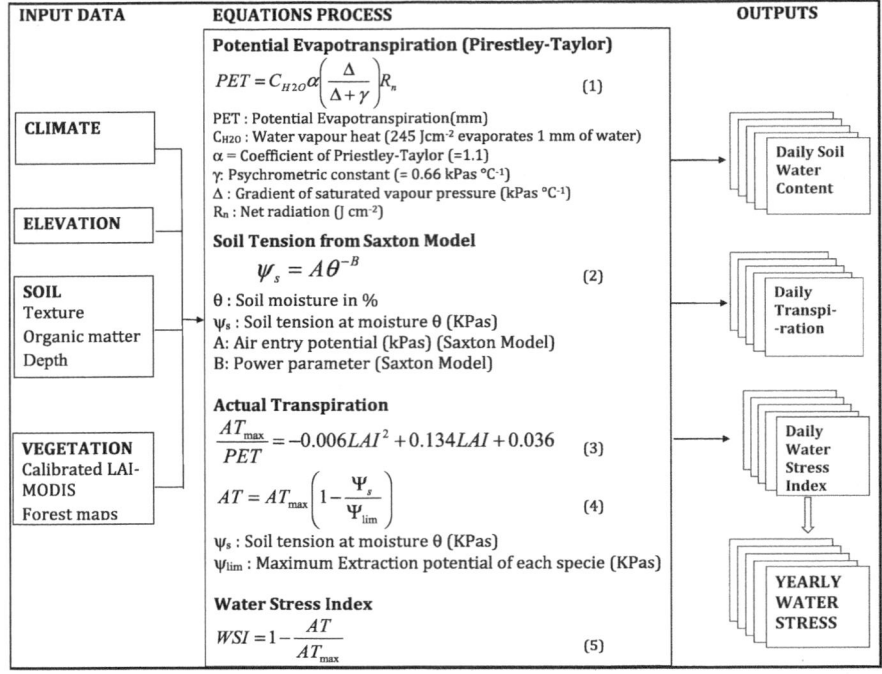

Fig. 12.4 Overview of the water balance model SIERRA

annually integrated WSID (Water Stress Index Days) was based on the summation of the days where WSI exceeds 0.3.

12.5.2 Water Stress Index Effect on Ecohydrologic Equilibrium

To depict the vegetation LAI dynamic effect on the distribution of the WSID, we made grid-based comparative analyses with simulated WSID using (LAI-MODIS-CALIB) and with unchanging LAI (LAI-FIX = 2.5) meaning that no vegetation dynamic is being integrated in the water balance model, which is in this hypothetic case driven only by climatic variables. Figure 12.5 represents the number of years where simulated yearly drought WSID exceeded 3 months within the 6-year period 2003–2008, using these two scenarios.

The hypothesis of ecohydrological optimality (described in Sect. 12.2) is based on the fact that there is a balance between the soil water reserve depending on field capacity and precipitation, and the leaf area [10, 34]. It assumes a mutual influence between optimized plant cover and maximum soil moisture (Eq. 12.1). Therefore,

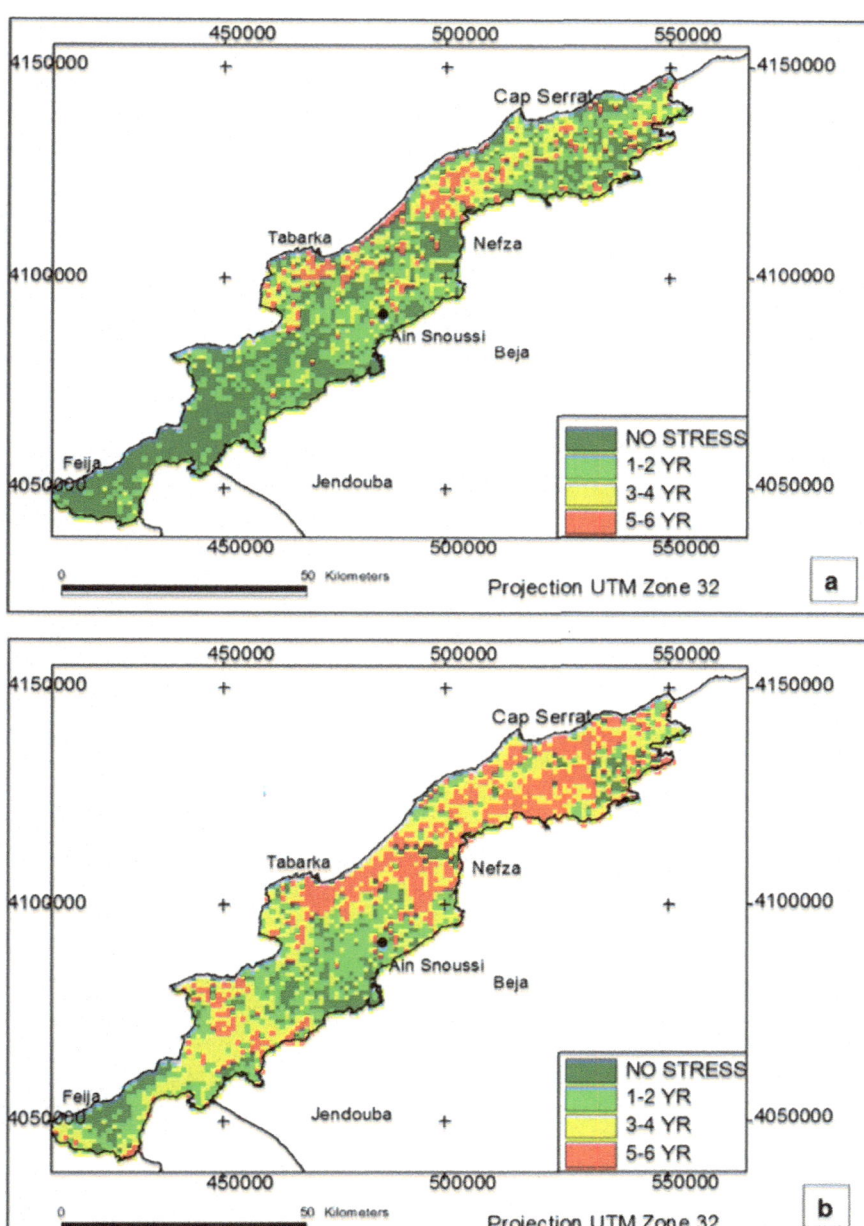

Fig. 12.5 Number of years of hydric stress (a) Case of LAI-MODIS-CALIB based simulations (b) Case of unchanging LAI [36] (Reproduced by kind permission of John Wiley and Sons Publishing)

plant species in water-limited environments must balance their optimum interception of light and minimize their transpiration to avoid drying out too quickly. Among the processes of functional adjustment of species to their environment, the modulation of the leaf area according to the drought is a process commonly identified under the Mediterranean climate [35]. Taking advantage of the long-term study integrating time-series satellite images integrated in a water balance model, we carried out spatial analyses based on modelled time-series maps of annual water stress in comparison with the characteristics of soil water availability, namely water depth determined from soil maps and water potential determined by the Saxton and Rowls model [33]. Spatial analyses were carried out to make the crossing between WSID and shallow/dry soils, and WSID and medium/high tension soils.

In the low-stress zones (WSID = 1 year over the simulated 6 years), the WSID simulated with the LAI-MODIS-CALIB are comparable between shallow soil and deep soil at the opposite to simulations made with unchanging LAI. The same trend is observed with soil water tension (Fig. 12.6a). This means that with an LAI that is hypothetically invariable in space (i.e. no adjustment between vegetation and soil), stress is strongly conditioned by the intrinsic characteristics of the soil. On the other hand, if we integrate the true variations of the LAI as captured by MODIS, we can find a lowly stressed vegetation that has conditioned itself on the soil underlying despite that this is a priori not favourable in terms of water availability (shallow soil or high retention). For example, Fig. 12.6a shows that with LAI-FIX, 37% of the vegetation with 1-year of water stress during the period 2003–2008 is found on shallow soil, whereas it increases to 41% if the simulations were done with LAI varying from year to year. This tendency is accentuated by considering the soil water retention criterion expressed by the soil tension (from 30 to 43%). These results are consistent with the fact that the vegetation adjusts to the underlying soil so that the moisture in this soil is the one that provides the possible water amounts, and therefore, the least stress to the vegetation. This confirms the hypothesis of the ecohydrological equilibrium at the scale of the Khroumirie ecosystem.

It is also interesting to observe from this analysis that for areas experiencing greater hydric stress duration (3 years and more, Fig. 12.6b, c), simulations with fixed or variable LAI show less significant differences in the proportions of areas stressed according to the soil water reserve. This suggests that successive years of water stress can be at the origin of the breakdown of the ecohydrological equilibrium. These results showed the potential of integrating time-series products such as LAI in water balance models for a better understanding of the ecohydrological ecosystem functioning.

12.6   Data Quality of MODIS

Despite the availability of a wide variety of online products among which we explored MODIS images, we noted that their use often requires corrections and calibrations with in situ measurements or with specific filtering tools for time-series data to

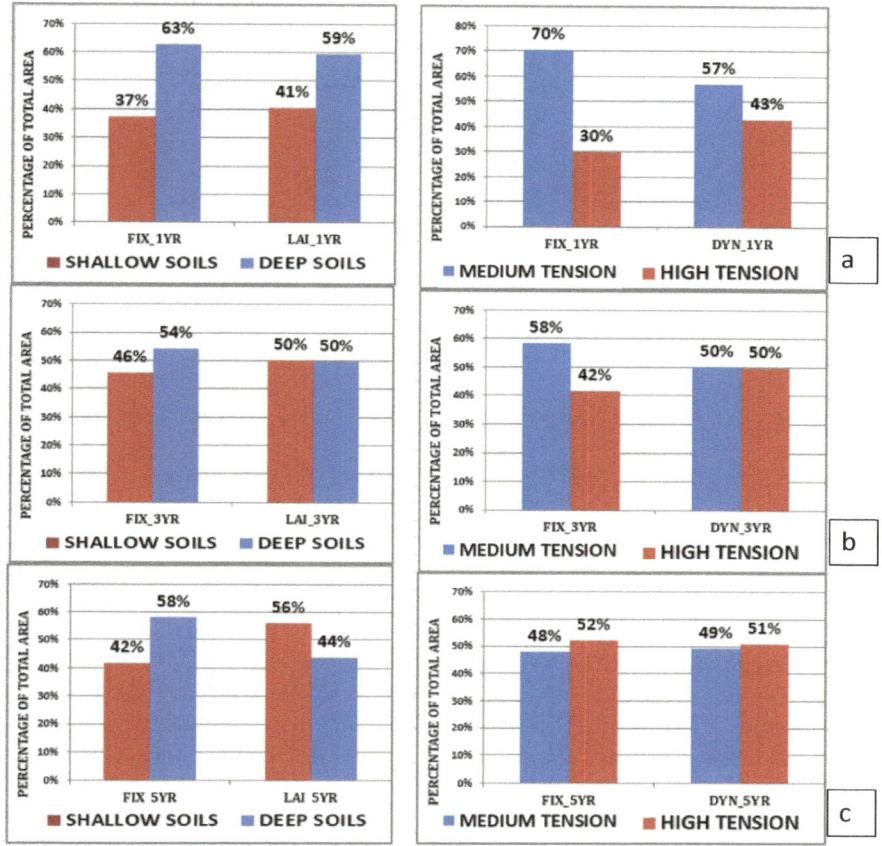

Fig. 12.6 Spatial proportions in % of areas with different annual hydric stress duration (**a** 1 year; **b** 3 years, **c** 5 years) according to soil depth and soil water potential

validate them before they are used in the calculation of ecohydrological indicators. The sources of errors in remote sensing products are due either to sensor failures or to algorithms used by data distributors (Space Agencies) which are generally tested for some regions of the globe and not for all ecosystems types. In the works synthesized above, we have found inaccuracies of the used products that we first proceed to their corrections before integrating them within the different models.

12.6.1 Calibration of LAI-MODIS

The LAI-MODIS product (MOD15A2) is a time series of LAI derived by algorithms based on BRDF (Bidirectional Reflectance Distribution Function). The algorithm used for calculating LAI from the BRDF over 8 days at a spatial resolution

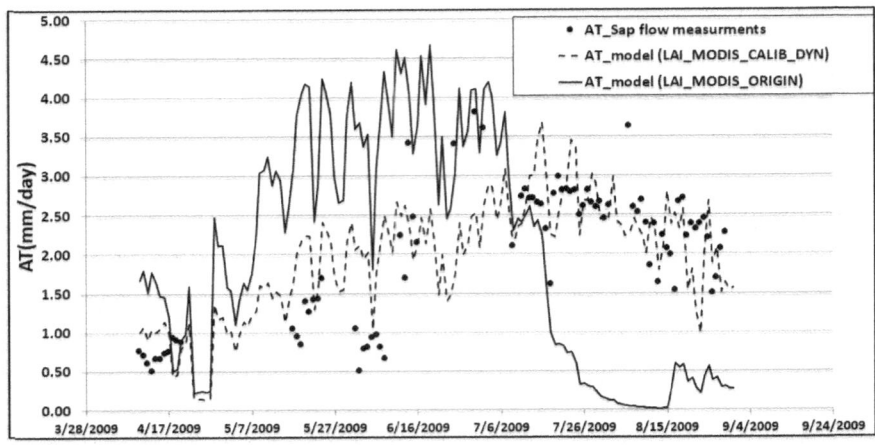

Fig. 12.7 (a) Daily actual transpiration from sap flow measurements (black dots) and model actual transpiration (AT) with LAI-MODIS original (full line) and LAI-MODIS calibrated (dotted line) for the year 2009 [36] (Reproduced by kind permission of John Wiley and Sons Publishing)

of 1 km [37] compares observed and modelled values of canopy and soil structures over different natural conditions. We compared the LAI-MODIS product with
LAI measurements carried out on about 30 plots in the Khroumirie ecorgion in
2006, and we observed significant disparities between the measured values and the
values resulting from the MODIS product. Therefore, a calibration of the product
LAI-MODIS was necessary to correct the values provided for this biophysical
variable.

The calibration methodology was based on the LAI measured in situ, and the NDVI
derived from SPOT images (details of calibration methodology in [36]). These corrections were necessary as the overestimation of non-corrected LAI-MODIS generates
important disparities between the measured and the estimated values of actual transpiration simulated by water balance model compared to the corrected values as
shown by Fig. 12.7.

12.6.2 Quality Assessment of MODIS Time-Series Images

Time-series images can be subject to errors due to gaps, clouds or sensor noise.
This is the case for MODIS images widely used in various researches whose authors
reported errors and gaps within this product [38–40]. Therefore, the adequate use of
this product must be preceded by corrections based on time-series images filtering
either by global or local fitting methods. The global methods fit the data to the long
time scale of observations such as Fourier filtering, whereas the local methods use

the surrounding value of time-series data either by median smoothing approaches [41] or by Savitzky-Golay filter approach [42].

In the study described in Sect. 12.4 where a statistical model was developed from time-series vegetation indices, TIMESAT tool was used to make the filtering of these images. This tool developed by [43] offers two types of filters: local model fitting based on minima and maxima and adaptive filtering.

In the first case, data are fitted to local model functions defined in intervals around maxima and minima in the time-series vegetation indices. The local fitting function uses two types of functions which are: the asymmetric Gaussian function and the double logistic function. The second method is known as the adaptive Savitzky-Golay filter [44] and uses a linear combination of nearby values in a window to fit locally a polynomial function by a least square method.

The details of this filtering process are described in [45]. Time-series smoothing result in the elimination of abrupt changes and data variations become more continuous with the elimination of spikes as shown in Fig. 12.8. The adaptive Savitzky-Golay filter compared to local filtering process produce variations in time-series that conserve local variations for all the tested vegetation indices.

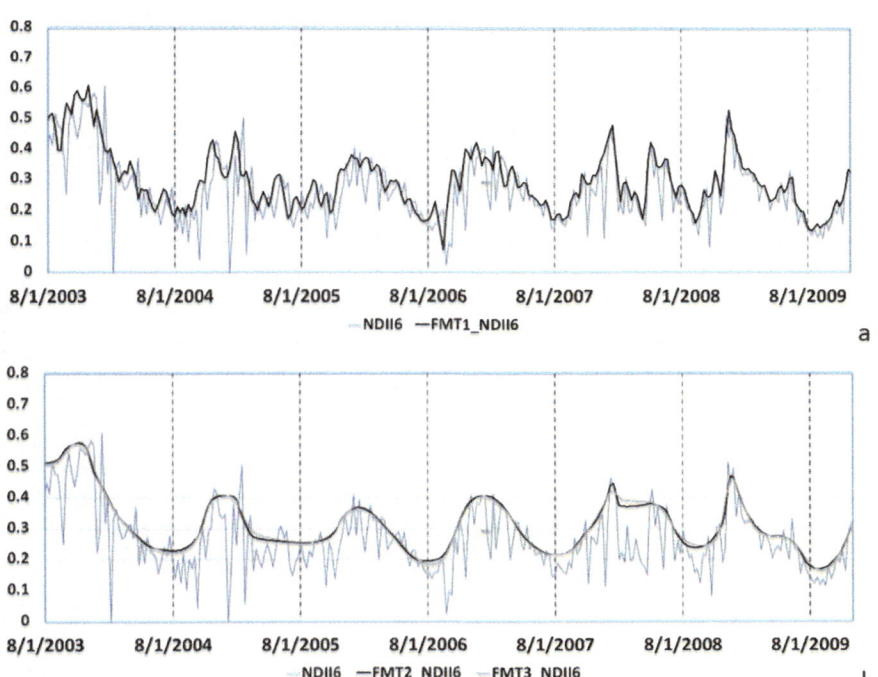

Fig. 12.8 Example of original and TIMESAT filtered time-series profiles of the vegetation indice NDII6. (a) FMT1: Savitzky-Golay adaptive filter; (b) FMT2: Asymmetric Gaussian function; FMT3: Double logistic function [45] (Reproduced by kind permission of Scientific Research Publishing Inc.)

12.7   Conclusions and Perspectives

The objective of this chapter was to present the contribution of spatial and temporal information to ecohydrological modelling, where time-series remote sensing is the essential data of this spatio-temporal information. The concepts of remote sensing integration in ecohydrological reasoning were first described, and various modelling case studies have been presented. The most salient results were related to the integration of remote sensing to ecohydrologic reasoning and the modelling of regionalized water balances. The potential of remote sensing data in modelling physical variables influenced by drought such as the canopy equivalent water thickness was showed through an approach based on a statistical modelling of the spatio-temporal dynamics of the canopy surface biophysical variable. Such ecohydrological indicators could be used in water and carbon budgets as well as in climate change models in limited water resources environments. For operational applications, the founding of this study could also bring useful information in forest management and prevention of fires.

On the other hand, we have demonstrated that remote sensing as a source of spatio-temporal vegetation dynamics is promising because the spectral data measured by the sensors integrate the behaviour of the soil-vegetation complex, which is the basis of the ecohydrological equilibrium hypothesis. In this case, computation of water stress maps over years, and their comparison with soil-water availability edaphic characteristics allowed the verification of the ecohydrological optimality hypothesis at the ecoregion scale using spatiotemporal variability of vegetation.

Ecohydrological assessment by remote sensing will be refined in the future by the availability of free products, especially LANDSAT and SENTINEL, which offer a good repetitively and high spectral and spatial resolutions.

References

1. Wilhite DAG, Glantz, MH (1985) Understanding the drought phenomenon: the role of definitions. Water Int 10: 111–120
2. Newman BD, Wilcox BP, Archer SR, Breshears DD, Dahm CN, Duffy CJ, McDowell NG, Phillips FM, Scanlon BR, Vivoni ER (2006) Ecohydrology of water-limited environments: a scientific vision. Water Resour Res 42:1–15
3. Nuttle W (2002) Eco-hydrology's past and future in focus. In: EOS: Transactions. American Geophysical Union (pp 205–212)
4. Van den Honert TH (1948) Water transport in plants as a catenary process. Discuss Faraday Soc 3:146–153
5. Eagleson PS (1978) Climate, soil, and vegetation: the expected value of annual evapotranspiration. Water Resour Res 14: 731–739
6. Thornthwaite CW (1948) An approach toward a rational classification of climate. Geogr Rev 38:55–94
7. Zalewski M, Janauer GA, Jolankai G (1997) Ecohydrology: a new paradigm for the sustainable use of aquatic resources. In: UNESCO IHP Tech. Document in Hydrology
8. Huxman TE, WBP, Scott R, Snyder K, Breshears DD, Small E, Hultine K, Pockman WT, Jackson RB (2005) Ecohydrological implications of woody plant encroachment. *Ecology* 86: 308–319

9. Rodriguez-Iturbe I (2000) Ecohydrology: a hydrologic perspective of climate-soil-vegetation dynamics. Water Resour Res 36:3–9

10. Eagleson PS (1982) Ecological optimality in water limited natural soil vegetation systems: 1. theory and hypothesis. Water Resour Res 18: 325–340

11. Hatton TJ, Salvucci GD, Wu HI (1997) Eagleson's optimality theory of an ecohydrological equilibrium: quo vadis? Funct Ecol 11:665–674

12. Chakroun H, Benabdallah S, Lili Chabaane Z (2015) Spatial decision support systems integrating ecohydrology in water limited resources regions. In: Advances in Environmental Research

13. Yebra M, DPE, Chuvieco, E, Riaño D, Zylstra P, Hunt ER, Danson M, Qi Y, Jurado S (2013) A global review of remote sensing of live fuel moisture content for fire danger assessment: Moving towards operational products. Remote Sens Environ 136:455–468

14. Ceccato P, Flasse S, Tarantola S, Jacquemoud S, Gregoire JM (2001) Detecting vegetation leaf water content using reflectance in the optical domain. Remote Sens Environ 77:22–33

15. Chakroun H, Mouillot F, Hamdi A (2015) Regional equivalent water thickness modeling from remote sensing across a tree cover/LAI gradient in Mediterranean forests of Northern Tunisia. Remote Sens 7: 1937–1961

16. Hunt ER, Li L, Yilmaz LT, Jackson TJ (2011) Comparison of vegetation water contents derived from shortwave-infrared and passive-microwave sensors over central Iowa. Remote Sens Environ 115:2376–2383

17. Tucker CJ (1979) Red and photographic infrared linear combinations for monitoring vegetation. Remote Sens Environ 8:127–150

18. Huete A, Didan K, Miura T, Rodriguez EP, Gao X, Ferreira LG (2002) Overview of the radiometric and biophysical performance of the MODIS vegetation indices. Remote Sens Environ 83:195–213

19. Huete A (1988) A soil adjusted vegetation index (SAVI). Remote Sens Environ 25:295–309

20. Qi J (1994) A modified soil adjusted vegetation index. Remote Sens Environ 48:119–126

21. Liu ZY, Huang JF, Wang FM, Wang Y (2008) Adjusted-normalized difference vegetation indice for estimating Leaf Area Indice of rice. Sci Agric 41:3350–3356

22. Gao BC (1996) NDWI A normalized difference water index for remote sensing of vegetation liquid water from space. Remote Sens Environ 58:257–266

23. Hunt ER, Rock BN (1989) Detection of changes in leaf water content using near and middle-infrared reflectances. Remote Sens Environ 30:43–54

24. Ceccato P, Gobron N, Flasse S, Pinty B, Tarantola S (2002) Designing a spectral indice to estimate vegetation water content from remote sensing data: Part 1. theoretical approach. Remote Sens Environ 82:188–197

25. Guerschman JP, Van Dijk AIJM, Mattersdorf G, Beringer J, Hutley LB, Leuning R, Pinpunic RC, Sherman BS (2009) Scaling of potential evapotranspiration with MODIS data reproduces flux observations and catchment water balance observations across Australia. J Hydrol 369:107–119

26. Rambal S, Damesin C, Joffre R, Méthy M, Lo Seen D (1996) Optimization of carbon gain in canopies of Mediterranean oaks. Ann For Sci 53:547–560

27. Kochendorfer JP, Ramırez JA (2010) 2010, Modeling the monthly mean soil water balance with a statistical-dynamical ecohydrology model as coupled to a two-component canopy model. Hydrol Earth Syst Sci 14:2099–2120

28. Mouillot F, Rambal S, Lavorel S (2001) A generic process-based simulator for Mediterranean landscapes (SIERRA): design and validation exercises. For Ecol Manage 147:75–97

29. Priestley CHB, Taylor RJ (1972) On the assessment of surface heat flux and evaporation using large scale parameters. Mon Weather Rev 100:81–92

30. Fisher JB, De Biase TA, Qi Y, Xu M, Goldstein AH (2005) Evapotranspiration models compared on a Sierra Nevada forest ecosystem. Environ Model Softw 20: 783–796

31. Garnier A, Breda N, Biron P, Villette S (1999) A lumped water balance model to evaluate duration and intensity of drought constraints in forest stands. Ecol Model 116:269–283

32. Campbell GS (1974) A simple method for determining unsaturated conductivity from moisture retention data. Soil Sci 117:311–314

33. Saxton KE, Rowls WJ (2006) Soil water characteristic estimates by texture and organic matter for hydrologic solutions. Soil Sci Soc Am J 70:1569–1578

34. Hoff C, Rambal S (2003) An examination of the interaction between climate, soil and leaf area index in a Quecus ilex ecosystem. Ann For Sci 60:153–161

35. Limousin JM, Rambal S, Ourcival JM, Rodriguez-Calcerrado J, Perez-Ramos IM, Rodriguez-Cortina R, Misson L, Joffre R (2012) Morphological and phonological shoot plasticity in a Mediterranean evergreen oak facing long-term increased drought. Oecologia 169(2):565–577

36. Chakroun H, Mouillot F, Nasr Z, Nouri M, Ennajah A, Ourcival JM (2014) Performance of LAI-MODIS and the influence on drought simulation in a Mediterranean forest. Ecohydrology 7(3):1014–1028

37. Knyazikhin Y, Martonchik JV, Myneni RB, Diner DJ, Running SW (1988) Synergistic algorithm for estimating vegetation canopy leaf area index and fraction of absorbed photosynthetically active radiation from MODIS and MISR data. J Geophyisical Res 103:32257–32275

38. Atzberger C, Vuolo F, Klisch, A (2015). Use of MODIS temporal signatures for regional-scale and cover mapping and crop status monitoring: activities at Buku University. In: IEEE Geoscience and Remote Sensing Society, the International Geoscience and Remote Sensing Symposium (IGARSS)

39. Choler P, Sea W, Briggs P (2010) A simple ecohydrological model captures essentials of seasonal leaf dynamics in semi-arid tropical grasslands. Biogeosciences 7:907–920

40. Yuan H, Dai Y, Xiao Z, Ji D, Shangguan W (2011) Reprocessing the MODIS leaf area index products for land surface and climate modelling. Remote Sens Environ 15:1171–1187

41. Reed BC, Brown JF, VanderZee D, Loveland TR, Merchant JW, Ohlen DO (1994) Measuring phenological variability from satellite imagery. J Veg Sci 5:703–714

42. Chen J, Jönsson P, Tamura M, Gu Z, Matsushita B, Eklundh L (2004) A simple method for reconstructing a high-quality NDVI time-series data set based on the Savitzky-Golay Filter. Remote Sens Environ 91:332–344

43. Jönsson P, Eklundh L (2004) TIMESAT—a program for analyzing time-series of satellite sensor data. Comput and Geosci 30: 833–845

44. Press WH, Teukolsky SA, Vetterling WT, Flannery, BP (1994) Numerical recipes in Fortran. Cambridge: Cambridge University Press

45. Chakroun H (2017) Quality assessment of MODIS time series images and the effect on drought monitoring. Open J Appl Sci 7:365–383

Part VI
RS and GIS to Assess and Monitor Dry, Arid and Wetlands

Chapter 13
Monitoring of Dryland Vulnerability by Remote Sensing and Geoinformation Processing: Case of Wadi Bouhamed Watershed (Southern Tunisia)

Najiba Chkir and Dalel Ouerchefani

Abstract Land degradation (LD) has become a crucial issue with both environmental and socio-economic implications. Natural forces, through periodic stresses of extreme and persistent weather events, and human use and abuse of vulnerable areas, jointly affect LD dynamics, creating negative feedbacks for the ecosystem equilibrium. Spatial assessment of environmental phenomena at regional scale involves the analysis and fusion of multiple, complex, multidisciplinary, and large-scale information. It is thus important to develop cost effective methodologies to assess and monitor dryland conditions. Remote sensing data and geoprocessing are currently widely tested for this purpose as repeatable and spatial cost-effective ideal tool. Meanwhile, standardized techniques and operational procedures still need to be developed to evaluate land degradation and desertification in the arid areas of Mediterranean regions. Changes in surface properties can be detected through remote sensing data analysis. The main sources of information for the large scale monitoring of soils and vegetation is nowadays derived from satellite imaging. Several indices based on visible near-infrared (VNIR) and short-wave infrared (SWIR) reflectance spectrum are used to produce qualitative and quantitative studies of land degradation and desertification through biological, geophysical and chemical properties description. The general objective of this chapter is to present an overview of dryland degradation and to discuss geo-information and remote sensing data analysis as a support tool for the assessment and monitoring of dryland vulnerability in Southeastern Tunisia. Land Use Land Cover (LULC) changes in the Wadi Bouhamed catchment during 1988–2000 and 2000–2011 periods have been evaluated by soil and vegetation radiometric indices (Normalized Difference Vegetation Index-NDVI; Brightness Index-IB) using LANDSAT TM et ETM + images. Data highlighted that desertification is extending downstream the watershed with a sand movement phenomenon. This is mainly explained by long drought events observed since 1988 enhanced by human

N. Chkir (✉)
Faculty of Arts and Humanities of Sfax, Geography Department, Laboratory of Radio-Analysis and Environment, University of Sfax, Sfax, Tunisia
e-mail: najiba_chkir@yahoo.fr

D. Ouerchefani
Laboratory of Eremology and Combating Desertification, Arid Regions Institute, Medenine, Tunisia

© Springer Nature Switzerland AG 2021
F. Khebour Allouche et al. (eds.), *Environmental Remote Sensing and GIS in Tunisia*, Springer Water, https://doi.org/10.1007/978-3-030-63668-5_13

practices. Starting from the '80, modification of agricultural activities intensification on the one hand and marginal land abandonment on the other had caused severe environmental impacts, including the increase in land degradation risk.

Keywords Desertification · Vulnerability · Land degradation indicators · Remote sensing

13.1 Introduction

Land degradation, in particular desertification, is a widespread phenomenon that threatens the future of our earth [1–5]. On a global scale, around 10–20% of drylands and 24% of the world's productive lands are degraded. Food, water, and energy security, climate change mitigation and adaptation are some of the crucial issues affected by land degradation. "It is estimated that 1.5 billion people across the world are already directly affected through reduced income or food security" [6]. According to the United Nations Convention to Combat Desertification (UNCCD), many of the pressures on land are driven by anthropogenic demands on land and land-based ecosystems, and unfortunately impact the most vulnerable human populations. The increase of world population and therefore the increase of land product needs are exacerbating the competition on natural resources, in particular on soil and water. Every year, around 24 billion tons of fertile farmland worldwide is lost due to soil sealing, erosion, and desertification [6]. Land degradation and related soil erosion can impact wide regions, causing dust storms, changing stream flows, polluting drinking water, and causing siltation in water bodies [1].

Africa is particularly vulnerable to land degradation and desertification (LDD), and it is the most severely affected region worldwide with about 45% of Africa's land area affected by desertification [7]. Unfortunately, soil degradation in drylands is irreversible [8, 9] and represent a severe and long-lasting disturbance that will prevent ecosystem recovery in the absence of comprehensive artificial restoration measures. Southern Tunisia as a part of the North African steppe, whose economic activity is based on agro-pastoralism, underwent significant degradation since the eighties. The combination of severe droughts and an exponential increase in livestock had a catastrophic impact on pastoral resources. The "alfa", an emblematic species of these arid lands is now disappearing [10, 11]. In this arid area, land degradation risk (LDR) is exacerbated by inappropriate human practices. Violent gully erosion induced by severe climate conditions related to long drought periods followed by torrential rainfall is enhanced by over grazing.

Satellite remote sensing presents a viable option to monitor land degradation. The advantage of satellites is that they provide spatially continuous [12], replicable and homogeneous information in a cost-effective manner and over large areas. Drought indicators can be derived from time series data what allows to determine long-term changes and pressures [13].

The objectives of this chapter are to present (1) an overview of methods used for the assessment of dryland degradation vulnerability in particular remote sensing and geoprocessing tools, (2) an overview of land degradation processes as observed in the watershed of Wadi Bouhamed as a part of dryland areas of Southern Tunisia, in relation with climate and human controlling factors, (3) an estimation of soil and water conservation techniques efficiency in the watershed of Wadi Bouhamed as assessed by geo-information data.

13.2   Global Overview of Dryland Degradation

Land degradation is a complex term that describes how the environment features such as soil, water, vegetation, rocks, air, climate and relief are evolving to worse state. Since 1994, the UNCCD has defined land degradation as a "reduction or loss, in arid, semi-arid, and dry subhumid areas, of the biological or economic productivity and complexity of rain-fed cropland, irrigated cropland, or range, pasture, forest, and woodlands" [14]. This definition is focused on the loss of land-based ecosystems services. Land degradation is therefore a major threat to global sustainable development by affecting biodiversity and stability of ecosystems at different levels. This destructive process is mainly observed as erosion and desertification.

13.2.1   Erosion

Erosion is a part of the geographical cycle. This process is the gradual shape of reliefs under the combined action of several factors such as wind, rainfall, runoff, temperature variations or human activities. Two types of erosion can be distinguished: geological or progressive erosion that occurs during long time under natural factors and accelerated erosion that is directly related to anthropogenic activities [15–19].

Erosion either geological or accelerated occurs according to three successive stages: detachment, transportation and then deposition.

- The *detachment* is caused by raindrops, runoff or by the wind. It depends on rainfall intensity, runoff rate, surface slope, detached particles weight, wind velocity and of rain distribution over the surface.
- The material *transportation* corresponds to the movement of sediments downstream, whether on the hillside or in the stream channel. Mediums required for material displacement can be running water, wind and gravity.
- The deposition is the process by which sediment settles out of the water or wind that is carrying it. Deposition can take place anywhere and damages can be as grave as those caused by the erosion itself.

13.2.1.1 Water Erosion

Water erosion comes in several different forms with various causes generally in relation with rainfall, runoff on a slope, and soil type. These various modes are classified as follow [20]:

Sheet erosion: Particles that are detached by raindrop impact are transported over a small distance in the surface flow without formation of elementary channels. This type of erosion removes a thin covering soil from large areas more or less uniformly during runoff. The sheet erosion impact on the field can remains until most of the productive topsoil has been lost; its existence is mainly reflected by high turbidity of streamflow.

Linear erosion: The linear erosion takes place along the channels which cut the surface according to diverse forms and sizes. It appears when the sheet erosion gets organized and digs well-defined but more and more deep forms. Rills are elementary channels that are small enough to not interfere with field machinery operations and that are filled as part of tillage operations. These temporary features cut the surface, detached and transport the sediments in surface runoff. On the other hand, gullies that are formed due to the confluence of many rills are an advanced stage where surface channels are eroded to the point where they become a nuisance factor in normal tillage operations. The sediments removed to the formation, enlargement and deepening of gullies are transported to the channel streamflow. Enlarged gullies become permanent topographic features that will tend to deepen and widen with every heavy rainfall. This gully erosion is the dominant process causing land degradation and soil losses in Tunisia.

Channel erosion: Channels are permanent topographic features formed due to the confluence of gullies, either with perennial or non-perennial streamflow. Channel erosion includes stream bed and bank erosion and is significantly more noticeable than sheet erosion. The erosion extent and intensity are directly related to the streamflow speed and energy and to the slope of the channel.

13.2.1.2 Aeolian Erosion

In most arid and desert areas, the wind has an important effect on the erosion of rocks by driving sand and soil particles when the soil or sand is not compacted or is of a finely granulated nature. Wind erosion can happen anywhere and anytime the wind blows.

Wind erosion is described by three types of soil particles movement (suspension, saltation and reptation or surface creep). These movements depend on both soil particle size and wind strength.

Saltation: Saltation is the primary means of soil movement. "The major fraction of soil moved by the wind is through the process of saltation" [21]. Fine soil particles (0.1 to 0.5 mm) are lifted into the air by the wind and drift horizontally across the surface by successive leaps increasing in velocity as they go. This process is therefore responsible of intense damage to both soil surface and vegetation.

Surface creep (reptation): This type of movement occurs when the large particles which are too heavy to be lifted into the air are rolled over the surface and meet particles that have been through saltation. This process concerns high density particles ranging between 0.5- and 2-mm size [23, 24].

Suspension: This process describes the movement of very fine dirt and dust particles when thrown into the air by the impact with other particles or by the wind itself. These particles can be carried very high and be transported over extremely long distances and high altitudes reaching 4000 m [24].

13.2.2 Desertification

The desertification indicates the process which leads to the degradation of lands and natural resources in particular the degradation of soils and vegetation, and in a more general way the loss of soil fertility. The desertification is a world phenomenon which is a high risk in dry zones.

According to the UNCCD [14], desertification is defined as a process of land degradation in arid, semi-arid and sub-humid areas due to various factors including climatic variations and human activities. In another way, desertification results in persistent degradation of dryland due to man-made activities that enhance climate conditions.

Three key desertification processes are observed in drylands [25]: (1) deterioration of vegetative cover due to overgrazing, wood cutting and burning; (2) wind and water erosion resulting from improper land management; and (3) salinization due to improper use of irrigation water. These processes are active in the desertification of Southern Tunisia.

The desertification is due to several natural factors bound to the severity of the climate and the drought. However, even if drought phenomenon is one of the desertification processes, it is not the dominant one. The desertification is mainly controlled by wind erosion; violent and whirling winds in dry and semi-arid zones cause the extension of affected zones and the decrease of vegetation cover.

The desertification is enhanced by human activities. The increase in population and urbanization induce an increasing food demand. Field productivity increase was made to the detriment of natural land cover, resources and environment [26]. Long-term inadequate practices of natural resources exploitation are some of the human-induced factors that exacerbated the risk of desertification. Through these practices, overgrazing, clearance extension, intensive livestock are the most important ones. Frequently, the satisfaction of short-term urgent needs for political issues associated with climate, demographic and economic crises enhance the impact of these practices and lead to desertification.

The desertification is a localized geographical phenomenon which, once introduced, tends to widespread proportionally with the land degradation rate. The risk of desertification is likely high when the anthropic destruction of the vegetal cover takes place, in particular when the natural environment is already ecologically fragile and climatically favorable to the acceleration of wind erosion processes [22].

Worldwide, desertification that constitutes an important constraint to socioeconomic development is affecting more than 81,000 sq km [27]. In Tunisia, most affected zones are located in the center and the south of the country. Southern Tunisian is submitted since several decades to an intensive desertification process and sand drift that strongly menace the natural ecosystems, reducing the agricultural and pastoral production and causing important exodus and migration movement of local population from farming areas to neighboring cities. The southeastern coastal plain of Jeffara is the most affected zone by land degradation and desertification.

13.2.3 Factors Affecting Land Degradation

The land degradation rate over a catchment depends upon various factors related to climate, soil, topography, soil cover and finally anthropic activities (Table 13.1).

Table 13.1 Factors affecting land degradation [28]

Factor	Parameter	Effect
Climate	Rainfall intensity Duration of rainfall Temperature Wind velocity	Splash erosion Flow erosion Weathering action Wind erosion
Soil characteristics	Soil mass characteristics (granulation, porosity, moisture content) Grain size and shape	Infiltration and runoff rates and hence erosion rate Particles transportation mode: Saltation, creep surface and suspension
Topography	Slope (orientation, degree and length)	Steeper slope: higher energy of flow, higher erosion and transportation rates
Soil covers	Vegetation/plant cover	Retardation of flow and erosion rates, protection from splash erosion, protection from deflation (wind detachment of particles)
Land uses or human activities	Agricultural practices, mining, roads, buildings construction, etc. Reservoirs	Increased erosion rates Sedimentation

13.2.3.1 Natural Factors Affecting Land Degradation

Natural hazards are the conditions of the physical environment which induce the existence of a high degradation risk.

Climate: The soil erosion is mainly influenced by the rainfall (amount, frequency, duration and intensity) and by the wind (direction, strength, and frequency of high-intensity winds). Hydric erosion is more widespread, and its impact is greater than that by the wind. The effect of climatic factors is enhanced by the drying-out of the soil. Climate change through its effect on rainfall intensity is likely to increase soil erosion rates and soil erodibility and will affect vegetative cover and patterns of land use [29].

Soil characteristic: Soil erodibility is the main aspect of soil properties that reflects the soil sensitivity to erosion. Soil with medium texture is the most vulnerable to erosion, especially in hillsides. Clayey and muddy soils, which are enriched in organic matter, have better resistance to erosive actions. This is essentially due to the stability of aggregates and favorable porosity system. In Tunisia, soils are mostly sandy-clay and highly vulnerable to erosion and therefore to land degradation.

Topography: The topography is the combination of the geological processes as well as the erosion. The erosion is mainly influenced by the form, the gradient, the length and the position of the slope. Although this powerful influence on erosion, the presence of erosion and heavy runoff on gentle slopes indicates that this phenomenon can occur without any need for a steep slope: the action of rain is enough [30].

13.2.3.2 Human-Induced Factors Affecting Land Degradation

Human induced factors that cause the land degradation are mainly due to inappropriate land use and unsuitable land management practices [31].

Inadequate tillage practices: Tillage is used to prepare the soil to crop production activities. It aims to make the structure and the texture of the topsoil suitable for sowing, germination and easy root penetration. The tillage practices are of vital importance in areas with limited water and soil resources. The use of mechanical tools made possible to extend farming areas into poor soils. However, most of weakly structured soils, with many loamy and fine-textured topsoils, are depredated within a few years with this techniques. In many cases, the tillage increases the surface water runoff causing increased water erosion particularly when the ploughing is made along the slope. Aeolian erosion is also increased by tillage when the upper layer of the soil is left bare. The runoff and aeolian erosions are also indirectly increased by the tillage that reduces the soil organic matter and the soil structural stability.

Clearing and overexploitation of natural vegetation: Clearing of natural vegetation is used to create large open landscape for livestock pasture, plantations of commodities, or settlements. however, the cut down of natural vegetation causes damage to habitat, biodiversity loss and aridity which quickly accelerate natural erosion.

Overgrazing: This phenomenon is observed in areas with intensive pasture. In most of the cases, animals exceed the average of the land availability for grazing. The removal of plant/vegetation material is repeated quickly and avoid the leaf/pasture mass to regrow. This situation is quickly reached in arid areas where natural vegetation is scarce and submitted to a long drought. On the other hand, the animals damage the soil by compacting the dry soil with their hooves. This can prevent grass from growing and slow down the percolation of water through the soil enhancing the runoff and therefore water erosion.

In Southern Tunisia, as in the plain of Jeffara, agriculture is historically based on olive trees, field crops and pasture depending on water points (wells, springs) with nomad tribes adapting their uses and needs to climate variation. However, since the nineteen's, land management strategies aim to transform the tribal nomad system and to encourage population settlement what likely modified farming practices and enhanced human pressure on the land.

13.3   Modeling and Mapping Land Degradation

Global assessment of land degradation is not an easy task. A wide range of method could be found in the literature, often with distinct disciplinary-oriented indices leading to different generated data among disciplines [32, 33].

Two types of approaches can be distinguished: global scale methods that consider the assessment of the land degradation over large areas and local scale methods that focus on the determination of land degradation parameters at the parcel scale.

Remote sensing had become the most common tool used for the study of land degradation (soil and vegetation) in arid areas [34–37] thanks to its advantages in providing dynamical, multi-temporal or time series land cover information with increasing spatial resolution images.

Recent advances in space and computer technologies provide the opportunity to process a big database such as storing and interpreting both spectral and spatial data including elevation, slope, aspect and relief of the earth environment. Hence, the integration of soil erosion models, field data and data provided by remote sensing technologies through the use of geographic information systems (GIS) are nowadays widely used to assess soils erosion risk [38–41]. Several studies focused on the spatial extrapolation of parcel results obtained by conventional methods on large areas [35, 42–44] using different algorithms of remote sensing tools.

Satellite data with its favorable synoptic view and repetitive covers offers the possibility of mapping, monitoring and estimating soil erosion and of detecting changes in erosion rates in relation with land management decisions [45–47]. Several studies have shown high efficiency of remote sensing for environmental issues in particular for spatial assessment of land uses and land degradation process [35, 42–44]. Remote sensing methods used to detect the evolution of land uses and land covers can be classified according to two main algorithms [48–50]: image transformations (indices calculation) or image classifications.

The focus of this chapter is to present land use and land cover change (LULCC) assessed by remote sensing time series to better understand how human decisions impact land surface.

13.3.1 Image Transformations

Basic mathematical treatments of the raw image bands referred as *image transformation* produce a new imagery that highlights particular features or properties of interest, better than the original input images. This treatment could be applied to multiple bands of data from single multispectral image or from two or more images of the same area acquired at different times (i.e. multitemporal image data). Basic image transformations are simple arithmetic operations to the image data [48 50, 51].

Image addition or subtraction: Arithmetic operations are performed on two or more georeferenced imageries referenced to the same coordinate system. These operations are applied to the same area using separate spectral bands on a single date or on multiple dates. This procedure is often used to identify changes that have occurred between images collected on different dates.

Image division or spectral ratioing: This transformation allows to highlight low-level variations in the spectral reflectance of various surface covers. The division of two different spectral bands data results on an image that enhances variations in the slopes of the spectral reflectance curves that may be masked by the pixel brightness variations in each of the bands [48].

Principal Components Analysis: Principal Components Analysis (PCA) is a statistical transformation technique that reduces the set of data defining a given phenomenon. The algorithm identifies uncorrelated components and orders them in terms of explained variance of the original set. For remote sensing data, it is common to find that the first two or three components of multispectral image bands are reliable to explain all of the original variability in reflectance values. PCA is traditionally used as a means of data compaction. PCA is also applied in environmental monitoring.

13.3.2 Image Classifications

The image classification refers to the computer-assisted interpretation of remote sensing images based on the detection of the spectral signatures of land cover classes [48]. The quality of the classification depends on the ability to reliably detect the distinctive signatures of the land cover classes of interest in the band set being used. The image classification can be supervised or unsupervised according to how it is performed.

Supervised Classification: The supervised classification is based on the spectral information of classes of interest such as land cover/uses classes as identified on the image. These are called training sites [52, 53]. Supervised classification accuracy depends on training sites numbers and representability [53]. The spectral signature of each class of interest is produced by the statistical characterization of the reflectance based either on simple indices such as the mean or the range of reflectance on each band, or on more complex indices such as detailed analyses of the mean, variances and covariances over all bands. The classification consists on the comparison of each pixel reflectance to spectral signature of each classe in order to decide to which one it can be attributed. Several statistical decision rules could be used but the most sophisticated and widely used is the Maximum Likelihood procedure.

Unsupervised classification: In contrast to supervised classification, unsupervised classification requires no advance information about the classes of interest. Cluster analysis will mathematically associate each pixel into the most prevalent natural spectral groupings, or clusters, with similar reflectance patterns. These clusters are later recognized in terms of landcover classes based on field knowledge. Unsupervised classification is faster than supervised one but could product some cluster grouping that are mathematically homogeneous without any natural representations.

Accuracy Assessment: A vital step in the classification process, whether supervised or unsupervised, is the assessment of the accuracy of the final images produced. In order to calibrate and therefore improve the classification procedure, the land cover for a set of sample locations in the field is compared to what was mapped in the image for the same location. The accuracy could be assessed for the entire study area as well as for individual classes. The confusion matrix produced by the iterative approach allows to identify cover types for which the accuracy is below the desired level due to either errors of commission (pixels mistakenly included in a particular class) or errors of omission (pixels mistakenly excluded from that class). The confusion matrix is used to refine the classification approach.

13.4   Study Area: The Wadi Bouhamed Watershed

The watershed of Wadi Bouhamed is situated in Southern Tunisia, a part of the North African steppe that underwent significant degradation in the eighties. The watershed is located at about 20 km south of the nearest city, Medenine, it extends between 634,830 and 663,995 of North latitude and between 3,653,282 and 3,700,613 East longitude (UTM-Carthage) (Fig. 13.1). This watershed is characterized by a fragile natural environment due to extremely irregular rainfall regime. Overexploitation of natural resources (water, soil, vegetation) by high human pressures increases the vulnerability of the watershed to land degradation hazard.

13.4.1   Climate Context of the Studied Area

Bioclimatic classification is based on two factors: (i) the water stress undergone by the vegetation directly related to the water climate balance between the vegetation

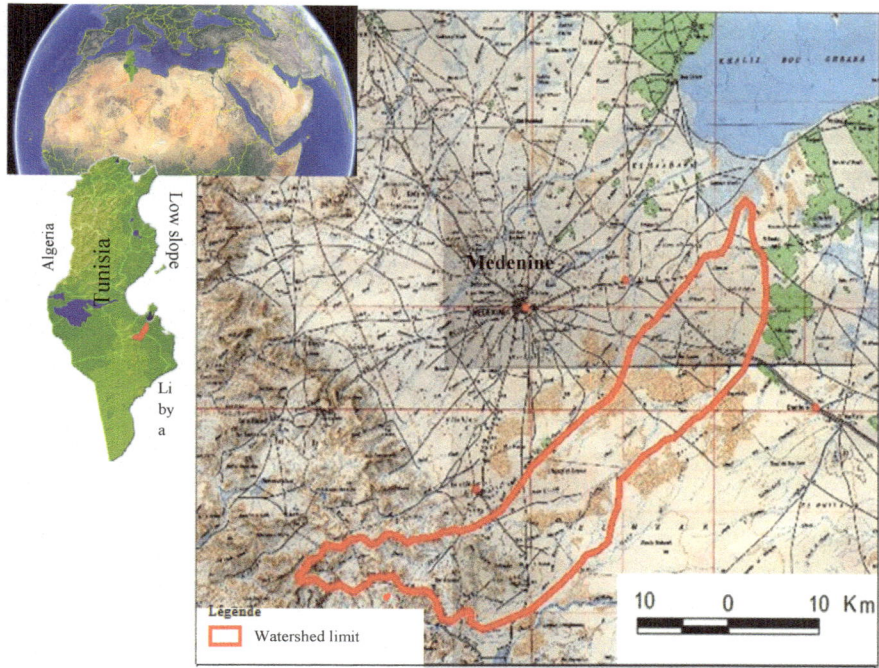

Fig. 13.1 Geographic delimitation of the Wadi BouHamed watershed on top of Medenine topographic map (Gabes sheet—/200 000)

needs and the climate supply; and (ii) the thermal winter constraint that limits the vegetation growth due to the cool condition. According to the Emberger classification, the studied area is situated under the lower arid bioclimate stage with mild winter. Precipitation is of low range with an annual mean under 200 mm/year and irregular. Low temperatures recorded during January and February vary between 5,5 et 8,7 °C and higher temperatures recorded during July and August can be over 40 °C. Violent and dry winds, mainly blow from the western saharian part of the region during the vegetation season, the steppe landscape is therefore marked by several of aeolian sandy forms.

13.4.2 Vegetation Cover

As a part of the presaharian region of Tunisia, the watershed of Bouhamed is submitted to severe climate conditions related to aridity that control the steppic specification of the vegetation (Table 13.2–Fig. 13.2). The vegetation is mainly composed of *Rhanterium Suavuolens* and *Artemesia campestris* groups that cover about 35% of the area. Ecosystems of steppic areas are often related to isohumic soil types [53]

Table 13.2 Vegetation species in the watershed of Wadi Bouhamed (2013)

Vegetation species	Areas	
	Hec.	%
Steppe with *Rhanterium Suaveolens* (Arfej) and *Ranterium Suaveolens* and *Artemesia campestris*	17,816	34.7
Steppe with *stipa grostiss pungens* (Sbat) and *Stipagrostis pungens* group	14,583	28.4
Steppe with *Artemissia herba-alba* (shih) and *Artemissia herba-alba* and *gymnocarpes deconder* groups	10,755	21
Steppe with *Stipa tenacissma* (Alfa;Halfa) and *Stipa tenacissima* and *Lygeum Spartum* groups	5223	10.1
Steppe with *Lygeum Spartum* (sparte;Halfa mahboula) and *Lygeum spartum* and *Erodium glaucophyllum* groups	1346	2.6
Steppe with *Anthyllis henoniana* (Gezzir) and *Anthyllis henoniana* and *Gymnocarpos decander* groups	650	1.2

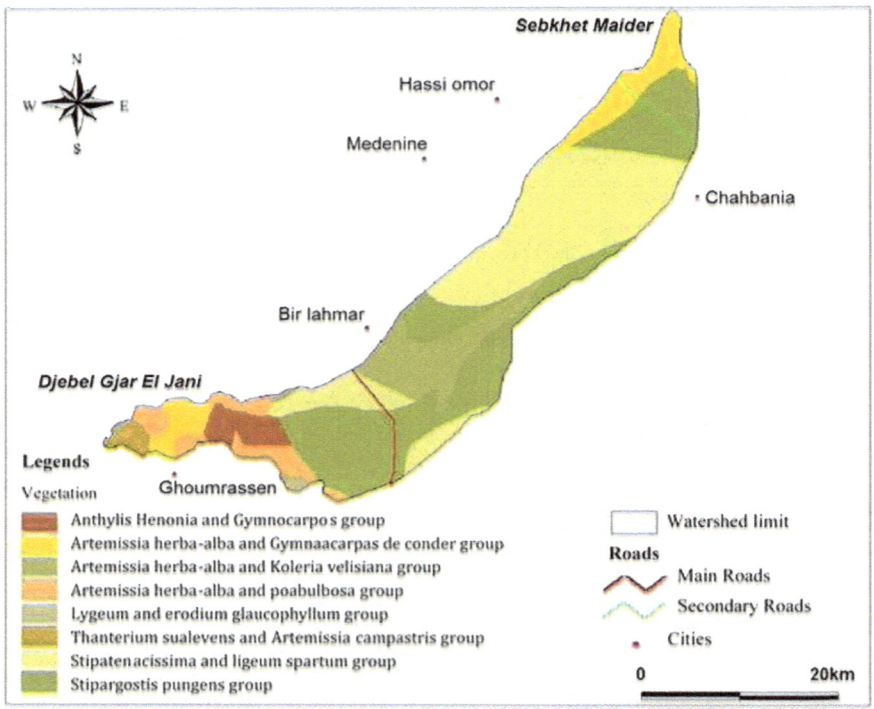

Fig. 13.2 Vegetation species map of the Wadi Bouhamed

with some carbonated or gypsy evaporite crusts found at limited depth in alluvial and aeolian deposits. These soils have light sandy to loam-sandy texture. *Artemesia Herba alba* and *Gymnocarpos Decander* groups cover 21% of the watershed. These species developed on regolisolic soils and poor loamy soils with carbonate nodules. *Stipa tenacissma* and *Lygeum Spartum* groups are observed on 10% of the watershed covering calcomagnesic soils and crusted soils [45]. In low areas where poor hydromorphic and salty soils are developed, vegetation is mainly made by *Ziziphus lotus* (jujubier), *Nitraria retusa* and *Lyceum arabicum* groups.

13.4.3   Slope

Four classes of slopes have been defined (low, medium, high and very high slope). Water erosion is remarkable in high and very high slope areas (61.5% of the watershed) located upstream of the watershed at Jebel Ghoumrassen (Fig. 13.3) indicating a high vulnerability to land degradation. Low and medium vulnerability areas are mostly located north of the watershed where infiltration is favored by sandy soils.

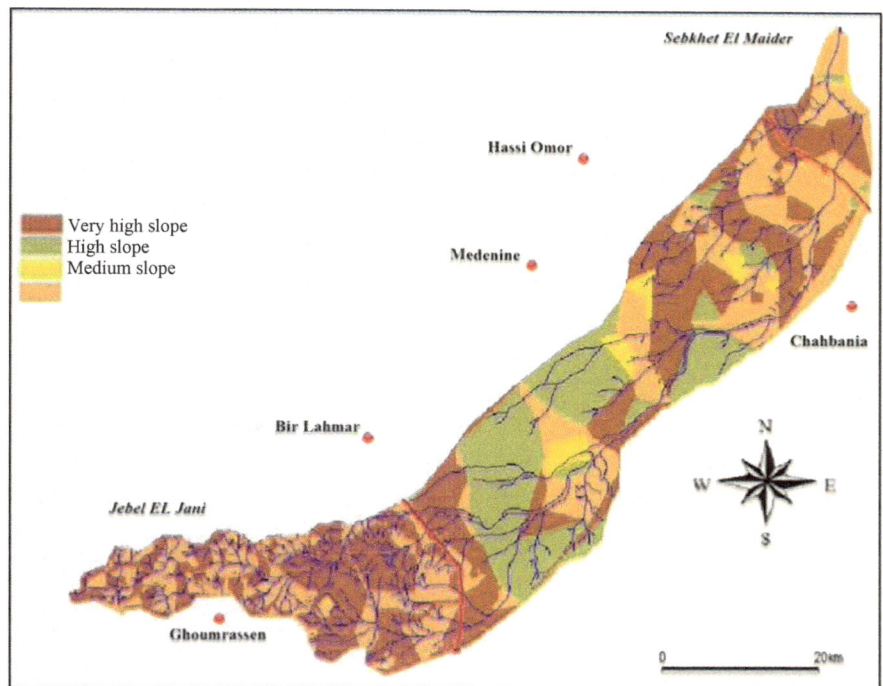

Fig. 13.3 Slope map of the Wadi Bouhamed watershed

13.4.4   Human Issues

The Wadi Bouhamed watershed is situated in a rural area where the population density is very low (1 inhabitant per squared kilometer). The local population are gathered in so-called *arouch*, the local name of tribes, «aouled Belgacem» and «aouled Hamad» being the most known ones. Each of these tribes has its own farming parcels. However, field trips have shown that most of the houses are abandoned due to the migration of the population [55].

Although a limited number of people, human activities had highly affected the Wadi Bouhamed watershed. Clearing and pasture are the main farming activities observed in this region. These activities degraded soil and reduced vegetation cover, what made the soil more exposed to rain and wind and therefore more likely to get washed or blown away. This situation had activated the desertification process as observed in some areas covered sandy desert soils (Fig. 13.4). Field trips carried out on the watershed highlighted several inadequate agricultural practices that contribute to water and wind erosion enhancement such as intensive exploitation by successive cropping along the year and tillage along the slope (Fig. 13.5).

However, the watershed is also situated in an area where national authorities have implemented several features for water and soil management.

Fig. 13.4 Sample of a desertification effect in Ben Guerdane (Southern Tunisia—2013)

Fig. 13.5 Sample of degradation in Ben Guerdane steppe due to tillage technics (2013)

13.4.5 Soil Conservation Management

Since the 1980s, large-scale watershed management and reconstruction programs have been implemented in key areas throughout Tunisia that had serious soil erosion problems. According the field observation, many different soil and water conservation features are spread over the watershed. Gabions are main systems implemented upstream the watershed in piedmont areas of Jebel Ghomrassen while palm windbreaks used to reduce the aeolian erosion and to fix sandy dunes are implemented in the valley. These features are mostly built by the local Soil and Water Conservation Office (750 hectares) but also by local farmers (about 150 hectares) when their fields are directly affected by erosion.

13.4.5.1 Traditional Soil Conservation Features: Jessours and Tabias

Traditional features called *jessour* and *tabia* are typical harvesting structures well known in Southern Tunisia and in particular in mountain areas under arid climate [56, 57].

Jessour is low narrow-based berms designed to intercept and temporarily pond concentrated runoff. The ponded water is then slowly released through infiltration and increase water available to crops. Jessour are mainly used for growing trees such as olive, almonds or palm trees (Fig. 13.6). A single unit is made of three components: (i) the impluvium is which the area drained by the conveyed runoff, (ii) the terrace is the cropping zone where soil is deposited through long-term sedimentation and (iii) the dyke that acts as a barrier to retain runoff and sediments at the terrace area [56–58].

Tabia structures are similar to jessour [56, 58] but they are built on longer distances (Fig. 13.7) along mountain foothills and piedmonts, perpendicular to slope gradient in order to divert, slow down and convey runoff.

Fig. 13.6 Jessour located upstream of Bouhamed watershed of (2013)

Fig. 13.7 Tabias structures in the Wadi Bouhamed plain (2013)

13.4.5.2   Civil Engineering Structures: Gabions

Gabions are retaining walls made by rectangular wire mesh baskets filled with rock and fixed on slopes and channels for erosion protection. They are widely used because of their flexibility in terms of design characteristics and fixation location. Gabions are particularly useful in stream and channel applications where there is both a need for a natural look and high amounts of water flow are expected [59]. In the studied watershed, 11 gabions structures are implemented along the Wadi Esseder and 7 structures along the Wadi Bouhamed. This type of structures is exclusively built by the local Soil and Water Conservation Office and covers 18 hectares. Gabion structures are implemented in order to reduce runoff amount and soil displacement, to decrease runoff energy and scouring risks, to protect the downstream structure from sediments and to reduce flooding intensity.

13.4.5.3 Aeolian Erosion Protection and Windbreak Systems

Wind erosion is one of the major land degradation factors affecting the south of Tunisia since the twentieth century [60]. Land Protection strategies initiated by the national authority are mainly based on erosion control structures in order to either prevent or control erosion risks. Aeolian erosion is prevented by four different techniques [61]:

- Increase the surface roughness to reduce the wind speed at the soil surface so that the wind is less able to move soil particles,
- Introduce irregularities in the landscape in order to avoid windstorms
- Implement field shelterbelts perpendicular to prevailing winds to reduce the wind speed. Field shelterbelts reduce the wind velocity for distances up to 30 times its height.
- Maintain a vegetation cover or a crop residue cover as long as possible to maintain soil moisture and stability.

The efficiency of these practices depends on local climate and soil conditions and on local agricultural practices such as crop rotation and tillage practices. However, these techniques are low-cost and high-performance techniques what allow an easy implementation by local farmers. In the watershed of Wadi Bouhamed, main wind erosion control structures are natural and man-made shelterbelts.

Man-made shelterbelts: Windbreaks systems made of palms are often fixed around agriculture areas on the top sandy dunes. However, some technical specific conditions are often not taken into consideration what reduce the longevity and the efficiency of most structures (Fig. 13.8). For example, palms should be spaced out at least from 2 to 4 cm to be crossed by the air otherwise the shelterbelt becomes an obstacle submitted to the wind power and destruction risks.

Natural shelterbelts: Windbreaks systems constituted by trees of different species (*Acacia ligulata, Acacia cyanophylla and Eucalyptus occidentalis*) are implemented along roads since 1965 and in some areas of the watershed (Fig. 13.9). These species are well-adapted to local climate conditions so that they do not need irrigation or maintenance. However, this practice has been abandoned.

Fig. 13.8 Palm windbreak on the top of sandy dunes in the watershed of Wadi Bouhamed (2013)

Fig. 13.9 Tree windbreaks implemented in the watershed of Wadi Bouhamed (2013)

13.5   Land Degradation Over the Watershed of Wadi Bouhamed

This part of the chapter aims at the identification of the status of land degradation in the Wadi Bouhamed watershed on the global scale.

13.5.1   Methods and Multi-sources Data

Multi-source (map-, space- and field-based) datasets were used to obtain both static and dynamic monitoring of land degradation in the watershed of Wadi Bouhamed. GIS software has been used to produce georeferenced thematic maps controlled by field points. Available maps have been digitalized to be used as a background to satellite images.

13.5.1.1 Field-Based Data

Soil properties measurements have been carried out on 15 sites (Fig. 13.10). For each site, in situ observations of several characteristics have been first defined: coordinates, geomorphology, vegetation cover, land use, neighboring, water and soil conservation structures. Two soil samples have been collected for each site: (i) «movable» soil collected at the surface, this sample is representative of the easily removed part of the soil either by water or by wind and (ii) "fixed" soil at a depth 5–10 cm, this sample is representative of the stable part of soil due to the compaction. Granulometry analyses has been determined in the Laboratoire d'Erémologie et Lutte Contre la Désertification of the Institute of Arid Regions of Médenine (IRA-Tunisia).

All soils are sandy to loamy sandy soils with sand fraction greater than 95% for "movable" soils and a light enrichment up to 5% in loam and silt for "fixed" soil.

13.5.1.2 Remote Sensing Data

Remote sensing data were seven-band Landsat5 TM. This study is based on the comparison of three Landsat5 TM scenes for different dates (17 March 1988, 18 March 2000 and 25 March 2011). Each band remote sensing data was input using ENVI image processing system. False Color Composites was created for the study area using the first three bands of Landsat5 TM (1, 2 and 3). In this combination

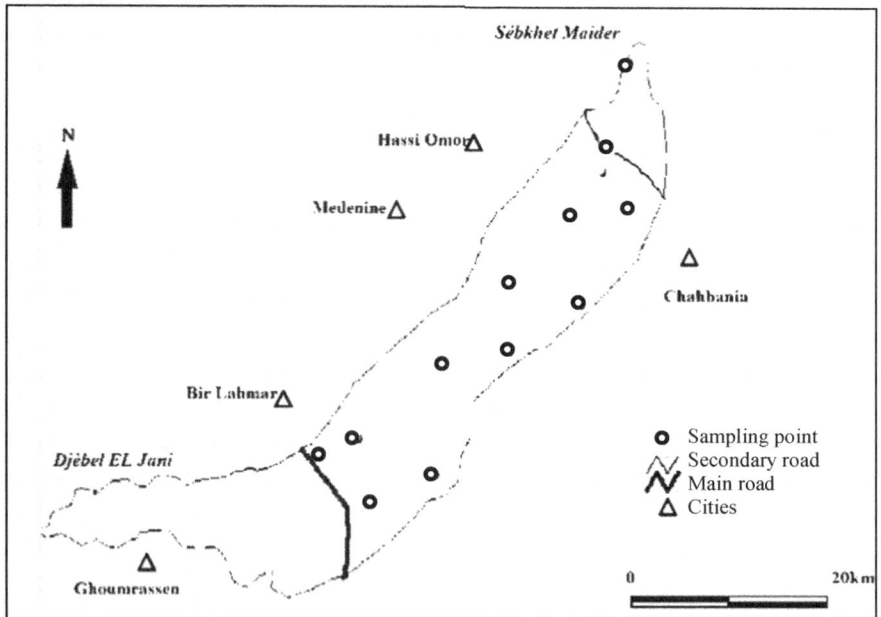

Fig. 13.10 Soil sampling map of the Wadi Bouhamed watershed

(RGB), the visual identification of natural landscape features is easier. The change detection is based on a radiometric index for soil (Brightness Index- BI) and the Normalized Difference of Vegetation Index (NDVI) for vegetation. Processing and UTM 32 N projections have been carried out after the extraction of the study area from the Landsat5 image using the watershed limits digitalized by GIS.

Pre-processing: Before the extraction of information and the main data analysis, radiometric or geometric corrections have been applied to the multispectral bands. Geometric corrections consist of correcting the geometric distortions of the image due to sensor-Earth geometry variations and of converting the data to real world coordinates on the Earth's surface [62]. These corrections have been carried out on the three Landsat images using a first-order polynomial rectification with a Nearest Neighbor Resampling. The scene of 1988 has been corrected on the UTM 32North projection system then other images have been corrected by the image to image method with 32 control points. Radiometric corrections consist of correcting the data for sensor irregularities and for sensor or atmospheric noise so that data accurately represent the reflected or emitted radiation measured by the sensor [62]. Radiometric corrections have been carried out on the multispectral bands using calibration data available in metadata files of each scene [63].

Classification processing: Supervised classification method was used in conjunction with the maximum likelihood classifier to sort image pixels. The classification was done on multi-temporal Landsat TM/ETM + data by creating training sites of LULC classes with distinct pixel values/signatures to ensure class separability. The classification accuracy is assessed by the total accuracy and by the Kappa statistic.

Transformation processing: Conventional vegetation and soil indices, respectively NDVI and BI, have been used to monitor the evolution of vegetation cover as an indicator of land degradation. Brightness index (BI) [64] is one of the most used spectral indices of land degradation in arid and semi-arid areas such as Tunisia, Chili, Spain, Morocco [65–69]. This index is calculated using visible, red and near-infrared radiations [66, 68, 70]. The Normalized Difference Vegetation Index (NDVI) is also the most used one for vegetation cover monitoring [71–73].

13.5.2 Land Uses Changes in Wadi Bouhamed Watershed as Inferred by Image Classification

The processing Landsat scenes created land uses/cover maps for each date. The classification should be performed given the land cover class themes defined according to the purpose of the study (Fig. 13.11). The classification system was referred to the land cover map issued by the Regional Authority for Agriculture Development (CRDA) completed by field knowledge. Land cover class for the study area was divided into seven theme classes as following: (1) Tree crop, (2) Bare soils, (3) Wheat crop, (4) Halomorphic zones, (5) Sandy zone, (6) Pasture zone, (7) Over-pasture zone (degraded).

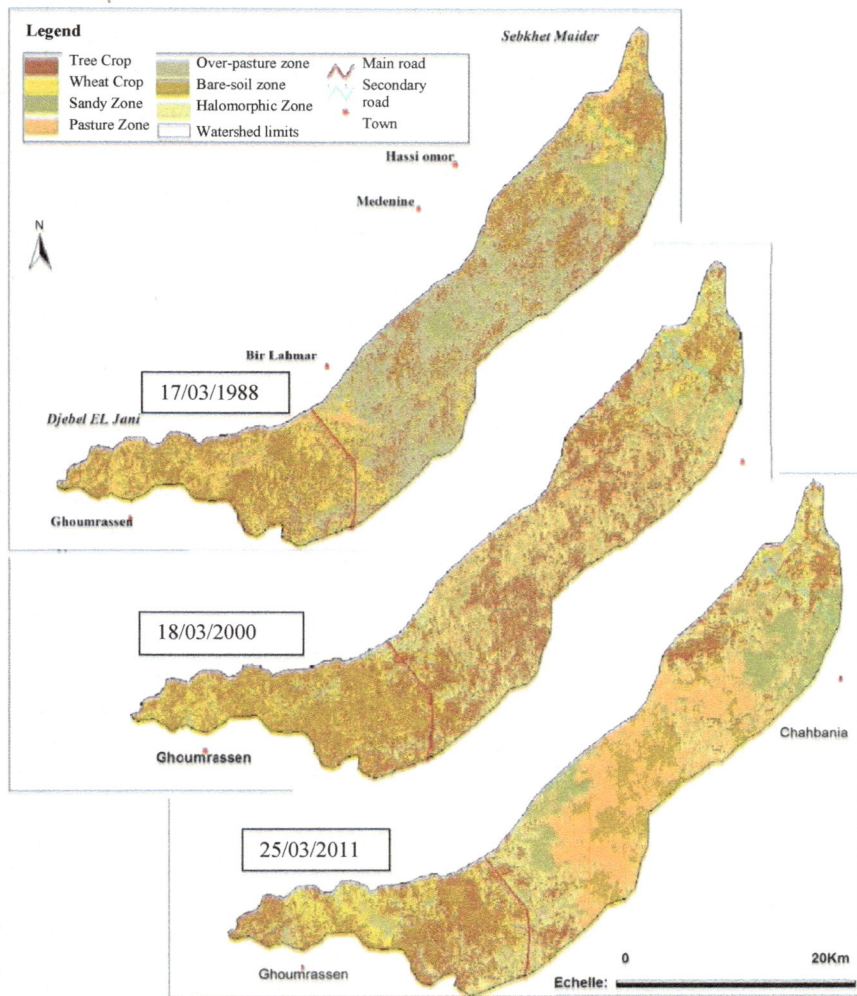

Fig. 13.11   Land use land cover changes on the Wadi Bouhamed watershed between 1988 and 2011

The total accuracies of the classification are high for the three dates, respectively 96/7, 81.8 and 91.1% indicating a good agreement between classified maps and thematic classes. Moreover, Kappa values indicate a strong agreement and a good accuracy for the classification of 1988 ($\hat{K} = 0.94$) and middle accuracy for 2000 and 2011 (respectively $\hat{K}$ equal to 0.77 and 0.70).

Following the confusion matrix for the three dates (1988, 2000 and 2011), thematic classes of the wheat crop, sandy zones, halomorphic zones are well identified by the classification. Other thematic classes such as tree crop, bare soil, pasture or over pastured zones have less accuracy what can be explained by the spatial resolution of

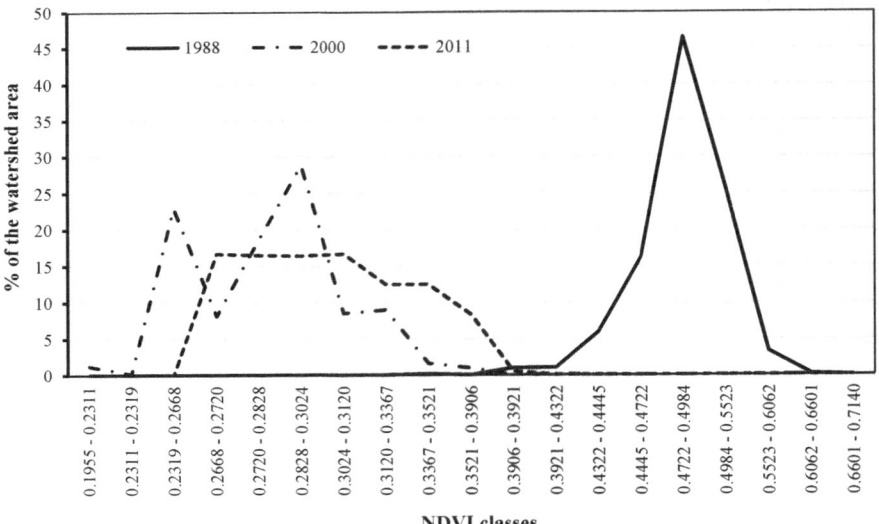

Fig. 13.12 Multitemporal variation of NDVI distribution of the Wadi Bouhamed watershed during the 1988–2011 preriod

the Landsat image (30 × 30 m). In fact, the confusion between tree crop zone and bare soil zone is likely due to the space interval between trees that made pixels of bare soil zone and tree crop zone with the same range of mean radiometric value [63].

The over pastured zones changed drastically between the three periods (Table 13.3). This thematic class that covered 34.6% of the watershed in 1988 decreased to less than 10% in 2000 and increased again to about 20% in 2011. This fluctuation is explained by climate conditions. Indeed, the 1988 year was a dry one during which the natural vegetation degradation has been enhanced by increased pasturing needs. This drought was followed by a relatively wet period explaining the regression rate of 25% of over pastured zone in 2000.

Although severe climate conditions during the studied period, tree crop zone areas remain of the same range, respectively 25, 32 and 27% for 1988, 2000 and 2011. This fact could be explained by the state soil management strategy that was based on olive trees and eucalyptus trees plants all over the Saharan border. The regression of wheat crop areas is explained by both local agriculture practices changes and by farmer abandonment of parcels.

It's remarkable that sandy zone areas have extended twice between 1988 and 2011. This is due to wind effect strengthened by the vegetation degradation that lets large opened areas.

Table 13.3 Distribution of thematic classes in 1988, 2000 and 2011

Thematic classes	1980		2000		Variation 1988–2000	2011		Variation 2000–2011
	Area (Hect.)	%	Area (Hect.)	%		Area (Hect.)	%	
1. Tree crop	12782.61	24,9	16711.11	32,6	7,7	13896.18	27,1	−5,5
2. Bare soils	5729.22	11,2	7244.01	14,1	2,9	4640.76	9,1	−5
3. Wheat crop zone	5451.48	10,6	3702.06	7,2	−3,4	3391.56	6,6	0,6
4. Halomorphic zones	4210.29	8,2	11340.72	22,1	13,9	8744.94	17,1	−5
5. Sandy zone	3100.41	6,1	3877.83	7,6	1,6	5852.97	11,4	3,8
6. Pasture zone	2231.73	4,4	3303.99	6,4	2	3521.16	6,9	0,5
7. Over-pasture zone	17732.7	34,6	5058.63	9,9	−24,7	11190.87	21,8	11,8
Total	51238.44	100	51238.44	100	–	51238.44	100	–

13.5.3 Vegetation Cover in Wadi Bouhamed Watershed as Inferred by NDVI

The Normalized Difference Vegetation Index (NDVI) indicates the vegetation density and photosynthetical activity and is one of the most commonly used vegetation indices. The NDVI values range from minus one (−1) to plus one (+1). Negative values indicate not vegetative areas such as water bodies, zero means bare soils with no vegetation and values close to +1 (0.8 − 0.9) indicate the highest possible density of green leaves.

The NDVI of the Wadi Bouhamed watershed calculated by ENVI software on Landsat images for the three dates (17 March 1988, 18 March 2000 and 25 March 2011) range from 0.022 to 0.714 (Fig. 13.12). No negative values have been obtained confirming that no water bodies are observed over the watershed. Lower values indicate to bare soils areas or areas with sparse vegetation.

Following the first image, more than 88% of the watershed have NDVI values ranging from 0.4445 and 0.5523. These relatively high values are related to steppe vegetation covering the plain of the watershed (Fig. 13.13a).

The image of 2000 has low NDVI values indicating a quasi-total degradation of the vegetation over the watershed. Vegetation cover revealed by 1988 image has disappeared and highest NDVI values are mainly related to halomorphic vegetation located around the wetland and along the coast at the watershed outlet. This situation is still observed in 2011 since in 2000, 95% of the watershed have NDVI ranging from 0.2319 and 0.3367 while in 2011, 78% of the watershed have NDVI ranging from

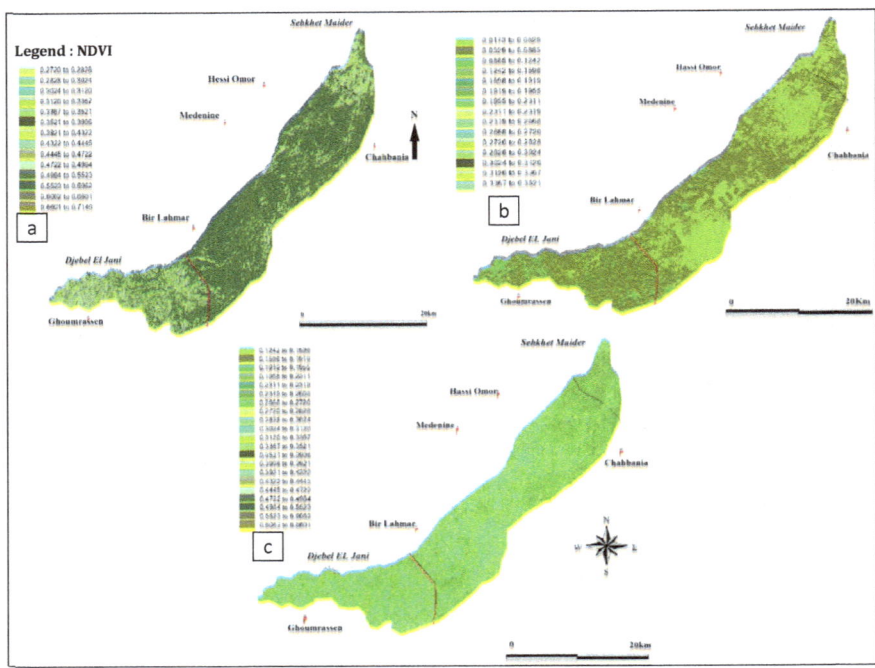

Fig. 13.13 NDVI maps of the Wadi Bouhamed watershed in (a) 17 March 1988 (b) 18 March 2000 and (c) 25 March 2011

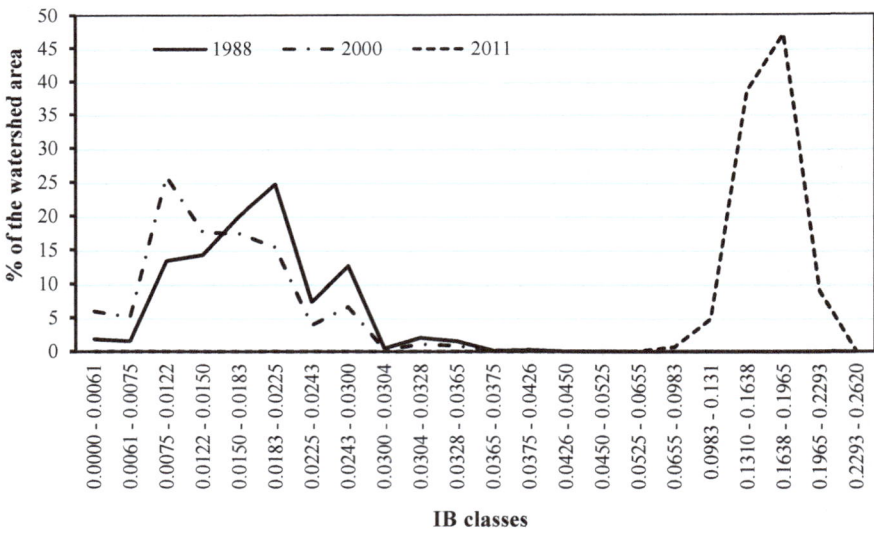

Fig. 13.14 Variation of brightness index (BI) during the 198–2011 period over the Wadi Bouhamed watershed

0.3367 to 0.3921. Vegetation is located in the plain and upstream of the watershed. Highest values of NDVI remains related to halomorphic vegetation downstream the watershed. The degradation of the vegetation is directly related to dry climate conditions and over pasture observed during this period.

The vegetation cover had declined during the 1988–2000 period due to both dry climate conditions and changes in agriculture practices since grain cultivation have been abandoned for tree farming that reflects lower NDVI (Fig. 13.13b, c).

13.5.4 Soils Distribution in Wadi Bouhamed Watershed as Inferred by Brightness Index

The Brightness index [64] describes the bare-soil reflection of sunlight. Soil brightness is a combined effect of soil type in particular clay grain presence, organic matter content, water content and outcropping salts at the acquisition date. All these characteristics are variable under arid areas by land degradation. The brightness index is related to the soil albedo [81]. The decrease in BI could indicate that soils have been wetted by rainfall or irrigation, that soil surface roughness has increased due to ploughing or erosion, or that vegetation cover has increased. The increase of IB corresponds to the inverse phenomenon and could be used to diagnose land degradation [47, 82, 83].

The brightness index calculated for 17 March 1988 (Fig. 13.8) range between 0 and 0.0983. Low values indicate dark soils what could be explained by relatively high water contents following rainy season and clay-sandy composition while high values indicate light soils mainly enriched in fine sands. The 1988 image taken during the rainy season indicates that soils are retaining water what could explain the relatively high NDVI values of this date and the growing up of vegetation. This is also the case of the brightness index calculated for 18 March 2000 (Fig. 13.14) ranging between 0 and 0.0983. However, NDVI of this image is lower probably since this rainy season followed a long drought period that constraint the growing up of the vegetation.

The brightness index calculated for 25 March 2011 (Fig. 13.14) range between 0.0375 and 0.1965. Highest values are explained by the sand drift that affected the major part of the watershed. Size grading of soil samples collecting during field trips showed that most of the samples are classified according to FAO as fine to medium sands what indicate a high vulnerability to aeolian erosion. On the other hand, over pasture practices enhance the brightness index due to the soil compaction by animals.

IB values variations during the studied period show that the watershed has been affected by land degradation (Fig. 13.15a, b, c). The entire watershed (99%) has higher IB values in 2011 compared to 1988 and 2000. This is due mainly to the sand drift all over the area and to the effect of over pastured that limited the vegetation cover and increased the soil compaction.

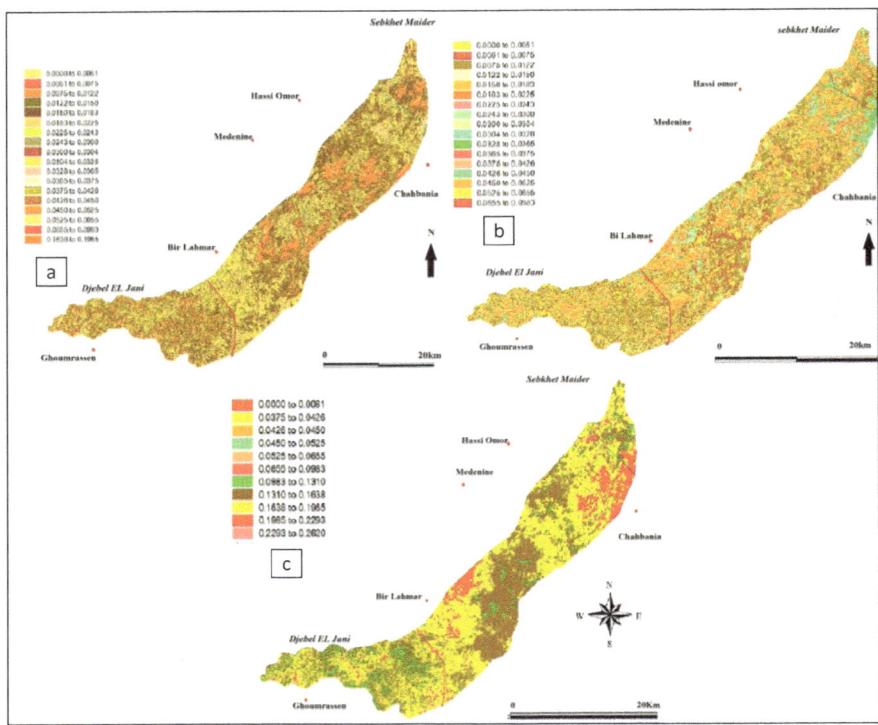

Fig. 13.15 Spatial distribution of BI maps of the Wadi Bouhamed watershed in (a) 17 March 1988 (b) 18 March 2000 and (c) 25 March 2011

13.6 Conclusions

Land degradation is a key ecological process that affects all the Tunisian country in particular its southern part. Even if less than 15% of the Tunisian population lives in these arid regions, the threat of desertification ranks among the most important environmental problems and has significant impacts on meeting human well-being. Due to severe climate and biophysical conditions, the main farming activities are based on clearing and overgrazing what break down a fragile equilibrium and tend to activate the desertification process. These regions located on the Saharan border are large areas where physical and human-induced factors of land degradation interact at local as well as at global scale. Remote sensing has long proven to be a time- and cost-efficient tool for monitoring change to desert environments; it is commonly used to monitor land uses/cover changes and to identify drylands degradation risks.

This research uses geographically approaches supported by remote sensing data to describe the Wadi Bouhamed watershed, considered as representative of southern Tunisia. This watershed is under arid climate conditions with very low thermal amplitude, violent winds and low precipitation not exceeding 200 mm a year.

This studied zone is characterized by sparse vegetation cover, quasi-sterile soils and high erosion risks due to the particular properties of the ground as fine texture, mainly sandy mineralogy, low structural stability and low organic content. This zone is also submitted to high human pressure due to its proximity to Medenine, the first city of South Tunisia. Multitemporal LANDSAT5 TM images (17 March 1988, 25 March 2000 and 25 March 2011) have been processed in order to map the vegetation cover and the soils types using respectively the Normalized Difference Vegetation Index (NDVI) and the Brightness Index (BI). The comparison of data obtained for the three dates allows land uses/covers between 1988 and 2000 and between 2000 and 2011. The main outcome of this research is both the decline of vegetation cover in favor of over pastured zones and the extension of sandy soil downstream of the watershed indicating a sandy drift phenomenon. This could be partly explained by the long drought period that affected the area. Furthermore, as in most arid regions, clearing and livestock as indicated by the decrease of grain cultivation areas are the main agriculture practices of local farmers what enhance the vegetation cover decrease and the soil properties degradation by overgrazing and compaction. The impact of the national strategy of water and soil management instituted since the 80 s is highlighted during the 1988–2000 period with the increase of tree crop areas but seems to have slowed down later. The research has proven that beyond complex and high-number methodologies and approaches that are available for remote sensing data processing thanks to computer sciences advances, simple spectral indices could be used to produce reliable maps and valuable information for dryland degradation monitoring.

13.7 Recommendations

The southern part of Tunisia is predisposed to soil erosion due to erodible soils together with poor farming practices. The protection and conservation of soil productivity is a major challenge in this area for the sustainable development of local population. The ability of decision-makers to implement cost-effective and efficient land management strategies depends on the availability of spatial data about both natural and human-induced factors that control land degradation. The research has proven that beyond complex and high-number methodologies and approaches that are available for remote sensing data processing thanks to computer sciences advances, simple spectral indices could be used to produce reliable maps and valuable information for dryland degradation monitoring. The research has also point to the need for a long-term monitoring of land covers in order to early identify mismanagement and inappropriate land uses and therefore guide the soil remediation efforts. Further refinement of national erosion assessment should be carried out using recent developments in the application of GIS and remote sensing techniques. More accurate erodibility maps capturing multitemporal variations of natural factors and produced at national, regional and local scale should be adopted by decision-makers as reliable decision support tools.

References

1. UNCCD (2013) A stronger UNCCD for a land-degradation neutral world. www.unccd.int
2. Dregne HE, Kassas M, Rozanov AB (1991) New assessment of the world status of desertification. Desertification Control Bull 20:6–19
3. UNCED (1992) Managing fragile ecosystems, combating desertification and drought. United Nations Conference on Environment and Development
4. Reynolds JF, Stafford Smith M (2002) Global desertification: do humans create deserts? In: Stafford-Smith, Reynolds, MJF (eds) Do humans create deserts? Dahlem University Press, Berlin, pp. 1–22
5. Sivakumar MVK, Stefanski R (2007) Climate and land degradation—an overview. In: Sivakumar MVK, Ndiang'ui, N (eds) Climate and land degradation. Environmental science and engineering (environmental science). Springer, Berlin, Heidelberg
6. EDL (2014) A global initiative for sustainable land management. The Economics Of Land Degradation. http://www.eld-initiative.org
7. EDL (2017) The costs of land degradation and benefits of sustainable land management in Africa. The Economics of Land Degradation. http://www.eld-initiative.org
8. Van de Koppel J, Rietkerk M, Weissing FJ (1997) Catastrophic vegetation shifts and soil degradation in terrestrial grazing systems. Trends Ecol Evol 12:352–356
9. Rietkerk M, Van de Koppel J (1997) Alternate stable states and threshold effects in semi-arid grazing systems. Oikos 79:69–76
10. Mtimet A (2001) Soils of Tunisia. In: Zdruli P, Steduto P, Lacirignola C, Montanarella L (eds) Soil resources of Southern and Eastern Mediterranean countries. Bari: CIHEAM, Options Méditerranéennes: Série B. Etudes et Recherches 34:243–262
11. Hirche A, Salamani M, Boughani A, Belala F, Essafi B, Gashut EH, Hourizi R, Grandi M, Ain Hamouda T (2017) Land degradation and restoration: The North African experiences. Geophysical Research Abstracts 19, EGU2017-11898
12. ESA (2017) Monitoring and evaluation tools to assess land degradation and environmental conditions. http://eo4sd.esa.int/files/2017/04/agri_2.pdf
13. Dewitte O, Jones A, Elbelrhiti H, Horion S, Montanarella L (2012) Remote sensing for soil mapping in Africa: an overview. Prog Phys GeographyProgress Phys Geogr 36:514–538
14. UNCCD (1994) United Nations convention to Combat desertification: article 1. http://catalogue.unccd.int/936_UNCCD_Convention_ENG.pdf
15. Starkel L (2005) Role of climatic and anthropogenic factors accelerating soil erosion and fluvial activity in Central Europe. Stud Quat 22:27–33
16. Marchetti M (2002) Environmental changes in the central Po Plain (northern Italy) due to fluvial modifications and anthropogenic activities. Geomorphology 44(3–4):361–373
17. Edward JA, Guillaume B, Besset M, Goichot M, Dussouillez P, Nguyen VL (2015) Linking rapid erosion of the Mekong River delta to human activities. Scientific Reports, vol. 5, Article n°14745
18. Sowa AHO, IbeSr KM (1992) The interaction of human activities and geological processes: a geo-environmental study in Southeastern Nigeria (Owerri urban area). J Afr Earth Sci (and Middle East) 14(4):539–544
19. Loczy D, Suto L (2011) Human activity and geomorphology. In: Gregory KJ, Goudie, AS (eds) The SAGE handbook of geomorphology. Sage, London, p 648
20. Blanco H, Lal R (2010) Water erosion. In: Principles of soil conservation and management. Springer, Berlin, p 256
21. Bagnold RA (1941) The physics of blown sand and desert dunes. Methuen, London, p 265
22. Khatteli H (1995) Aeolian erosion in arid and Saharan Tunisia: analysis of processes and research of remediation ways. Ph-D. Gent-Belgique University, p 170 (French)
23. Leon L (1988) Basic Wind Erosion processes agriculture. Ecosystems and Environ 22(23):91–101

24. Breshears DD, Whicker JJ, Johansen MP, Pinder JE (2003) Wind and water erosion and transport in semi-arid shrubland, grassland and forest ecosystems: quantifying dominance of horizontal wind-driven transport. Earth Surf Proc Land 28:1189–1209
25. Dregne HE (2002) Land degradation in drylands. Arid Land Resour. Manage 16:99–132
26. Hill J, Schütt B (2000) Mapping complex patterns of erosion and stability in dry Mediterranean ecosystems. Remote Sens Environ 74:557–569
27. Mainguet M, Dumay F (2006) Fighting against aeolian erosion: a way to fight against desertification: CSFD Thematic Folders. n°3. April 2006. CSFD/Agropolis, Montpellier, France, p 44 (French)
28. Subramanya K (2008) Engineering hydrology. Ed. Tata. McGraw-Hill, New Delhi
29. Bullock P (2005) Climate change impacts. In: Encyclopedia of soils in the environment, pp. 254–262
30. Roose E (1996) Land husbandry—components and strategy. FAO Soils Bulletin n°70
31. Gabriels D, Cornelis WM (2009) Human-induced land degradation. In: Land use, land cover and soil sciences: land use planning, vol 3. EOLSS Publications, p 290
32. Lal R, Blum WEH, Valentin C, Stewart BA (1997) Methods for assessment of land degradation. CRC Press, Boca Raton, 576 pp
33. Eswaran H, Lal R, Reich, PF (2001) Land degradation: an overview. In: Bridges EM, ID Hannam, LR Oldeman FWT Pening de Vries, SJ Scherr and S Sompatpanit (eds) Responses to land degradation. In: Proceeding 2nd International Conference on Land Degradation and Desertification, Khon Kaen, Thailand. Oxford Press, New Delhi, India
34. Escadafal R (1989) Characterization of arid soil surface with in situ observation and remote sensing: case of Tataouine region (Tunisia). Ph-D. Paris VI Univ., 317 pp (French)
35. Dubovyk O (2017) The role of Remote Sensing in land degradation assessments: opportunities and challenges. Eur J Remote Sens 50(1):601–613
36. Godert WJ, Mantel S (2001) The role of GIS and remote sensing in land degradation assessment and conservation mapping: some user experiences and expectations. Int J Appl Earth Obs Geoinformation 3(1):61–68
37. El Baroudy A (2011) Monitoring land degradation using remote sensing and GIS techniques in an area of the middle Nile Delta, Egypt. CATENA 87(2):201–208
38. Fernandez C, Wu JQ, McCool DQ, Stockle CO (2003) Estimating water erosion and sediment yield with GIS, RUSLE and SEDD. J Soil Water Conserv 58(3):128–136
39. Gitas LZ, Douros K, Minakou C, Silleos GN, Karydas CG (2009) Multi-temporal soil erosion risk assessment in N. Chalkidiki, "using a modified USLE raster model". EARSeL eProceedings 8:40–52
40. Ganasri BP, Ramesh H (2016) Assessment of soil erosion by RUSLE model using remote sensing and GIS: a case study of Nethravathi Basin. Geosci Front 7(6):953–961
41. Panagos P, Borrelli P, Meusburger K (2015) A new European slope length and steepness factor (LS-factor) for modeling soil erosion by water. Geosci 5:117–126
42. Dwivedi RS, Kumar AB, Tewari KN (1997) The utility of multi-sensor data for mapping eroded lands. Int J Remote Sens 18(11):2303–2318
43. Bou Kheir R, Girard M., Shaban A, Khawlie M, Faour G, Darwich T (2000) Contribution of remote sensing for the modeling of water erosion of soils in the coastal region of Lebanon. Télédétection 2(2):79–90 (French)
44. Bachaoui B, Bachaoui EM, El Harti A, Bennari A, El Ghmari A (2007) Mapping of water erosion risk: case of the Moroccan High Atlas. Télédétection 7(4):393–404 (French)
45. Dengiz O, Yakupoglu T, Kaskan O (2009) Soil erosion assessment using geographical information system (GIS) and remote sensing (RS) study from Ankara-Guvenc Basin, Turkey. J Env Biol 30(3):339–344
46. Singh A (1989) Digital change detection techniques using remotely-sensed data. Int J Remote Sens 10:989–1003
47. Coppin PR, Bauer ME (1996) Digital change detection in forest ecosystems with remote sensing imagery. Remote Sensing Reviews 13:207–234

48. Kumar M, Singh RK (2013) Digital image processing of remotely sensed satellite images for information extraction. Conference on Advances in Communication and Control Systems 2013 (CAC2S 2013)
49. Dubois JM, Pham TH, Bonn F (2007) Methodological approach for detecting changes in a fragmented environment using medium spatial resolution images: application to a coastal region in Vietnam. Revue Télédétection 7(1):303–323 (French)
50. Richards J, Jia X (2006) Remote sensing digital image analysis. Springer, Berlin, Heidelberg
51. Lambin EF, Geist HJ, Lepers E (2003) Dynamics of land-use and land-cover change in tropical regions. Annales Review of Environmental Resources 28:205–241
52. Jensen JR (2004) Introductory digital image processing —a remote sensing perspective. Prentice Hall, Upper Saddle River, NJ, 316 pp
53. Masoud A, Koike K (2006) Arid land salinization detected by remotely-sensed landcover changes: A case study in the Siwa region. NW Egypt J Arid Environments 66(1):151–167
54. Chaieb M (1989) Influence of soil water content on some vegetal species in the tunisian arid zone. Ph-D. Univiserty of Sciences and Technologie. Languedoc, Montpellier, p 292 (French)
55. Di Bartolomeo A, Fakhoury T, Perrin D (2010) Consortium for applied research on international migration—migration Profi le Tunisia. European University Institute
56. Bonvallot J (1986) Tabias and jessour of South Tunisia. Agriculture in marge zones. Cahier de l'ORSTOM, série Pédologie XXII(2):163–171 (French)
57. Missaoui H (1996) Soil and water conservation in Tunisia. In: Pereira LS, Feddes RA, Gilley JR, Lesaffre B (eds) Sustainability of irrigated agriculture. NATO ASI Series (Series E: Applied Sciences), vol 312. Springer, Dordrecht
58. Ben Mechlia N, Oweis T, Masmoudi M, Khatteli H, Ouessar M, Sghaier N, Anane M, Sghaier M (2009) Assessment of supplemental irrigation and water harvesting potential: methodologies and case studies from Tunisia. ICARDA, Aleppo, Syria. iv + 36 pp
59. Peyras L, Royet P, Degoutte G (1991) Flow and energy dissipation in gabion stepped weirs. La Houille Blanche 1:37–47
60. Floret C, Le Floch E, Pontanier R (1993) Agriculture and desertification in arid zones of Northern Africa. Agriculture in Mediterranean zone. Soils of the Mediterranean zone: uses, management and perspectives. Zaragoza: CIHEAM, Cahiers Options Méditerranéennes 1(2):39–51 (French)
61. Khatteli H (1993) Inventory and technical evaluation of fighting actions against silting in six governorates of South Tunisia. Revue des Régions Arides 5:59–90 (French)
62. Schowengerdt RA (2006) Remote sensing: models and methods for image processing. Elsevier Academic Press, Cambridge, p 560
63. Ouerchefani D, Dhaou H, Essifi B (2013) Application of atmospheric, radiometric and geometric correction on LANDSAT TM et ETM + images of SouthEastern Tunisia. Revue des Régions arides 32(3):89–110. P25 (French)
64. Escadafal R, Bacha S (1996) Strategy for the dynamic study of desertification. In: Escadafal R, Mulders MA, Thiombiano L (eds.) Proceeding of the International Symposium AISS, Ouagadougou, Burkina Faso, 6–10 February 1995. Monitoring of Soils using remote sensing and GIS, Editions ORSTOM. pp. 19–34 (French)
65. Cailleau D, Mougenot B (1996) Identification of soil degradation using remote sensing in the Sahel region of Niger. In: Escadafal R, Mulders MA, Thiombiano L (eds) Monitoring of Soils using remote sensing and GIS. Paris: ORSTOM, pp 169–179 (French)
66. Haboudane D, Bonn F, Royer A, Sommer S, Mehl W (2002) Land digital geomorphometic attributes. Int J Remote Sens 18:3795–3820
67. Belghith A (1997) Spectroscopic, satellite and integrated study of the ecosystems degradation in arid conditions (pre-Saharan Tunisia)—Interest of optical and microwave data. Ph-D, Paris VII. 264 pp (French)
68. Bennari A, El-Harti A, Haboudane D, Bachaoui M, El-Ghmari A (2007) Integration of spectral and geomorphometric variables in a GIS for mapping areas exposed to erosion. Revue Télédétection 7:393–404 (French)

69. Chikhaoui M, Bonn F, Merzouk A, Lacaze B, Alami M, Mejjati AM (2007) Soil degradation mapping using aster data approaches. Revue Télédétection, 7(1–2–3–4):343–357 (French)
70. Maimouni S, Bennari A, El-Harti A, El-Ghmari A (2011) Potentials and limits of the spectral indices to characterize the degradation of soils in semi-arid environment. J Can Dent Assoc 37(3):285–301 (French)
71. Séguis L, Puech C (1997) Method for determining radiometric invariants suitable for the semi-arid landscape of West Africa. Int J Remote Sensing 18(2):255–271 (French)
72. Rouse JW, Hass RH, Schell JA, Deering DW (1973) Monitoring vegetation systems in the great plains with ERTS. Third ERTS symposium, NASA SP-351. 1:309–317
73. Hunter RE, Richmond BM, Alpha TR (1983) Storm-controlled oblique dunes of the Oregon coast. Bull the Geol. Soc. America 94(12):1450–1465
74. Huete A, Post DF, Jackson RD (1984) Soil spectral effect on space vegetation discrimination. Remote Sens Environ 15:155–165
75. Huete A, Jackson RD, Fost DF (1985) Spectral response of a plant canopy with different soil backgrounds. Remote Sens Environ 17:37–53
76. Huete A, Jackson RD (1987) Suitability of spectral indices for evaluating vegetation characteristics on arid rangelands. Remote Sens Environ 23:213–232
77. Qi J, Chehbouni A, Huete A, Kerr YH, Sorooshian S (1994) A modified soil adjusted vegetation index. Remote Sens Environ 48:119–126
78. Qi J, Huete A (1995) A feedback based modification of the NDVI to minimize canopy background and atmospheric noise. IEEE Trans. Geo. & Remote Sensing 33:457–465
79. Pech RP, Davis AW, Lamacraft RR, Graetz RD (1986) Calibration of Landsat data for sparsely vegetated semi-arid rangelands. Int. J Remote Sensing 7:1729–1750
80. Schmidt H, Karnieli A (2001) Sensitivity of vegetation indices to substrate brightness in hyper-arid environment: the Makhtesh Ramon Crater (Israel) case study. Int J Remote Sensing 22(17):3503–3520
81. Lillesand PTM, Kiefer RW (2000) Remote sensing and image interpretation, 4th edn. Wiley, New York, p 724
82. Mougenot B, Cailleu D (1995) Identification of soil degradation using remote sensing in the Sahel region of Niger. Proceeding of the International Symposium AISS (working groups RS and DM), Ouagadougou, Burkina Faso, 6–10 February, pp 169–179 (French)
83. Escadafal R (2012) Long-term observation of arid environments by satellites: feedback and perspectives. In: Requier-Desjardins M, Ben Khatra N, Nedjraoui D, Wata Sama I, Sghaier M, Briki M (eds.). Environmental monitoring and development. Achievement and perspectives: Mediterranean, Sahara and Sahel Montpellier: CIHEAM/OSS-Options Méditerranéennes: Série B. Etudes et Recherches, 68:41–69 (French)

Chapter 14
Assessing Tunisian Oasis Dynamics Using Earth Observation and Landscape Metrics: Case of Djerid and Nefzaoua Regions

Faiza Khebour Allouche, Ibticem Abidi, Eric Delaître, Mohamed Saeid Desouky Abu-hashim, Dalel Ouerchfeni Bousaida, Safa Hamad, and Ribh Riahi

Abstract Future sustainable management of oasis is related to good selected archives of historical dynamics, assessment of actual stats by using earth observation data, and GIS tools. In this study, we combined remote sensing and landscape metrics to monitor the dynamics of Tunisian oasis landscapes, in particular, to characterize land use in the Nefzaoua and Djerid regions. The methodology is based on the analysis of NDVI maps and then through the metrics index analysis using Patch Analyst for oasis class. The application covers a period of thirty-six years from 1979 to 2015 from the processing of Landsat images series. Results reveal an unequally distributed oasis progression in space, with a slightly different rate over the period

F. Khebour Allouche (✉)
Laboratory of Phytopharmacy and Weed Science, Higher Institute of Agronomy-Chott Meriem, University of Sousse, ISA CM BP 47, 4070 Sousse, Tunisia

GREEN-TEAM Laboratory (LR17AGR01), Higher Institute of Agronomic Sciences-Chott Meriem, University of Sousse, Sousse, Tunisia

I. Abidi
High Institute of Agronomic Science-Chott Meriem, University of Sousse, ISA CM BP 47, 4070 Sousse, Tunisia

E. Delaître
UMR-S ESPACE-DEV, IRD Montpellier, 911 Avenue Agropolis, 34394 Montpellier, France
e-mail: eric.delaitre@ird.fr

Mohamed Saeid Desouky Abu-hashim
Soil Science Department, Faculty of Agriculture, Zagazig University, Zagazig, Egypt

D. Ouerchfeni Bousaida
Institute of Arid Regions-Medenine, Route Djorf Km 22.5, Médenine, Tunisia

S. Hamad
Regional Commissariat for Agricultural Development—Tozeur, Route Nafta Avenue Farhat Hached, 2200 Tozeur, Tunisia

R. Riahi
Regional Commissariat for Agricultural Development-Kebili, Rue Saleh Ben Youssef, 4200 Kelibia, Tunisia

© Springer Nature Switzerland AG 2021
F. Khebour Allouche et al. (eds.), *Environmental Remote Sensing and GIS in Tunisia*, Springer Water, https://doi.org/10.1007/978-3-030-63668-5_14

285

considered. The increased oasis area was mainly due to uncultivated land cover. However, the analysis of spatial metrics suggests a transition from an accelerated fragmentation phase of the oasis landscape, to an expansion phase by the continuous spreading of existing oasis surfaces. A novelty of this study is that the results observed for landscape metrics can be correlated with a human dimension. The use of Geographical Information System based metric information system and spatial analysis with Inverse Distance Weighted interpolation enabled the future mapping of oasis extension in South Tunisia.

Keywords Earth observation · NDVI · Metrics · Oasis · Landscape dynamics · Management

14.1 Introduction

Tunisian oasis forests cover about 40 000 ha, with diverse and intensive production systems. According to the world bank in 2014, there are three types of oases in Tunisia: (i) Coastal oasis (17.3% of the total area of the oasis); (ii) Mountain oasis (5.8%); and (iii) Saharan oases (76.8%). These Oasis are two categories, "traditional" and "modern" oases (or irrigated schemes/perimeters). These Oasian environments have rich and complex biotic, abiotic, and socio-economic characteristics. The Oasis dynamics are directly related to global socio-environmental changes. However, in Tunisia, this ecosystem is facing multiple problems, including climatic deterioration, urbanization, and human population in excess of the ecosystem's carrying capacity, causing a real crisis under the influence of an inappropriate development model. Thus, these ecosystems are altered due to adjacent human activity and excessive land use. They constitute a heritage that must be preserved for future generations. Hence, there is an urgent need to develop methodologies for assessing and monitoring oases from multiple perspectives and especially by using remote approaches. Monitoring, evaluating, and predicting the spatiotemporal changing patterns of oases is essential for managing oases for sustainability [1]. Authors have treated different topics related to diversity agro-systems [2], sustainability [3], irrigation system [4], hydro-geochemical processes [5]. However, all of these studies require intensive field surveys. Hence, to monitor the complexities of oasis change of long duration effectively, multi-type, time-serial remote sensing data are needed [6]. This study, therefore, approaches oasis dynamics from the landscape scale. We quantified and qualified landscapes structure, accounting for the global and local spatial autocorrelation of the data. We mapped NDVI and oasis patches (spatial resolution 5 km × 5 km) using Landsat images to address the following questions:

– Can landscape metrics be used to characterize oasis structure?
– Can quantitative landscape indexes reflect the socio-economic status and dynamics of oases?
– With metric used, can we propose a scenario of future oasis extension?

14.2   Materials and Methods

Figure 14.1 illustrates the sequence of treatments required for the application of the proposed method. Two important steps are considered: the image processing for 8 images Landsat, Two images Landsat 1 MSS, one image Landsat 3 MSS, three images Landsat 5 TM, one Landsat 7 ETM + and one Landsat 8 OLI-TIRS, and the calculation of landscape metrics tested for only one Landsat image.

14.2.1   Study Area

Figure 14.2 shows the location of Nefzaoua and Djerid regions placed on southern Tunisia between 33–34° N and 7–9° E. At present, Tunisian oases collectively cover 1.9% of the total tree area of the country, with 43,700 ha of oases supporting a total of nearly 5.2 million date palms. They are located mainly in Kébili (57%), Tozeur (29%), Gabès (10%) and Gafsa (3%). The development of oases was revived in the 1970s and 1980s with the implementation of a vast program rehabilitating 20,000 ha of ancient oases, and creating 4500 ha of new oases. Further oasis development occurred during the 1990s through illicit extensions (drilling of boreholes). In the four southern governorates, there are 210 oases, representing 9% of the total irrigated land. The oasis area has increased from about 16,700 ha in 1974 to about 41,700 ha in 2010, due to water resource development programs supported by the Tunisian Government and private developers. Oases are home to about 950,000 people (equivalent to 10% of the total Tunisian population). Traditional oases have been established on private

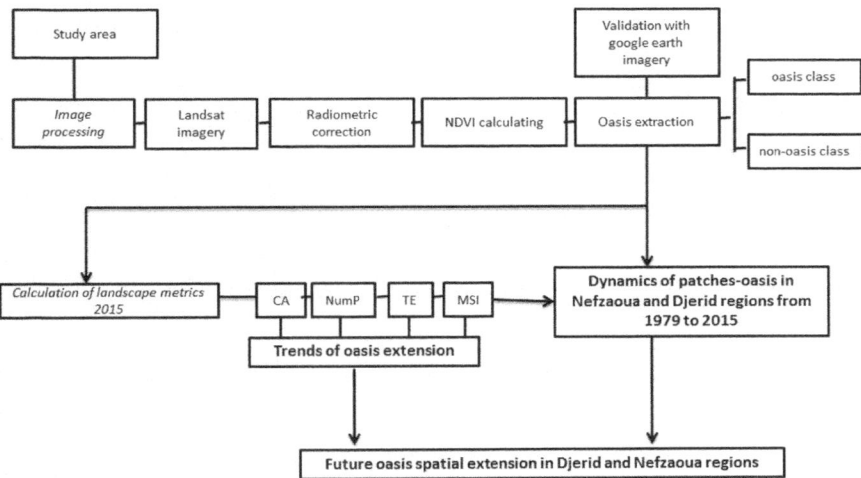

Fig. 14.1 Flow shart for future oases spatial extension in Djerid and Nefzaoua regions, Tunisia

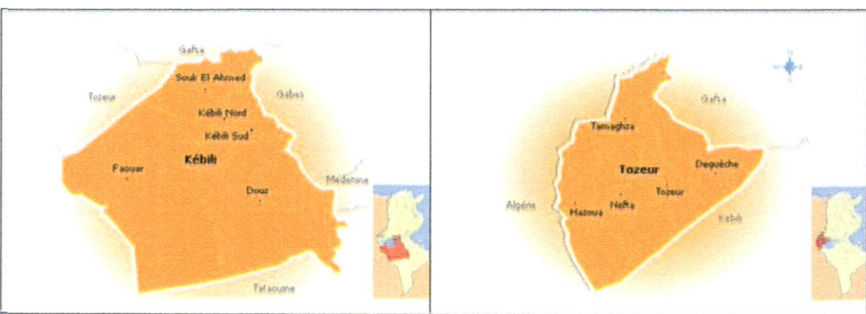

Fig. 14.2 South Tunisia, showing the location of the study areas (@*Google earth*)

properties, whereas the extensions concern state lands such as the case of Tozeur oasis and a local collective farming in Kebili [7].

14.2.2 Image Processing

Remote sensing techniques have become standard for LULC (Land Use and Land Cover) change detection analysis at local, regional and global scales [8–15]. Remote sensing plays a key role in acquiring broad-scale environmental information for conservation planning [16], and particularly Landsat imagery provides a sufficiently high spatial and temporal resolution to map land cover in a way suitable for identifying small habitat patches and corridors in many areas of the world [17]. Landsat images with adequate spectral properties have been used previously for landscape change studies in mountain areas with similar characteristics and can provide better information on LULC changes than point data collected by on-site instruments during in situ surveys Landsat imagery [18].

Eight cloud-free Landsat images were selected, covering 36 years (Table 14.1), and relative radiometric normalization was used to remove radiometric distortions and make the images comparable [19]. We then used these normalized images to calculate Normalized Difference Vegetation Index (NDVI) to measure gains and losses in oasis extent over time [20, 21]. We calculated NDVI by applying Eq. (1) using Envi 5.3.

$$\text{Round}\left(\left(\frac{\text{(float (B1) - B2)}}{\text{(B1 + B2)}} \times 10000\right) + 10000\right) \tag{1}$$

where B1: PIR (near infrared) and B2: R (red), using the software's Round and Float settings.

After interpretation of the NDVI values per pixel, the final interval retained after thresholding for oases extraction was on the order of (0.17; 0.67)]: below the NDVI value of 0.17 and above 0.67, no pixel corresponded to an oasis. Table 14.2 shows the different thresholding values used for the extraction of oases, using "Google Earth" imagery for verification for the years 1984, 1995, 2009, and 2015. Two classes were obtained after thresholding: (i) an oasis class encompassing all pixel values

Table 14.1 Dates of Landsat TM acquisition

Year	Satellite	Sensor	Path	Row	Date of acquisition
1972	Landsat 1	MSS	206	037	29-08-1972
1973	Landsat 1	MSS	206	037	07-02-1973
1979	Landsat 3	MSS	206	037	13-05-1979
1984	Landsat 5	TM	192	037	02-07-1984
1995	Landsat 5	TM	192	037	14-05-1995
2003	Landsat 7	ETM+	192	037	10-04-2003
2009	Landsat 5	TM	192	037	28-11-2009
2015	Landsat 8	OLI-TIRS	192	037	29-11-2015

Table 14.2 Different thresholding values used for the extracting of Oases

Year	Value of thresholding
1979	0.17
1984	0.18
1995	0.17
2009	0.17
2015	0.18

in the range [(0.17; 0.67)]. This class was coded [(1)] and (ii) a non-oasis class encompassing all pixel values outside the range [(0.17; 0.67)] and was coded [(0)].

14.2.3   Calculation of Landscape Metrics

To generate samples, we overlaid a grid 5 km × 5 km onto the image of the two land cover classes (oases and non-oases) for 2015. The grid generated 1014 cells (Fig. 14.3). From the 1014 cells, twelve cells were selected within the two study areas for the landscape metrics analysis.

Landscape metrics can be used to quantify the individual pattern to detect the spatiotemporal pattern of landscape change [22–26] and as indicators for the interaction between spatial pattern and ecological processes [27]. Some landscape

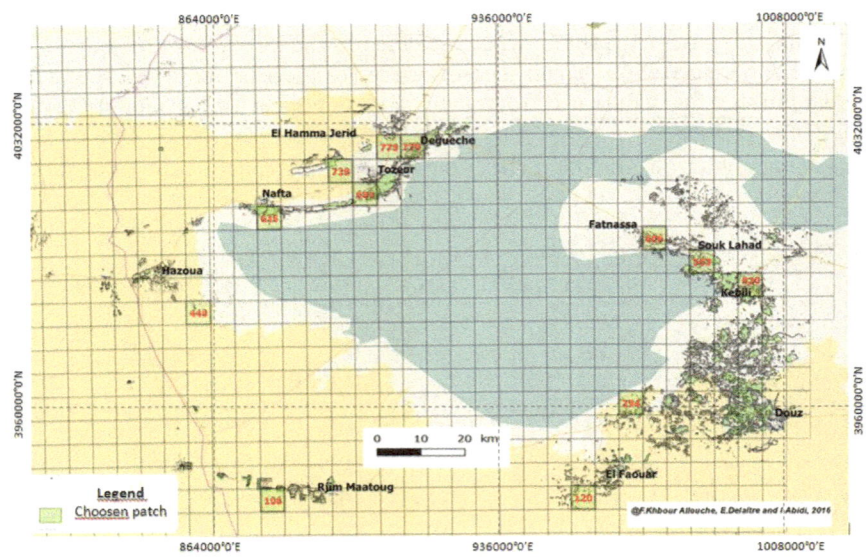

Fig. 14.3 Geographical location of 12 patches study area

metrics have also been used to reflect a landscape's socio-economic status [28]. Five commonly-used landscape metrics that characterized the changes in cover and shape of oases are used: Class Area (CA), Number of Patches (NumP), Mean Patch Size (MPS), Total Edge (TE), and Mean Shape Index (MSI) [29]. These metrics were calculated using Fragstats 3.2 [30]. To interpret trends in these metrics as a response to human activities, we used data collected from the Regional Agricultural Development Commissions of Tozeur and Kebili and from Southern Development Office in Tunisia (2015). Interventional studies involving animals or humans, and other studies require ethical approval must list the authority that provided approval and the corresponding ethical approval code.

14.3   Results

14.3.1   Shape Dynamics of Oases

Over the 36-year time span, oases have increased in number, size, and perimeter in both study areas. In 1979, oases were much more concentrated in the North of Nefzaoua (center of Kebili, Souk Lahad, and Fatnasa) but, became less dense on the south one. For the Djerid region, oases occupied the center of Tozeur rather than El Hamma Djerid. In 1984, there was a major change in the Djerid region: trends in oasis extent increased in the center of Tozeur and in Hamma Djerid, with a new appearance of oases in Nafta, Hazoua, and Dgeuche. However, in the Nefzaoua region, oases were continuously increasing throughout Kebili, Souk Lahad, Fatnassa, and Douz. In 1995, we observed an increase of oasis density with newly irrigated perimeters in many areas, such as El Faouar, Rjim Maatoug, and Hazoua. In 2009, there was a continuous increase of oases, particularly in Rjim Maatoug and Hazoua. Recently, in 2015, irrigated areas were still increasing in surface and number in the study area but more remarkably in Nefzaoua (Fig. 14.4).

At the patch level, Fig. 14.5 shows the spatiotemporal dynamics within each of the twelve grid cells in the two study areas. We detected different chronologies of oases occurrence among the grid cells. For example, in cells 294, 563, 606, 520, 728, 684, oases were already present since 1979 and had been increasing through 2015. Other oases have appeared in 1984 such as in cells 776 and 635, in 1995 in cells 120 and 775, and in 2009 for cells 106 and 449.

14.3.2   Shape Dynamics of Oases

Figure 14.6 shows the changes in oases area (CA) at Nafzaoua and Djerid regions. Between 1979 and 2015, CA of oases in Nefzaoua and Djerid was increasing. This indicates that at the end of this period, the oases surface is increasing, linked to

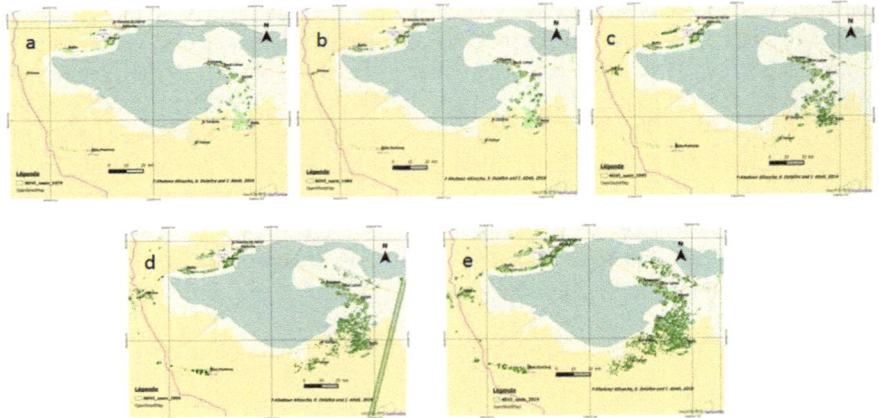

Fig. 14.4 Evolution of NDVI-oasis in Nefzaoua and Djerid regions in **a** 1979, **b** 1984, **c** 1995, **d** 2009 and 2015

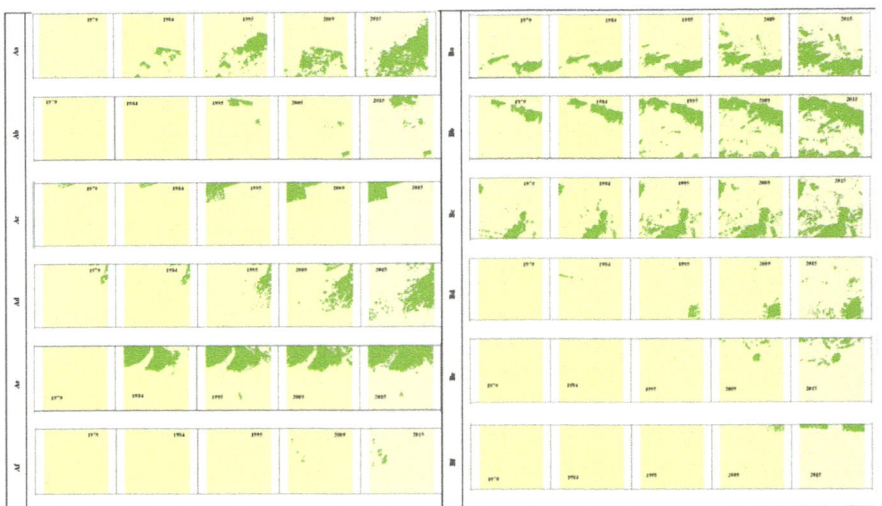

Fig. 14.5 Dynamics of patches-oasis in Nefzaoua and Djerid regions from 1979 to 2015 Aa. Degueche (776), Ab. Hamma Djerid (775), Ac. New Tozeur (728), Ad. Old Tozeur (684), Ae. Nafta (635), Af. Hazoua (494), Ba. Fatnasa (606), Bb. Souk Lahad (563), Bc. Kebili (520), Bd. El Dergine (294), Be. Faouar (120), and Bf. Rjim Maatoug (106)

the increase of oasis surfaces to the detriment of uncultivated land. However, patch surface evolution was marked by a slight increase in North nefzaoua, and an evolutionary expansion was observed between 1995 and 2009 in the South, there was a great change in the structure of oasis landscape. In the Djerid region, the variation was

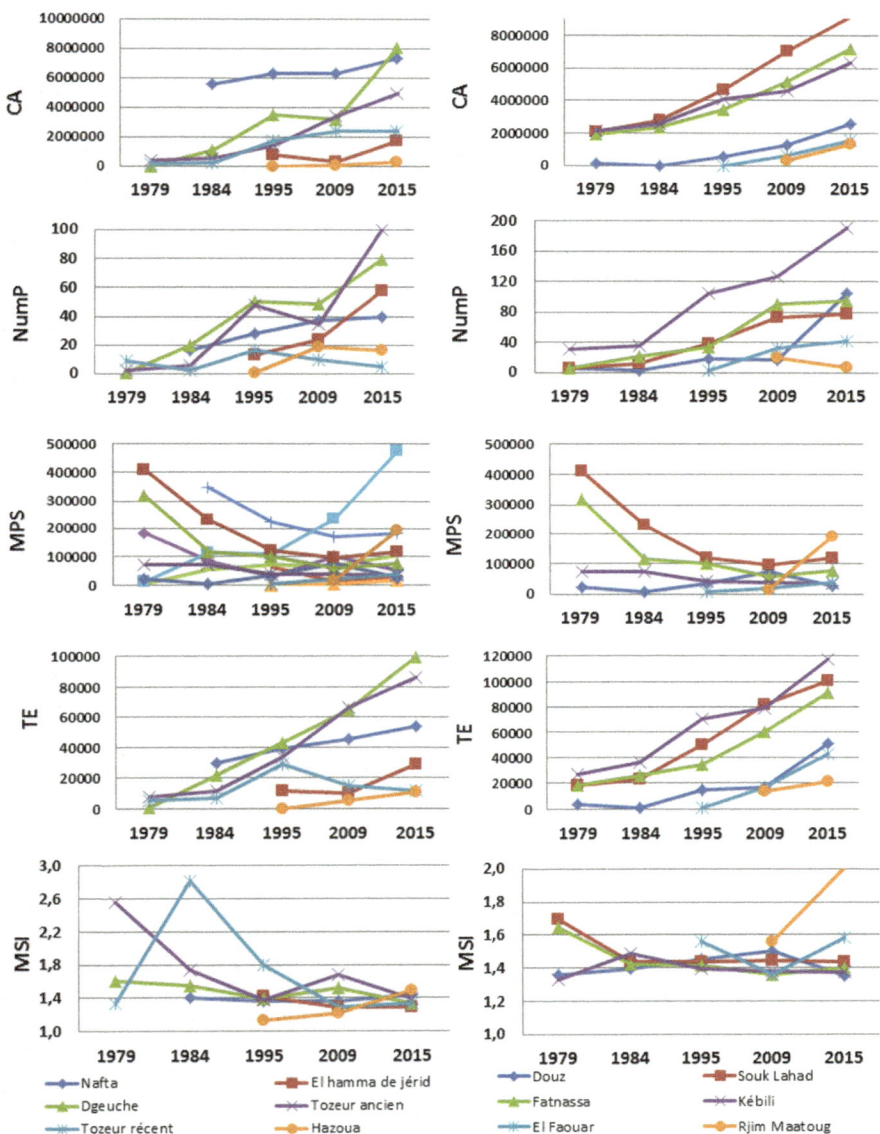

Fig. 14.6 Variation of Class Area (CA), Number of Patches (NumP), Total edge (TE) and Mean shape index (MSI) for Djerid (A) and Nefzaoua (B) regions from 1979 to 2015

more miscellaneous. CA increases continuously between 1979 and 2015 in North Djerid and between 1984 and 1995 in the South West one.

The variability of the oasis fragmentation process based on patch number (NumP) presents an increasing general trend of the curves is all the study area. However, decay of NumP is noted on patches located in West Nefzaoua (Hazoua) and South Djerid (Rjim Maatoug) since 2009. NumP is correlated to CA positively on the North side and negatively on both the South and on the West. At the opposite, between 1979 and 2015, there was a decreasing MPS in two regions, except for some patches such as Hazoua and Rejim Maatoug. On the other hand, MPS is correlated to CA negatively on the North side and positively on both the South and West. NumP and MSI metrics show that at Nefzaoua there are a continuous oasis fragmentation in the North (Kebili, Fatnassa and Souk Lahad), represented by a diffusion process whereby the extension of oasis surfaces was carried out through the creation of new oases, unconnected to existing oases in 1995. Yet, the increased connectivity in the South-West (Rjim Maatoug) was produced by the creation of new oases close to existing ones. However, In the Djerid region, NumP and MPS indicate that the North side (Dgeuche patch), was characterized by continuous connectivity between oases. More in the West side (Nafta, El Hamma Djerid, and Hazoua). These two metrics show a high degree of fragmentation produced between 1984 and 2009 due to the creation of a large number of new oases far from those already existing, with increased connectivity after 2009 as new oases began to connect existing ones.

The Total edge index (TE), strongly correlated to CA, increases between 1979 and 2015 in two regions but some peaks are observed in curves on 2009 especially in Dergine, Kébili, and Old Tozeur. In the Nefzaoua region, this pattern increases continuously from 1979 to 2015. This indicates that environmental conditions are more variable in the North (Kébili, Souk Lahad, and Fatnassa) than in the others. Nevertheless, the high MSI value in Nefzaoua region is noted in the South (Rjim Maatoug) in 2015. This indicates that at Rjim Maatoug, Fatnassa and Souk Lahad, there are many oases with complex shapes while in the rest of the cells, the oasis shapes are simple. This implies that in Nefzaoua Northside (Fatnassa and Souk Lahad), oases shapes were complex but then were simplified, and the opposite is true for the South (Rjim Maatoug). On the spatial scale, it is distinguished that the largest values of Dejrid all contours (TE) are marked in the patches located in North (Dgeuche and old Tozeur), so the ecological conditions are more variable than in the rest. However, two high values of MSI are marked in 1979 and 1984, respectively in old Tozeur and new Tozeur, while the remaining patches have low values of MSI. This implies that at old Tozeur level the patch form was complex then it became simple while, for new Tozeur the form was simple in 1979 then it became complex in 1984 and returned to simple until 2015.

14.4   Discussion

Through the result of CA pattern, it can be said that over the last 40 years, the extent of oases in Nefzaoua and Djerid has expanded, to the detriment of uncultivated land. This result is reflected by the growing human population, hence the effect of the human land use on the landscape structure. The number and area of Nefzaoua oasis are continuously increasing, in fact, according to the Ministry of the Environment (2012), the proliferation of drill holes and illegal surveys, especially in the North of Nefzaoua, has increased the oases from 15,837 ha in 2001 to 23,856 ha in 2008. This strong extension is reflected in water withdrawals that largely exceed resources. This has caused groundwater retreatments and consequently an increase in water salinity, and therefore problems in the oasis patches detected by other metrics.

Similarly, the same Djerid Oasian dynamic exists, but it has been a decline in some irrigated perimeters in 2009, particularly in the North of Djerid (El Hamma Djerid and Dgueche patches) due to problems of resource depletion and salinity rise blocking geothermal crops channels. Indeed, the loss of some oases especially at Tozeur patches can be explained by several factors such as desertification, oasis fragmentation, uncontrolled urbanization, pollution, over-exploitation of irrigation water and consequent deterioration of the irrigation water quality. Recently, the increase in the number of patches during the last period, particularly in 2015, is reflected by the progress of agricultural investment projects in Kébili (43 projects), El Faouar (40 projects), Souk Lahad (22 projects), Degueche (15 projects), Tozeur (28 projects), Nefta (9 projects) and Hazoua (3 projects) [31, 32].

NumP and MPS were correlated throughout the study period in Dgueche, El Faouar, and in a different period in Hamma Djerid, Nafta, Souk Lahad, Fatnasa, and Kebili. However, these indexes are inversely correlated for most studied regions and more particularly between 1979 and 2009, reflecting increased fragmentation and decreased connectivity among oases. The oasis fragmentation has been driven by increasing populations, and the persistence of positive inter-delegation migratory balances, resulting in the creation of new private irrigated schemes to Rjim Maatoug and Hazoua between 1995 and 2009 proved by the negative correlation between CA and NumP found in West Nefzaoua and South Djerid. Then, the positive or negative correlation of NumP and MPS indicates the kind of ownership structure, from example, the positive correlation indicates collective lands in the Nefzaoua region, and the negative one marks the importance of state land in the Djerid.

NumP is correlated to CA but north side positively and south and west negatively. The correlation between CA and fragmentation metrics proved the increase of number and area of Oasis between 1979 and 2015 in our study area, which is too characterized by an important fragmentation. The positive one is explained by the increase of population and of the number irrigated perimeters and water exploitation, and projects but the negative correlation is explained by the reduction of traditional oases characterized by old plantations, high density of trees (400 trees/ha), highly fragmented individual plantations, and low yields [33].

TE pattern increased in Nefzaoua, reflecting variability in the landscape structure of oases over time. While in Djerid, more precisely Degueche, Tozeur, Nafta, and Hazoua, the environment has been more stable, explained by the desertion of oases, especially after the revolution. The dynamics of oases are partly the result of population movements, and there is a positive migration balance during the last period, particularly in Kebili, Faouar, Degueche, and Hazoua having positive TE. The highest increases were recorded for Nefzaoua during the 1979–1984 and 1995–2009 periods respectively at the level of Dgeuche (57.65%) and Hazoua (41.17%). This implies that at these periods, the oasis environment has become rich and variable in plant species, which has led to a richness of biodiversity and consequently to ecological variability. This variability is probably explained by the practice of various crops in stages at within oases plots. On the other hand, the small increases recorded in Nafta indicate a monotony in the oasis environment, and this can be explained by the practice of monoculture. Furthermore, TE decreases were observed over the period 1995–2015 in Tozeur (-0.5% over the period 1995–2009 and -0.22% over the period 2009–2015) and El Hamma Djerid (-0.18 over the period 1995–2009). This can be explained by the loss of some irrigated perimeters. Too, El Hamma Djerid has experienced a regression on the surface of oases (-0.6%). This loss of biodiversity can be explained by the impact of the market economy, which has favored the exaggerated extension of Deglet Nour dates at the expense of other varieties. Therefore, the values of these indexes are more variable in Djerid, but the majority are positive. The biggest increases were recorded over the periods 1984–1995 and 1995–2009 respectively in El Dergine (21.33%) and El Faouar (20.25%). This implies that at these periods, the oasis environment has become rich and variable in plant species, which has led to a richness of biodiversity and consequently to ecological variability. This variability is probably explained by the practice of various crops in stages at within oases plots.

MSI was highly variable, with increased complexity reflecting the geometrical shapes of Kebili patches, explained by the importance of private PPI (4984 ha) and the cultural diversity especially to vegetable crops. More simplified configurations were observed in Tozeur and Degueche patches, reflecting the degradation of these oases following overexploitation between 1979 and 1995. The renovation and rehabilitation of oases between 1995 and 2009 increased configuration complexity, but the simplicity reappeared after the revolution following the farmers' desertion. This last event is also observed at Dergine.

Oasian landscape dynamics can be deduced by using CA pattern variation but to understand more and characterize this variation, the use of other metrics was interesting. Thus, to better draw the story of these oases, the interpretation of NumP, MPS, TE, and MSI trends using social and economic information is more effective.

Figure 14.7 shows that the scenario of management dealing with the future oasis extension is based on the kind of correlation found between each pair of metrics: NumP/MPS and MSI/TE. To achieve this objective, two hypotheses have been put forward: (i) Hypothesis 1: there will be the possibility of the future extension of the oases if the correlation is positive between the two pairs of metrics. (ii) Hypothesis 2: expansion trends are medium to small if correlations are negative for both

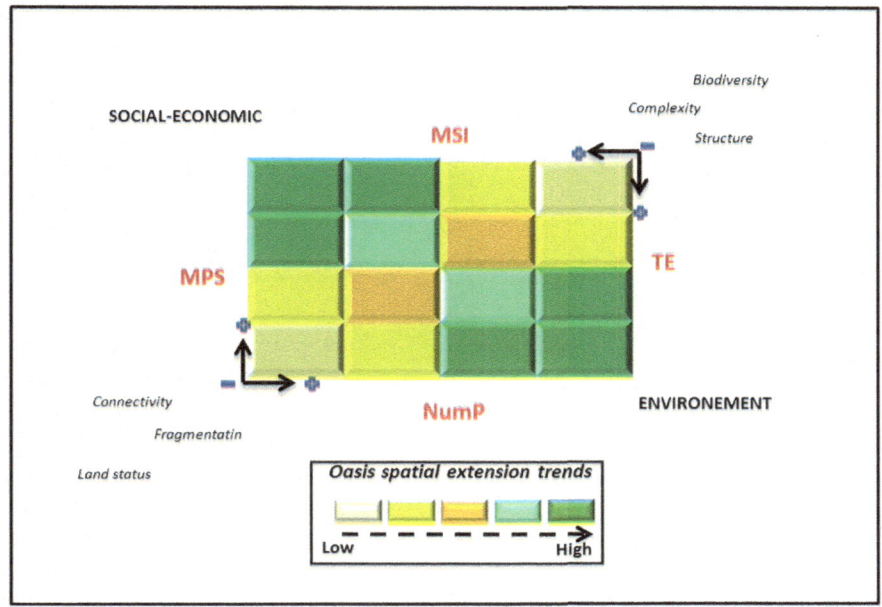

Fig. 14.7 Trends of oasis extension in correlation with NumP-MPS and MSI/TE correlation metrics

pairs of metrics. For mapping future management of an oasis in the study area, GIS interpolation is applied by using IDW tool from the center of each patch.

Figure 14.8 shows a test of spatial generalization of oasis extensions of the study area from the patch studied. A negative gradient oriented southwest east is deducted, and the risk of extension in the southwest of the study area is greater, especially in Hazoua and Rejim Maatoug. Side El Faouar and the new tozeur, these risks are means and decrease in Souk Lahad and Tozeur. This result demonstrates the utility of using metrics to study the dynamics of oasis forests. Two important results must be taken into account and developed for the applications of natural, agricultural or urban landscapes. The first one shows the environmental and anthropogenic changes deduced in this landscape by these indices. The second result underlines that the hypotheses emitted on the correlation of metrics from the patch help us to predict the future development of the oases. According to [34], the result is related to 2016–2020 strategic management. The decision makers can predict extensions of the oases, particularly border posts in Hazoua, in Tozeur Governorate to fix populations and upgrade utilizing existing facilities for better trading with cities in Algeria and Libya.

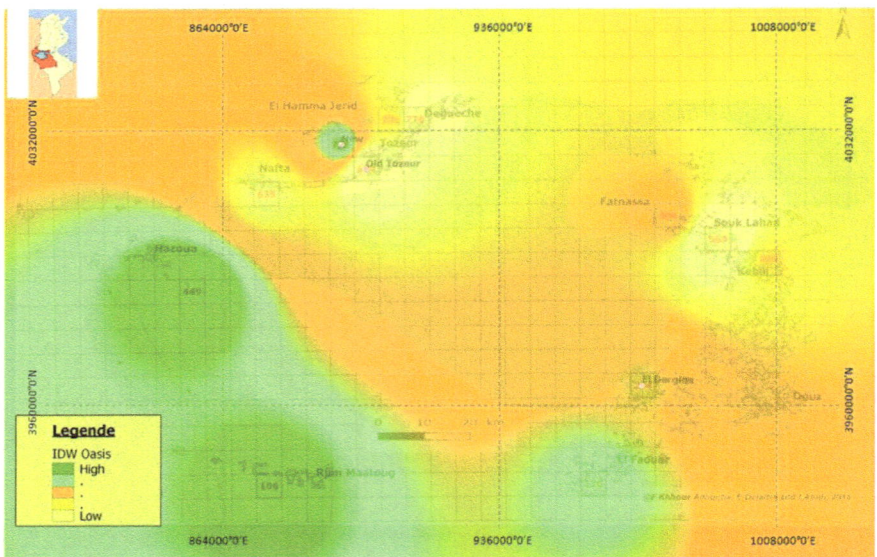

Fig. 14.8 Map of the future oases spatial extension in Djerid and Nefzaoua regions

14.5   Conclusion

The present study used a set of landscape metrics to characterize the landscape-level dynamic of oases in Tunisia. The first novelty of this study is the connection of oasien metric results to the oasis-related human dimensions Including land use, management, and regeneration.

The extension of oasis coverage mainly reflected increasing populations in the region and the rising numbers of agricultural investment projects, particularly on private land and the rent of publicly owned land. Indeed, the correlation between MSI and TE indexes reflected a variability of the oasis structure and increasing complexity, driven by increasing resident populations and a large population movement (net migration). In other areas, the lack of correlation between these two indices reflected problems of landscape degradation, desertification, or exhaustion of natural resources, thus causing migration of the population to other areas. The correlation of NumP and MPS indicated a significant fragmentation of oases and an important dynamics of the landscape structure. This case was mostly present where there were large areas of collective or State lands (governmental lands) and greater cultural diversity, generating areas of forage, fruit, and vegetables. In contrast, the reverse correlation of these metric indexes indicates a more dominant private land status characterized by high oasis plot size and monoculture. The second result, based on the first novel one is that with the kind of correlation between each pair of metric and using GIS interpolation, we can map the future oasis extension and propose scenarios for sustainable management of these oases in the long term. Overall, population growth, rapid urbanization, and improved living standards have led to a significant increase in water consumption in the region, some of which increased oasis area, but may drive water shortages and competition in the future (thus leading to oasis loss). Diversification of economic activities has also contributed to increased pressure on the region's water resources. These spatially explicit results can help decision makers locate landscape scale risks of degradation and indicate at the patch level, which agricultural investment projects can contribute to regional sustainability.

14.6 Recommendations

- Multi-scale time-series investigations of satellite images on patch-mosaic, gradient, and graph network models can go beyond simply describing dynamics. This chapter recommends the application of other types of metrics to exploit the presence of other metric-human dimension relationships.
- Test this relationship with other types of landscapes in order to model this relationship metric-human dimension. Applicable model in the detection of changes that will be used in planning strategy development.

References

1. Yuchu X, Gong J, Sun P, Gou X (2014) Oasis dynamics change and its influence on landscape pattern on Jinta oasis in arid China from 1963a to 2010a: integration of multi-source satellite images. Int J Appl Earth Obs Geoinf 33:181–191
2. Benaoun A, Elbakkey M, Ferchichi A (2014) Change of oases farming systems and their effects on vegetables pecies diversity: case of oasian agro-systems of Nefzaoua (South of Tunisia). Sci Horti 180:167–175
3. Mekki I, Frederic J, Marlet S, Ghazouani W (2012) Management of groundwater resources in relation to oasis sustainability: the case of the Nefzawa region in Tunisia. J Env Manag 121:142–151
4. Pun M, Mutiibwa D, Li R (2017) land use classification: a surface energy balance and vegetation index application to map and monitor irrigated lands. Remote Sens 9(12):1256. https://doi.org/10.3390/rs9121256
5. Hadj Ammar F, Chkir N, Zouari K, Hamelin B, Deschamps P, Aigoun A (2014) Hydrogeochemical processes in the Complexe Terminal aquifer of southern Tunisia: an integrated investigation based on geochemical and multivariate statistical methods. J Afr Earth Sci 100:81–95
6. Zhang Y, Chen Z, Zhu B, Luo X, Guan Y, Guo S, Nie Y (2008) Land desertification monitoring and assessment in Yulin of Northwest China using remote sensing and geographic information systems (GIS). Environ Monit Assess 147:327–337
7. JICA. Project on Regional Development Planning of the Southern Region in the Republic of Tunisia. Final Report. Part 2. Regional Development Plan of the Southern Region. Ed: MDICI & ODS. 2015, 643
8. Wickware GM, Howarth PJ (1981) Change detection in the Peace-Athabasca Delta using digital Landsat data. Remote Sens Environ 11(I):9–25

9. Avery TE, Berlin GL (1992) Fundamentals of Remote Sensing and Airphoto Interpretation, 5th edn. Macmillan Publishing Company, New York, pp 127–142
10. Jaiswal RK, Saxena R, Mukherjee S (1999) Application of remote sensing technology for land use/land cover change analysis. J Indian Soc Remote 27(2):123–128
11. Chandrasekar N, Cherian A, Rajamanickam M, Rajamanickam GV (2000) Coastal landform mapping between Tuticorin and Vaippar using IRS-IC data. Indian J Geomorphol 5(1&2):115–122
12. Alam SMN, Demaine H, Phillips MJ (2002) Landuse diversity in south western coastal areas of Bangladesh. Land 63:173–184
13. Jayappa KS, Mitra D, Mishra AK (2006) Coastal geomorphological and land-use and land-cover study of Sagar Island, Bay of Bengal (India) using remotely sensed data. Int J Remote Sens 27:3671–3682
14. Santhiya G, Lakshumanan C, Muthukumar S (2010) Mapping of land use/land cover changes of Chennai coast and issues related to coastal environment using remote sensing and GIS. Int J Geomatics Geosci 1(3):563–576
15. Mujabar PS, Chandrasekar N (2012) Dynamics of coastal landform features along the southern Tamil Nadu of India by using remote sensing and Geographic Information System. Geocarto Int 27(4):347–370
16. Pettorelli N, Laurance WF, O'Brien TG, Wegmann M, Nagendra H, Turner W (2014) Satellite remote sensing for applied ecologists: opportunities and challenges. J Appl Ecol 51:839–848
17. Baby S (2015) Monitorig the coastal land use land cover changes (LULCC) of Kuwait from space borne Landsat sensors. Indian J Geo-Mar Sci 44(6):1–7
18. Kawakubo FS, Morato RG, Nader RS, Luchiari A (2011) Mapping changes in coastline geomorphic features using Landsat TM and ETM imagery: examples in southeastern Brazil. Int J Remote Sens 32(9):2547–2562
19. Hall EC, Strebel DE, Nickeson JE, Goetz SJ (1991) Radiometric rectification: toward a common radiometric response among multidate, multisensor images. Remote Sens Environ 35:11–27
20. Mancino G, Nolè A, Ripullone F, Ferrara A (2014) Landsat TM imagery and NDVI differencing to detect vegetation change: assessing natural forest expansion in Basilicata, southern Italy. iForest 7:5–84
21. Vorovencii I (2014) Assessment of some remote sensing techniques used to detect land use/land cover changes in South-East Transilvania. Rom Environ Monit Assess 186:2685–2699
22. Patton DR (1975) A diversity index for quantifying habitat "edge". Wildl Soc Bull 3:171–173
23. Forman RTT, Gordron M (1986) Landscape Ecology. Wiley, New York
24. Gardner RH, Milne BT, Turner MG, O'Neill RV (1987) Neutral models for the analysis of broad-scale landscape pattern. Landsc Ecol 1:19–28
25. Schumaker NH (1996) Using landscape indices to predict habitat connectivity Ecology 77(4):1210–1225
26. Tang J, Wang L, Zhang S (2005) Investigating landscape pattern and its dynamics in Daqing. China Int J Remote Sens 26:2259–2280
27. Yang X, Lo CP (2002) Using a time series of satellite imagery to detect land use and land cover changes in the Atlanta, Georgia metropolitan area. Int J Remote Sens 23:1775–1798
28. Kong F, Yin H, Nakagoshi N (2007) Using GIS and landscape metrics in hedonic price modeling of the amenity value of urban green space: a case study in Jinan city. China Landsc Urban Plann 79:240–252
29. McGarigal K, Barbara JM (1994) Fragstats—Spatial pattern analysis program for quantifying landscape structure. USDA Forest Service, Pacific Northwest Research station; General Technical Report PNW-GTR, Portland, OR
30. McGarigal K, Cushman S, Neel M, Ene E (2002) FRAGSTASTS: Spatial pattern analysis program for categorical maps. Computer software program produced by the authors at the University of Massachusetts, Amherst
31. ODS-Kebili (2016) http://www.ods.nat.tn/upload/CHIFKEBILI.pdf

32. ODS-Tozeur (2016) http://www.ods.nat.tn/upload/CHIFTOZEUR.pdf
33. World Bank (2014) Tunisia oases ecosystems and livelihoods project (TOELP) 114 p
34. Ministry of Local Affairs and the Environment (2017) The traditional oases of Tunisia. Document synthesis. World Bank Group & fem, 46 p

Chapter 15
Contribution of GIS in the Environmental Monitoring of a Tunisian Biosphere Reserve (Bou Hedma National Park)

Olfa Riahi

Abstract With 17 national parks and 25 nature reserves, Tunisia is a country characterized by great environmental diversity that adopts an environmental policy whose ultimate goal is to ensure the sustainability of natural resources. From this perspective, the study of the ancient and recent dynamics of the environment is a key element in predicting its evolution in the future and the risks they face. This type of study leads to proposals for environmental protection measures. In this context, the study and monitoring of the environmental dynamics of Bou Hema National Park is integrated. The latter is characterized by its economic, geoscientific, historical and cultural value. It is home to the last acacia raddiana forest north of the great Sahara thanks to which has required the classification with UNESCO as a biosphere reserve. The environment of Bou Hedma National Park represents a legacy resulting from a long and complex evolution as evidenced by several morphostructural, chronostratigraphic and anthropogenic indices. These were used for the study and analysis of environmental dynamics. Understanding old environmental evolution is key to understanding and assessing the trend of current dynamics and to estimating sensitivity to potential environmental degradation. Thus, our research has shown that from Pleistocene to the beginning of the historical period, the dynamics of the geomorphological landscape were guided by the interaction of climatic fluctuations, lithology and the morphometry of the watersheds. Currently, environmental dynamics continue to show a trend of erosion resulting in stony paving covering almost the entire pediment. In addition, the environment in Bou Hedma National Park is more sensitive to the potential degradation of soil quality than to the potential degradation to water erosion. In order to limit sensitivity to the degradation of Bou Hedma National Park, it is essential to changes in land-use methods and techniques. This modification could allow for the reduction of soil degradation by water erosion and thus the optimization and sustainability of this natural resource.

O. Riahi (✉)
Lr CGMED (LR 99ES02), University of Tunis, 94 Boulevard 9 Avril 1938, 1007 Tunis, Tunisia
e-mail: olfariahi@gmail.com

High Institute of Human Sciences, University of Jendouba, Avenue de UMA, Jendouba North BP. 104, 8189 Jendouba, Tunisia

F. Khebour Allouche et al. (eds.), *Environmental Remote Sensing and GIS in Tunisia*, Springer Water, https://doi.org/10.1007/978-3-030-63668-5_15

303

Keywords Environmental monitoring · Biosphere reserve · Sensitivity · Bou
Hedma · Tunisia

15.1 Introduction

Tunisia is a country characterized by great environmental diversity. It has 17 national
parks and 25 nature reserves [1] of which 8 sites are of interest to Unesco [2]. With
this in mind, Tunisia adopts an environmental policy whose ultimate objective is to
ensure the sustainability of natural resources. The study of the ancient and recent
dynamics of the environment is a key element in understanding current dynamics and
predicting their evolution in the future, and risks on the other potential degradation.
[3]. The subjected region to this research is part of the pilot observatory of Hadej-Bou
Hedma (Tunisia). The latter is part of the Long-Term Ecological Monitoring Network
Observatories (ROSELT network), implemented by the Observatory of the Sahara
and the Sahel [4]. The purposes of this network were to understand the interactive
relationship between the population and its environment and to develop decision-
support tools. This is based on environmental mapping, on monitoring of the different
ecological and geomorphologic factors, and on modelling of agro-pastoral population practices [3]. Our research, contributes to the improvement of basic knowledge
of the environment of Bou Hedma National Park and in particular on the ancient,
recent, current and potential evolution under the influence of climate change and the
theme. Moreover, Bou Hedma National Park belongs to a region that is a typical
example of a landlocked region surrounded by mountainous terrain almost everywhere. The cumulative effect of the region's enclave and limited resources can further
degrade an already difficult and constraining environmental situation. This offers
only unfavourable conditions for the preservation of the already vulnerable environmental balance. The rupture may have materialized by the degradation of vegetation
and soils, and consequently by the acceleration of the rate of water erosion. This
led to the desertification of certain areas of the region and Bouhedma National Park
and eventually threatened the future of the local population. Thus, it is a question
of seeking to propose measures to be taken to ensure a rational and sustainable
management of environmental potential in the park.

15.2 The National Park of Bou Hedma: A National
and International Environmental Value

It derives its importance from the existence of the last dense to sparse Acacia raddiana
forest (Acacia tortilis subsp. Raddiana) in Tunisia and [5]. Elsewhere in (north of the
Great African Sahara), these species exist as sparse individuals [5]. This relic forest
of Bou Hedma has been classified as a Tunisian biosphere reserve by UNESCO
since 1977 and proposed by the Tunisian State for a future ranking on the World

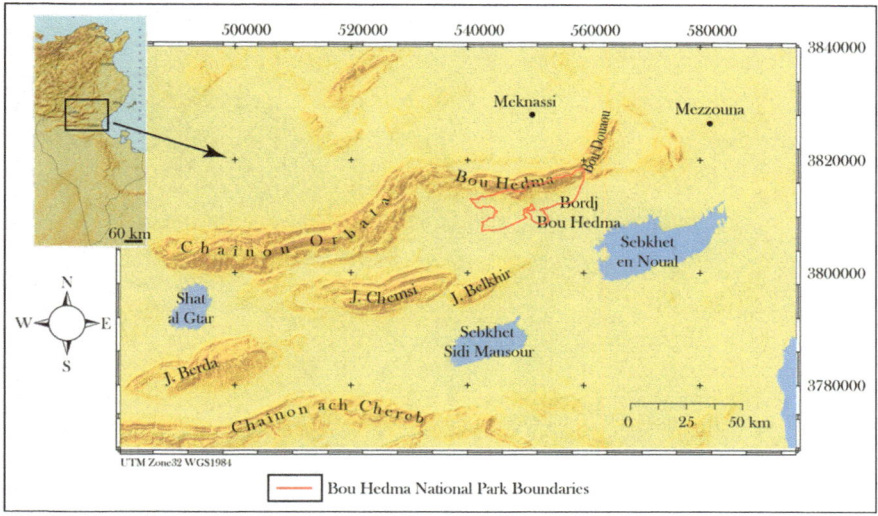

Fig. 15.1 Location of Bou Hedma National Park. *Source* Shuttle Radar Topography Mission (30 m) and [3]

Heritage List [6]. Besides, this relic forest is at the origin of the creation of Bou Hedma National Park (Fig. 15.1) in 1980 [3]. In reality, this date is that of the creation of Bou Hedma National Park under its current configuration [7]. However, in reality, this park was established over several stages. In 1936, The Bey Ahmed Bacha (see Vocabulary), had announced a part of Bled Talah as a beylicale hunting area [8] without assigning it a clear boundary. In 1957–1958, 700 ha were set aside and delimited by a plantation hedge. In 1970, the Directorate-General of Forests (under the Ministry of Agriculture) undertook efforts to reconstitute the natural biotope around Bordj Bou Hedma, leading in 1978 to the creation of a first enclosed area (Cutting by the Tunisian Forest Directorate) around Bordj [4] the plot called Bou Hedma [9], the ZII zone (the Hadej plot) and the ZIII (Fig. 15.2) zone named the Belkhir plot [9]. In this framed area, the buffer zones cover 2704.38 ha [9]. They were divided into three parcels. The first extends from the line of Jebel el Mech crests in the north to the bottom of the Piedmont of Jebel Bou Hedma in the south. The second extends from the western limit of the ZIII to about 1.6 km west of the wadi bed in Neguig. In the south, it is inserted between the western boundary of the ZIII, the eastern boundary of the ZII which is a temporary area of occupation. The third is located in the west of the IBA boundary. The areas of "temporary anthropogenic occupation" represent 2610.31 ha. These are two plots operated by the local population in cereal cultivation, arboriculture and market gardening.

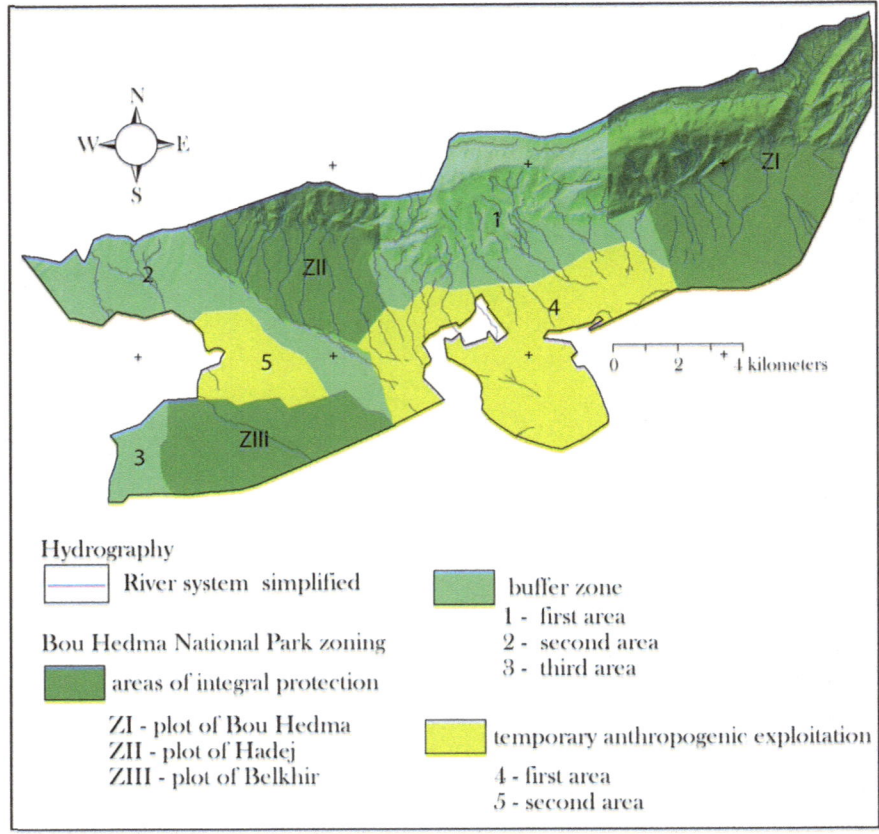

Fig. 15.2 Bou Hedma National Park zoning. *Source* Park zoning [3 and 9]

15.3 Biophysical and Land Use Characteristics

Bou Hedma National Park (Fig. 15.1) is located between 34°32'4.98"
N'; 9°41'14.83" E; 34°23'18.98" N and 9°23'22.47" E, a surface area equal to
16,448 ha [9]. It spreads, on the one hand, over a part of the southern slopes of
the chain of Hadej Bou Hedma, and, over the north-eastern part of Bled Talah on
the other hand. In addition the Bou Hedma link shows a corrugated line running in
an East-West direction, the Bou Hedma link culminating in 872 m. From its huge
aspect, the ridge line is strongly established in the landscape of the study area. It joins
the summits of the Jebels Hadej, from Noughis to Bou Hedma. The northern slope
is straight to slightly convex. Its slope values are comrised between 15 and 25%.
The southern slope is likely marked by a concave profile and by strong slopes which
are always above 50% and can reach the vertical pente at the level of the corniches.
On the other hand, the plateau of Bled al Talah ensures the transition between the

southern slopes of the Bou Hedma Massif (to the north) and the plain of Bled al Talah and the Sebkhet en Noual (to the south). In its central part, this plateau is aware of its minimum length. It does not exceed 2 km at this point. At the same time, it slightly extends equally to the West and to the East, where it reaches its maximum development. Its general deviation is from north to south with slope values that are between 3 and 6%. On the other hand, the north-eastern part of the plain of the village al Talah lies between the plateau of Bou Hedma in the north and the kodiat Oum al Far in the south. The boundary between the Piedmont and the plain is imprecise in both eastern and western parts. However, it is precise in the central part. Its overall elevation is equal to 55 m with slope values that are often less than 3%. The flatness of this plain is disturbed by the hill of Kodiat Oum el Far. Of a subcircular shape and whose altitude does not exceed 119 m on a hydrographic level, Bou Hedma National Park is part of five watersheds. They are from East to West of those of Bou Hedma, Oum el Far, Bou Zalâ, Hallouf-en Neguig, el Mech-Hadej, and aitha. These are all tributaries of order 2^1 which belong to the Sebkhet in Nual watershed. The flow is seasonal except for a few sections of streams belonging to the watersheds of the Wadi of el Mech-Hadej and Bou Hedma that have a perennial flow related to the activity of the sources. Thus, the study of hydromorphological (Fig. 15.3) characteristics[2] (watershed morphometry and watershed characteristics) showed that:

- the watersheds of Oum el Far and Bou Zalaâ have a hydrographic behaviour marked by the fastest concentration of stormwater and the highest peak flood flow values. These characteristics give the water system draining these watersheds the most important transportation regulation in the Bled Talah region.
- the watersheds of the wadis Bou Hedma and el Mech-Hadej are narrow, drained by a winding, dense to moderately river system, This results in a hydrographic behaviour marked by the longest concentration times and therefore the least important peak flood flow. In contrast, the volume of runoff is important which attributes to the wadis draining the watersheds of Bou Hedma and el Mech-Hadej the most important transport regulation capacity.
- the watershed of the wadi el Hallouf-en Neguig is elongated with a very dense hydrographic water system. This results in hydrological behaviour, marked by runoff concentration rates and average flow values.

Soils in Bou Hedma National Park are crude mineral soils (on the slopes and top of Piedmont), poorly evolved (below Piedmont), calcomagnesimorphs, and isohumic soils in the plain [12]. They show a low organic matter content against a high limestone and/or gypsum content. This is associated with coarse textures on the feet of the

---

[1] According to the classification of streams de Hack [10]. THi smethodology is a classification ascending whose hierarchy begins at the outfall. It consists of starting the classification from the outfall of the watershed. The main course is of order 1 over the entire watershed. All the tributaries leading into it are of order 2 and their tributaries are of order 3 and so on until the last upstream tributary.

[2] L'Hydromorphology is the study of the processes that guide the dynamics of streams or river dynamics, the functioning of the hydrological compartment and the resulting forms (Malavoi and Bravard [11]).

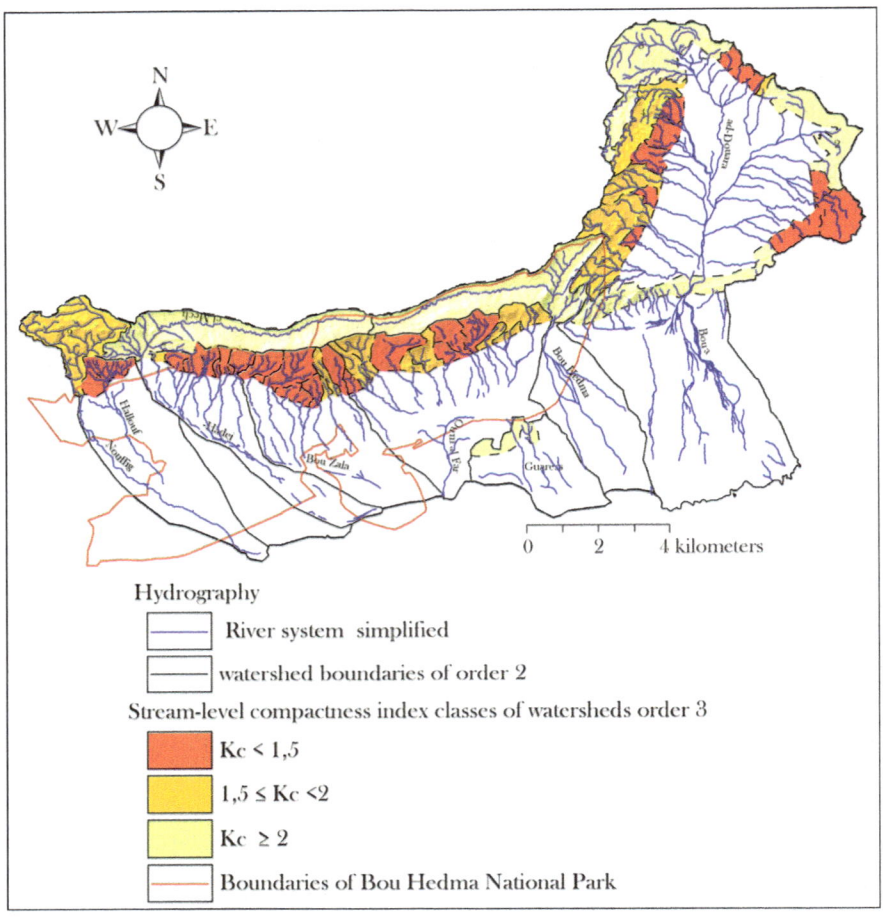

Fig. 15.3 Gravelius index of five watersheds of Bled al Talah

clay-silty links and textures in the plain [13–15]. Soils that can be grown in annual crops are very small in space. These are the poorly developed alluvial-bearing soils of the Oum el Far wadis and part of the Talah Plain [3 and 15]. Those that have good potential for arboriculture are the poorly evolved and isohumic soils on the cones of the wadis in Neguig and al-Hallouf, the wadi Bou Hedma and the wadi el Boua [15]. The medium-quality soils for annual crops mainly concern the poorly developed soils of the plain area between the Kodiat Oum el Far and the south and the pediments of the Bou Hedma chain. The non-cultivable soils that can be used as rangelands represent almost half of Bou Hedma National Park. These are crude mineral soils, low-grade soils, and calcomagnesimorphic soils [14 and 13].

15.4   Time Scales and Methodology

15.4.1   Time Scales

Throughout its history, Bou Hedma National Park, like all Tunisia basically, has witnessed several changes, both climatic and socio-political. In terms of climate changes, fluctuations have relayed arid and less arid periods [16–18] hence, alternating between phases of crises with and of stability. In addition, at the political and social level, the study area has experienced the succession of several civilizations. This succession is accompanied by different modes of use. This transition from one civilization to another, particularly during the period of Antiquity and the Middle Ages, is generally carried out by crises of occupation always accompanied by changes in tenure practices and land distribution. In an attempt to identify time scales that meet the objectives of our research, we have tried to conduct a quick analysis of History (both natural and entropic) of the Bled Talah region in which the current Bou Hedma National Park is located. This is the aim of looking for signs of physical change and patterns of land tenure practices and land distribution that have influenced environmental dynamics. The aim is to identify throughthe History of the occupation of the study area significant events and crucial periods for the understanding of the evolution environmental units of Bou Hedma National Park (climate change and/or a change in use pattern).

Finally, we tried to track, estimate, and specialise the history of the evolution of the study area and to integrate it in the form of data layers in a GIS geographic information system.

Moreover, since the Lower Pleistocene, the region has not experienced any major tectonic movement [17]. Thus, the final configuration of the large Morpho-of the Bled Talah region was established by the last major tectonic movements of the Lower Pleistocene, and since that time the large subdivisions of the watersheds of the region of Bled Talah remained overall unchanged [3]. Moreover, the anthropogenic occupation (Fig. 15.5) of the region of Bled Talah seems to be very old as evidenced by several archaeological evidences.[3] It was, in ancient and medieval times; a strategic passage linking Tripolitan with Byzacene and Zeugitaine [19]. It maintained its strategic importance until the Islamic period [21]. It has experienced threefold key socio-political changes that appear to be most influential on environmental dynamics. The first is that of the beginning of the Roman era with which the region experienced a great anthropogenicization manifested through a great effort of development (especially hydro-agricultural). The Romans have multiplied the means of production to make fresh pastures and to exploit new lands thanks to the progress

---

[3]Capsian and Neolithic sites described by Ballais [16] mausoleums of alignments and foundations, Roman baths, basilicas, remnants of localities, settlements and Africo-Roman farms decorated with mosaics, lime kilns, cisterns, Gsar Grewech's ….

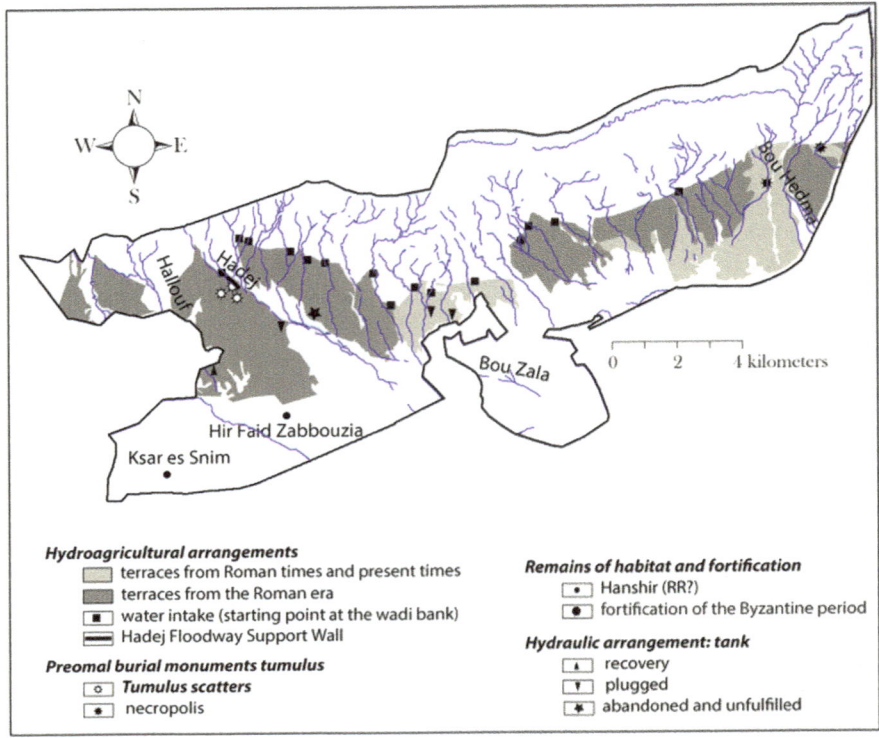

Fig. 15.4 Archeological Antiquity's sites of Bou Hedma National Park

of agricultural hydraulics [18]. They have influenced the environment through deforestation, clearing, hydraulic and hydro-agricultural development,[4] and the realization
of roads… (Fig. 15.4).

As a result, we believe that this period marks the beginning of man's influence on
environmental dynamics in the Bled Talah region. The second crucial change took
place from the late Antiquity and the High Middle Ages. This period begins with
the gradual transition from settlement and security to nomadism and insecurity. This
shift was accompanied by a breakdown in the maintenance of hydro-agricultural
facilities. The socio-political shift certainly influenced the environmental dynamics
of the Bled Talah region. The third crucial socio-political change in the region of Bled
Talah has taken place in modern times up to the present day. In contrast to the Middle

---

[4]These are water and soil development systems. Each system consists of two types of structures:
a main diversion channel and secondary channels and a set of growing terraces. The diversion
channels generally occupy the upstream portions of the Piedmont and allow the flow of flood water
to the parcels located a little further downstream of these structures. The Crop Terrace Systems are
dry stone stoves covered with earth fill. They are located downstream of the diversion channels and
received runoff and sediment from them [7].

Ages, the region of Bled Talah has experienced a trend towards settlement encouraged and/or imposed by the French authorities (during the occupation) and Tunisian authorities (after independence). The settlement of the population is accompanied by the implementation of land's cultivation and the change of the economic regime through the transition from a nomadic socioeconomic system based on pastoralism to a sedentary regime that changed considerably the environmental dynamics of Bou Hedma Park. The environmental dynamics of Bou Hedma National Park are greatly changed by a nomadic economic nomad based on pastoralism to a sedentary regime.

Thus, the study of the environmental evolution of the region of Bled Talah is structured around four periods/stages constituting the time scale. The first concerns the period from the Lower Pleistocene to the first century AD (Roman colonization). The second covers the Roman period. The third begins in the High Middle Ages and extends till the beginning of French colonization. And the fourth concerns the present era. In addition, the identification of the first period is based on the analysis of geological studies such as that of Boukadi [20] and geomorphologic works such as that of Ballais [16] and Ben Ouezdou [17] which in turn indicate that the last major tectonic movements in the region occurred in the Lower Pleistocene [17]. In sum, the time-scale periods are broken down as follows:

- Old

  - From lower Pleistocene till the beginning of the historical period
  - Roman times (from the first century AD with the foundation of the Colonia Iulia Concordia Carthago until its fall)

- Late-High Antiquity Middle Ages to Colonial Period
- current
- potential

15.4.2 Methodology

During our work, we followed several steps and methods depending on the state of the research (Fig. 15.5). Data collection consists of collecting all the information needed to analyze environmental dynamics in Bou Hedma National Park. The data collected is subsequently spatialized and integrated into a GIS database. Thus, the main objective of Multi-source Data Processing and Exploitation is to seamlessly shape the data (descriptive, map, field surveys, statistics, history, etc.) collected and spatialize it in order to to create a GIS database. This is in order to allow processing control, verification of results, export and import between multiple interfaces as well as updating data whenever necessary. Subsequently, the study of the sensitivity of the geomorphological landscape of the Hadej-Bou Hedma region to potential degradation is to identify geomorphological landscape units according to the degree of vulnerability. This approach is based on the study of biophysical and anthropogenic factors in action in the study area. The sensitivity of each unit depends on the interaction of the above factors and the result. In addition, the study of sensitivity to

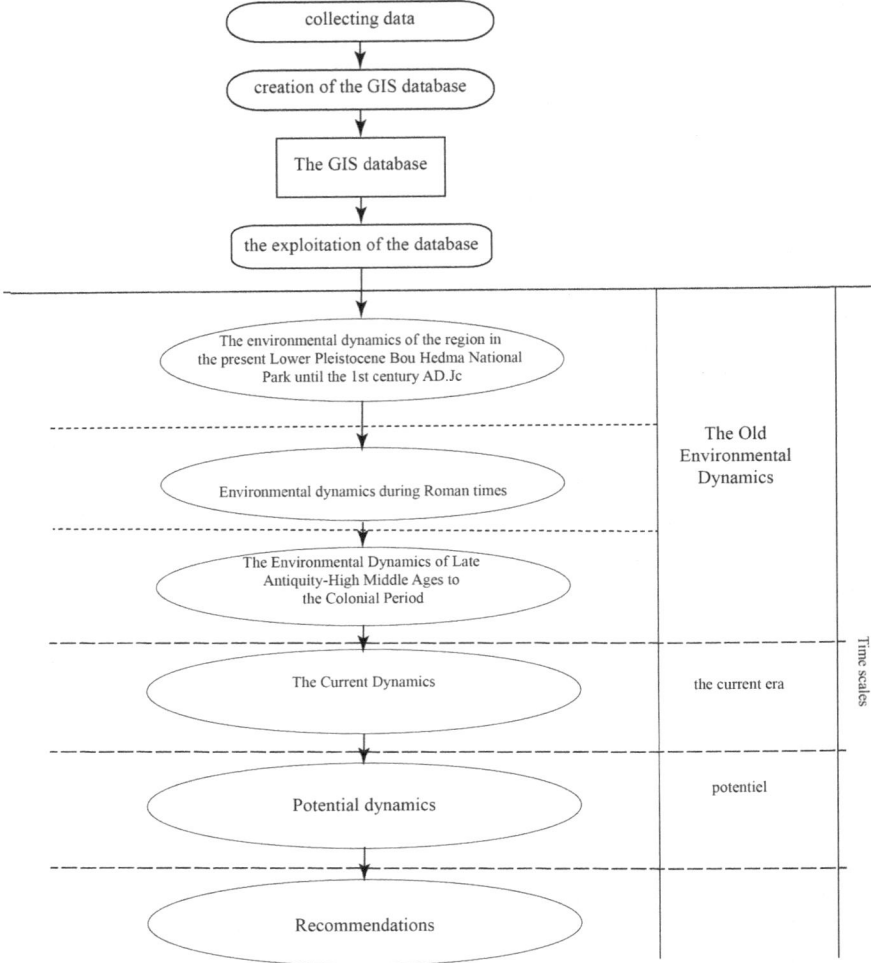

Fig. 15.5 The steps of realisation of the research

potential degradation involves two stages: automatic concatenation and classification. The first operation is to combine the different layers of data relating to each factor in action in Bou Hedma National Park. Classification consists of a typology based on a manual operation assisted by an automatic grouping of values (which results from the concatenation of the data layers).

15.5   The Old Environmental Dynamics of Bou Hedma National Park

15.5.1   The Environmental Dynamics of the Region in the Present Lower Pleistocene Bou Hedma National Park Until the 1st Century AD.Jc

By spatializing and using historical data as well as issues related to field works, two (Fig. 15.6) types of dynamics are identified during this period.

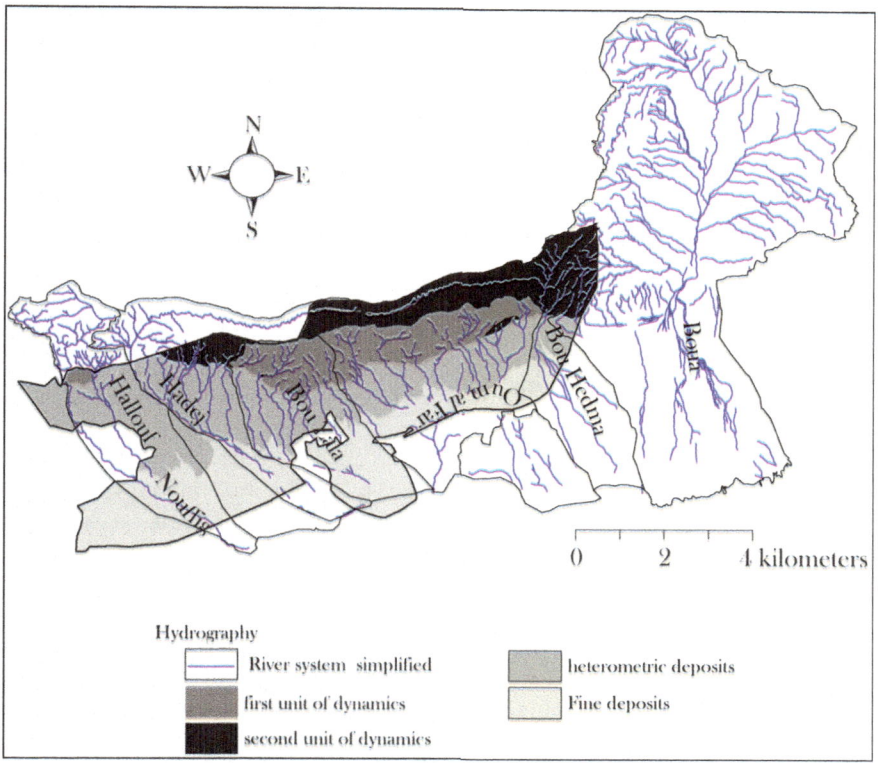

Fig. 15.6 Pleistocene environmental dynamics of Bou Hedma National Park

15.5.1.1   The First Environmental Dynamics Unit in Bou Hedma National Park

These are the park areas belonging to the watersheds of Bou Zalâ, Oum el Far and el Hallouf-en Neguig. Their impluviums are characterized by the highest density resulting in the fastest rate of runoff concentration, and therefore, the highest peak flood flows and the most important skill of wadis to carry important size. However, the reduced sizes of impluviums influence the volume collected. Downstream, these characteristics are reflected in the accumulation of series of heterometric coalescent waste cones; however, with reduced sizes and extensions.

15.5.1.2   The Second Dynamic Environmental Unit in Bou Hedma National Park

To this category of dynamics belongs the watersheds of the wadi Bou Hedma and that of the wadi el Mech-Hadej. The headwaters of these watersheds are narrow. This means that the rate of concentration of runoff is lower and the peak flood flow is modest compared to the watersheds belonging to the first unit of environmental dynamics. However, the volume of runoff is the largest. Downstream, this dynamic resulted in the implementation of cone terraces with encroached heterometric roots.

15.5.2   Environmental Dynamics During Roman Times

During this period, the area under consideration appears to be witnessing a major change in tenure choice. Human impact becomes more and more important, as it can be seen from the ancient structures and sites attributed to this period. In order to estimate the magnitude of human influence since the historical period on environmental dynamics in the present Bou Hedma National Park, we have mapped and integrated the spatial distribution into a GIS, on the Piedmont, remnants of ancient earthworks systems and floodwater diversion canals. It is obvious that the implementation of these types of structures contributes to the stabilization of the environment, especially in the case of systems of cultivation terraces. These represent an obstacle to runoff before having an erosive power and trap the sediments, thus giving rise to a deposition dynamic. As a result, man's intervention in the study area during the Roman period gave rise to the appearance of an area dominated by a dynamic deposition of fine materials on the strips of cones and those of cones-terraces (Fig. 15.7).

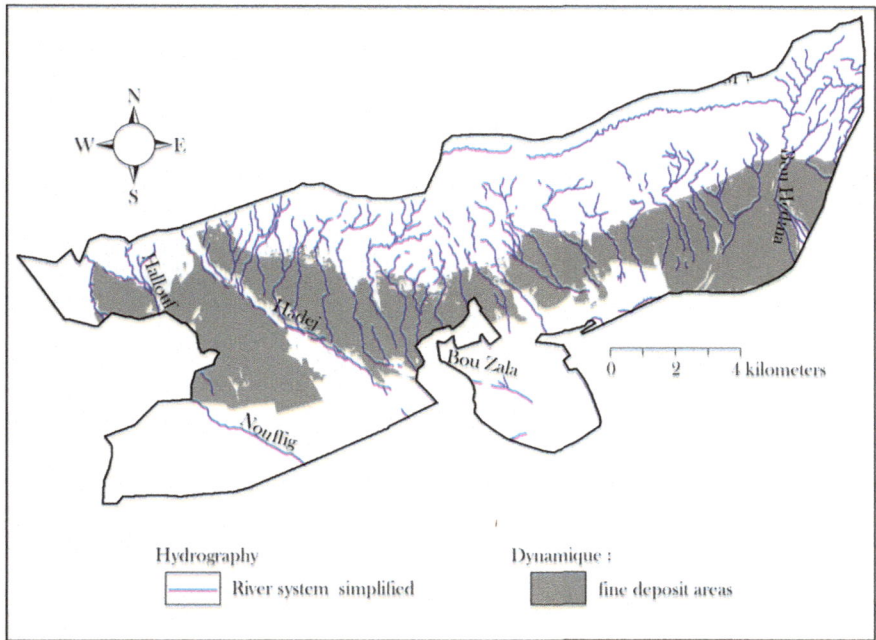

Fig. 15.7 Dynamics during Roman times of Bou Hedma National Park

15.5.3 The Environmental Dynamics of Late Antiquity-High Middle Ages to the Colonial Period

At the beginning of this period, the studied area experienced a gradual abandonment of the ancient hydro-agricultural developments due to socio-economic shift under the effect of insecurity. The way of occupation has shifted from a sedentary regime to one of nomadism. The latter lasted a long period until the beginning of the colonial period [21]. On the other hand, the progressive maintenance failure of these structures led to the destruction of the system of cultivation terraces, and the triggering of water erosion notably by the opening of ravines followed the break of the benches. Consequently, this transforms the areas previously protected by water and sediment retention planning into an erosion zone. In this way, the extracted particles are subsequently transported downstream; hence, the implementation of the historic terrace [17].

15.5.3.1 The Most Eroded Units

The most eroded parts of the present Bou Hedma National Park belongs mainly to the catchment areas of the al-Hallouf-en Neguig, Bou Zalâ, Oum el Far and part

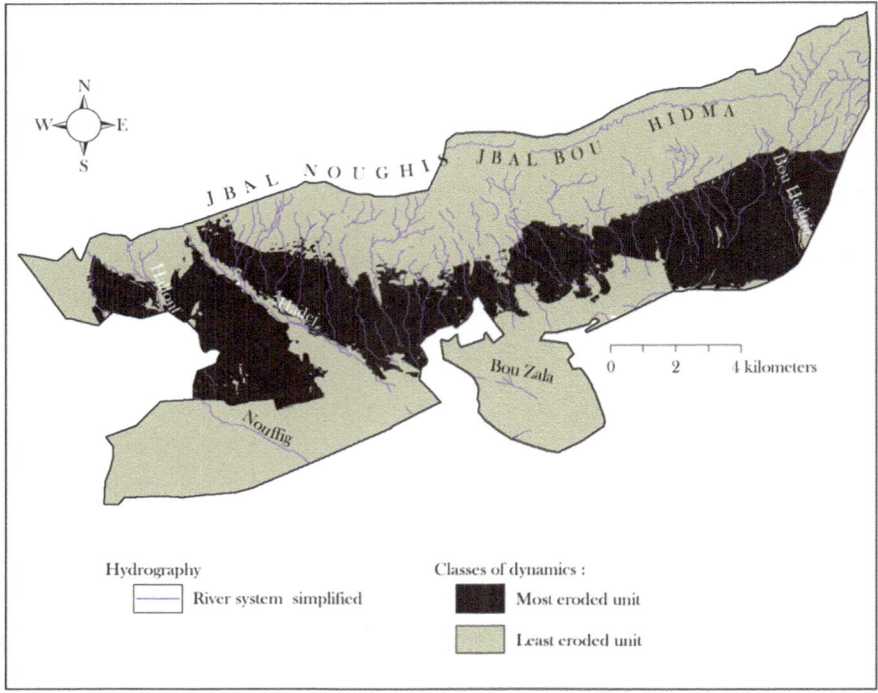

Fig. 15.8 The Environmental Dynamics of Late Antiquity-High Middle Ages to the Colonial Period of Bou Hedma National Park

of the non-hierarchical drainage area. The erosion in these parts is explained by the intervention of the anthropogenic factor. In fact, they correspond to areas with abandoned ancient hydro-agricultural developments that are covered by loose and thin deposits in the terraces of crops (Fig. 15.8).

15.5.3.2   The Least Eroded Units

These parts should be the site of a depository dynamic under the influence of two factors. First of all, it lacks arrangement for neglected antique amenities. Secondly, its position downstream of the previously developed parts favours the deposition of particles removed from them.

15.6   The Current Dynamics

At present, the dynamics in Bou Hedma National Park are mainly due to area, regressive, and lateral erosion (Fig. 15.9). On the top of the pediments, the area erosion exploits the entire surface generating the formation of a reg of dissociation and locally rocky pavement coarse.

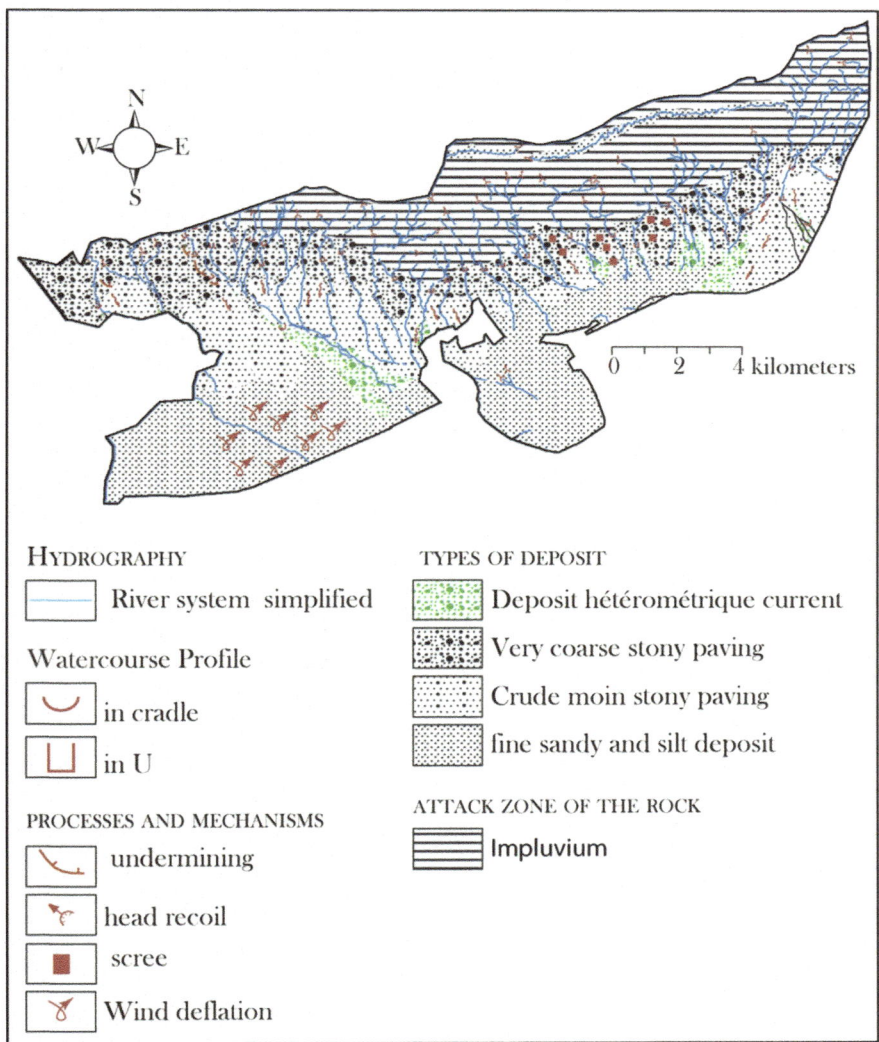

Fig. 15.9   The Current Dynamics of Bou Hedma National Park

Further downstream, a combination of process dynamics of opening, deepening and recoiling gully heads arecommon. This dynamic is related, firstly, to the fine and loose nature of the filling and, secondly, to the abandonment of recent hydro-agricultural developments. In the line of the trough-profile wadis, a tendency of bank retreat due to the combined dynamics of lateral and linear erosion. The banks of cradle-shaped wadis tend to evolve by landslide (Fig. 15.10). At the level of the plain, two different dynamics are encountered. The first is a tendency to deposition. The second is an attack dynamic. In addition, on the northern margins, in contact with Piedmont, the plain of Talah is the site of a coarse deposit of active cones. The magnitude of the deposit, its heterometry, and its location depend on the characteristics of the floods that are responsible for their installation (Fig. 15.9). Further downstream, a dynamic of gully erosion ravines setbacks is remarkable in fine salty and sandy deposits. Opening, deepening, and recoiling heads are the most active processes.

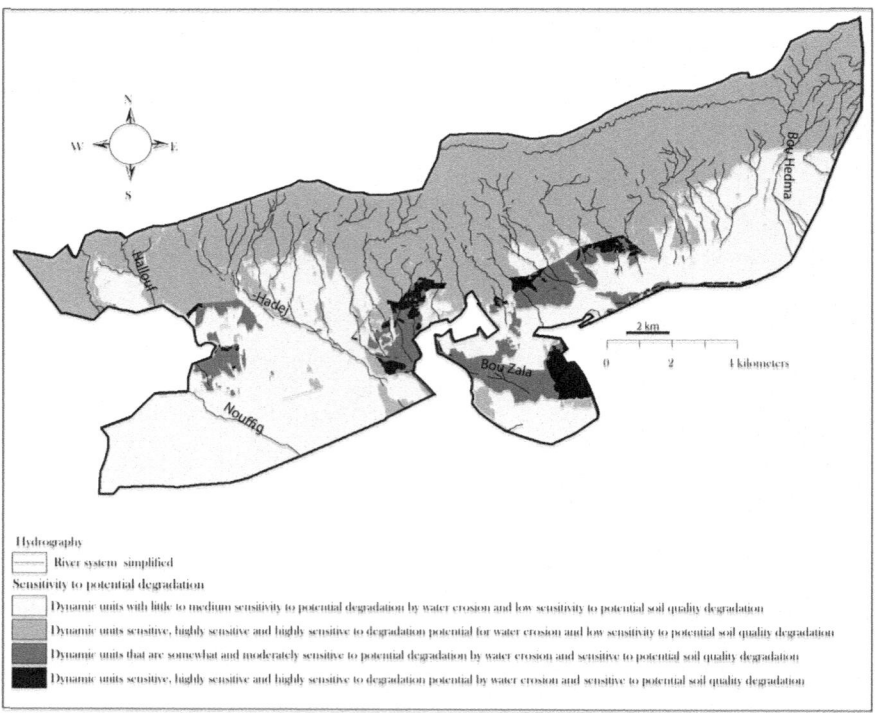

Fig. 15.10 Potential environmental dynamics of Bou Hedma National Park

15.7   Potential Dynamics

The potential environmental dynamics of Bou Hedma National Park would be guided on the one hand by the interaction of physical factors and on the other by the anthropogenic factor. On the basis of the classification of soils in relation to their potential, we have distinguished between units with low dynamics sensitive to potential soil quality degradation and others sensitive to degradation potential of soil quality. The first type of unit corresponds to the areas of integral defenses and buffers of Bou Hedma National Park on soils of good quality for shrub crops and non cultivable soils that can be used as cultivation grounds. The second type of unit corresponds to areas of temporary anthropogenic exploitation whose usesare inadequate to the resources and potentials offered by the soil. These are non-cultivable soils that can be used as a range. However, they are used for crops (field crops, arboriculture, and market garden crops).

- little sensitive to water erosion, which is the result of the interaction of several factors protecting the potential of the environment (slope between 0 and 3%, perennial vegetation) to which is added the existence of anti-erosion schemes
- Moderately sensitive to water erosion. These are cultivation areas on outcrops which are generally not resistant to water erosion covered by seasonal vegetation. In addition, there is a hydrographic behaviour marked by the most important transport competence in the study area.
- Sensitive to water erosion. They benefit from perennial vegetation dominated by tree species and are not affected by tillage techniques. The slope values are moderately high (between 3 and 6%). Resistance of outcrops is variable, and anti-erosive arrangements are still lacking.
- Very sensitive to water erosion, due to their sensitivity to the cumulative effect of fairly large slope values (between 6 and 15%). and moderately to weakly resistant outcrops. From there, the runoff can carry out a work of water erosion whether linear, regressive or lateral despite the presence everywhere of anti-erosive schemes locally accompanied by reforestation, and the durability of the dominant vegetation and the defensive.
- Highly apparent to water erosion. They have slope values ranging from 6 to 15% and seasonal vegetation cover associated with an almost total absence of anti-erosion schemes. The interaction of these factors increases the risk of potential degradation by water erosion regardless of the hydromorphometry regime and the resistance of outcrops (Fig. 15.10).

Thus the crossing of the two types of dynamics shows the existence of 10 units of dynamics classified in 4 categories according to their sensitivity to potential degradation.

15.7.1 Environmental Units with Accelerated Dynamics, Very Accelerated and Highly Accelerated to Potential Degradation Guided by Water Erosion and Sensitive to Potential Degradation of Soil Quality

These environmental units have the highest degrees of sensitivity to both potential degradation by water erosion and degradation of soil quality. These are the units:

- sensitive to potential soil quality degradation and sensitive to potential degradation by water erosion;
- sensitive to potential soil quality degradation and very sensitive to potential erosion degradation;
- Sensitive to potential soil quality degradation and highly sensitive to potential erosion degradation.

These environmental units are characterized mainly by the absence of all types of hydro-agricultural structure and inadequate anthropogenic usage of the potentialities of the soils. This mainly promotes soil fertility degradation and area erosion on interfluves. In order to limit the sensitivity to potential degradation of these environmental units in Bou Hedma National Park, several hydro-agricultural structures appear to be taken to the litho logical and orographic conditions.

15.7.2 Environmental Units that Are Sensitive, Very Sensitive and Highly Sensitive to Potential Degradation by Water Erosion and not Sensitive to Potential Degradation of Soil Quality

This category includes the environmental units of Bou Hedma National Park:

- low sensitivity to potential soil quality degradation and sensitive to potential erosion degradation;
- Low sensitivity to potential soil quality degradation and very sensitivity to potential erosion degradation;
- low sensitivity to potential soil quality degradation and highly sensitive to potential erosion degradation.
- In these environmental units, sensitivity to degradation by water erosion is more important than sensitivity to soil quality degradation.

15.7.3 Low and Medium Environmental Units Sensitive to Potential Degradation by Water Erosion and Sensitive to Potential Degradation of Soil Quality

The units are:

- sensitive to potential soil quality degradation and not sensitive to potential degradation by water erosion;
- sensitive to potential soil quality degradation and moderately sensitive to potential erosion degradation.

On the other hand, units classified in this category are mostly equipped with anti-erosive fittings. It is essentially the system of tabias (see Vocabulary) and seats of total retention. However, this type of hydro-agricultural development was committed to failure. This is due, on the one hand, to the absence of outlets and, on the other hand, to the lack of arrangement of the facilities by the peasants as well as by the state.

15.7.4 Low and Medium Environmental Units Sensitive to Potential Degradation by Water Erosion and not Sensitive to Potential Degradation of Soil Quality

Environmental units classified in this category have the lowest levels of sensitivity to both potential degradation by water erosion and degradation of soil quality which are the following (Fig. 15.11):

- low sensitivity to potential soil quality degradation and low sensitivity to potential degradation by water erosion;
- Low sensitivity to potential soil quality degradation and moderate sensitivity to potential erosion degradation.

15.8 Proposed Solutions to Limit the Potential Sensitivity of the Environment to Degradation

The analysis of sensitivity to potential degradation is a decision-making tool that allows us to propose some basic elements for the development of a possible management plan for Bou Hedma National Park. Accordingly, it can guide the right decisions on priority environmental dynamics based on the type of potential degradation they present. In addition, it allowed us to draw up a map of development choices.

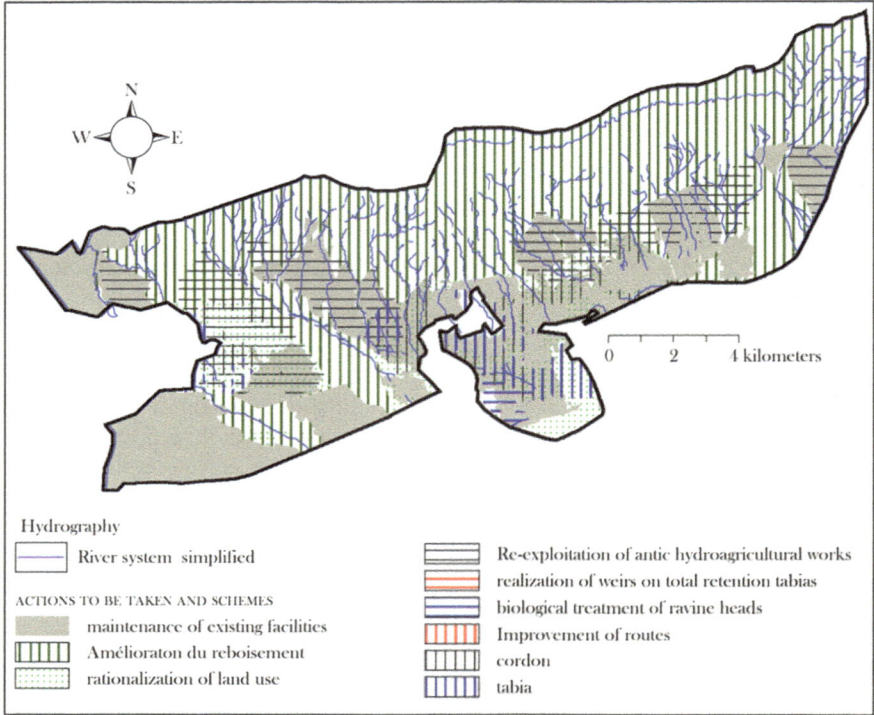

Fig. 15.11 Actions to be taken for the fight against sensitivity to degradation of Bou Hedma National Park

From this perspective, several measures can be considered. This is a set of social-economic and technical aspects which aim to optimise the human use of the potential of the environment while preserving and insuring the sustainability of the various environmental values of Bou Hedma National Park.

In order to limit sensitivity to the potential degradation of Bou Hedma National Park, it is essential to establish a policy to guide the cultivation, techniques, and methodologies of anthropogenic exploitation of arable land. This policy would lead to (Fig. 15.11):

- The rationalization of land use, particularly at the level of the temporary area of occupation.
- Changes in land-use methods and techniques. This modification could allow for the reduction of soil degradation by water erosion and thus the optimization and sustainability of this natural resource.

15.8.1 Environmental Units with Accelerated Dynamics, Very Accelerated and Highly Accelerated to Potential Degradation Guided by Water Erosion and Sensitive to Potential Degradation of Soil Quality

Thus, it is advisable to carry out jessours (see Vocabulary) through oued in order to retain runoff, dry stone cords, and tabias on interfluves in an attempt to retain diffuse runoff and sediments (Fig. 15.10).

15.8.2 Environmental Units that Are Sensitive, Very Sensitive and Highly Sensitive to Potential Degradation by Water Erosion and not Sensitive to Potential Degradation of Soil Quality

In these units, it makes sense, therefore, to target the slowdown of water erosion processes. In this perspective, it helps to:

- improve reforestation;
- Install dry stone cords on interfluves, etc.
- Re-exploiting ancient hydro-agricultural structures to the extent of promoting the regeneration of natural herbaceous vegetation which in turn requires two types of interventions:
- The first is to re-exploit the water intakes (see Vocabulary) that are currently perched in relation to the current levels of the wadis beds requirement. For example, dredging at the level of the wadi banks and levelling on the interfluve. The re-exploitation of the ancient water intakes will allow the redirection of the runoff on the interfluves, which will improve the routes.
- The second is the restoration of earthworks systems.

15.8.3 Low and Medium Environmental Units Sensitive to Potential Degradation by Water Erosion and Sensitive to Potential Degradation of Soil Quality

Thus, in these units:

- preserve and rehabilitate existing facilities;
- transform the total-retention benches into partial-retention benches by creating outlets on the body of these works;
- stabilize and revegetate the bodies of the seats;
- dredging of the reservoir of el Boua's new hilly lake

15.8.4 Low and Medium Environmental Units Sensitive to Potential Degradation by Water Erosion and not Sensitive to Potential Degradation of Soil Quality

In this category, it is necessary to carry out:

- biological treatment of gully heads in the plain (temporary use area) to combat head recoil;
- partial retention seats
- and to preserve and rehabilitate existing facilities;

15.9 Discussion

The special importance of Bou Hedma National Park in several respects (ecological, geological, environmental, etc.) has made it the subject of several multidisciplinary studies. However, the majority of these relates to the ecological and bio-vegetal aspect, particularly in the areas of defensation, studies concerning the preservation and development of soil and water resources are very rare with the exception of some general works and articles such as that of Zahar [22], Alaya et al. [23], Akrimi et al. [24] Floret et al. [25 and 26] and Abdelkebir [27]. They do not address environmental dynamics based on a multi-criterion approach that takes into account anthropogenic practices in a poor and fragile natural environment such as that of the Hadej-Bou Hedma region. This justifies the interest of carrying out our research based on a closer-scale monitoring of the ancient, recent and current environmental dynamics of Bou Hedma National Park as well as the analysis of the sensitivity to the potential degradation of this last. This approach is appropriate to be used as a protocol for the study of environmental dynamics in other national parks in arid climate.

15.10 Conclusion

Thus, the interaction of bio-physical and anthropogenic factors in Bou Hedma National Park results in two types of sensitivity to potential degradation. The first is the potential degradation of soil quality. The second is degradation by water erosion. Furthermore, Sensitivity to potential soil quality degradation in Bou Hedma National Park is the result of inadequate anthropogenic land usage. This is the temporary anthropogenic part of Bou Hedma National Park which is at risk of potential degradation of soil quality. Indeed, the sensitivity to degradation by water erosion has shown that more than the/4 of the area of Bou Hedma National Park are sensitive, very sensitive, and extremely or highly sensitive.

15.11 Recommendations

- The use of RADAR images especially for the spatial characterization of floods.
- Furthermore, using other multispectral and hyperspectral satellite image, accompanied by other satellite-based indices might result in more reliable and precise maps.

Acknowledgements This research was carried out as part of a research agreement between several institutions for which I thank on behalf of their directors. This is, on the one hand, the Faculty of Human and Social Sciences in Tunis (FSHST); the Laboratory For Geomorphological Mapping Of Environments, Environments and Dynamics (CGMED) and, on the other hand, the Institute for Research for Development (IRD ex-ORSTOM), Tunisia's Network of Ecological Surveillance Observors in Long Term (ROSELT network; Sahara and Sahel-OSS Observatory) and the Institute of Arid Regions (IRA Médenine).

Vocabulary

Bey title carried by the rulers of Tunisia in the Husseinite dynasty (1705–1957) vassals of the Ottoman Empire in Tunisia

Jessour (singular: jesser) construction (made of dry stones) to retain runoff and alluvials built across a wadi

National park a relatively large territory that has one or more ecosystems generally little or not transformed by human exploitation and occupation where plant and animal species, geomorphological sites and habitats are of special interest from a scientific, educational and recreational point of view, or where natural landscapes of great aesthetic value exist (Chapitre III of Tunisian Forest Code, law n° 88-20 of April 13, 1988 In. official website of the National Observatory of Agriculture Tunisia 1)

Nature reserve is a small site for the purpose of maintaining the existence of individual species or groups of natural, animal or plant species, as well as their habitat and conservation of migratory wildlife species. of national or global importance (Chapitre III of Tunisian Forest Code, law n° 88-20 of April 13, 1988 In. April 13, 1988 In. official website of the National Observatory of Agriculture Tunisia 1)

Tabia and banquette constructions (in Ground embankment) retention of runoff and alluvials built on interfluves following the levels curves

Water intake is a runoff diversion channel linked to water and sediment retention structures (Tabia and banquette)

References

1. http://www.onagri.nat.tn/atlas
2. http://www.roselt-oss.teledetection.fr/
3. Riahi O (2017) The dynamics of geomorphological landscapes and recent evolution of an arid natural environment: the region of Hadej-Bou Hedma (south Tunisia). Ph.D. thesis from the Faculty of Human and Social Sciences in Tunis (FSHST), University I. Tunisia
4. ROSELT/OSS C (2004) Biodiversity study at the Haddej pilot observatory in Bou Hedma, Tunisia. ROSELT Collection / OSS,CT n°7, Montpellier
5. Le Floc'h E, Grouzis M (2003) Acacia raddiana, a tree of arid zones with multiple uses. In A tree in the desert, Acacia Raddiana. IRD-Editions
6. http://whc.unesco.org/en/tentativelists/5384/
7. Decree no. 80-1606 of 18 Dec 1980 on the land title 36 S2 Sfax
8. Ferchichichi (2011) Evaluation of the legal and institutional framework relating to ecotourism and protected areas in Tunisia. CUICN, Tunisia
9. Karem A (2001) The role of national parks and nature reserves in conserving biodiversity. In Review of Arid Regions Tunisia
10. Hack JJ (1996) Studies of longitudinal stream profiles in Virginia and Maryland U.S. Geological Survey Professional Paper. USA
11. Malavoi J-R, Bravard J-P (2010) River hydromorphology elements. Paris: Collection "understand to act" of the National Office of Water and Aquatic Environments, France
12. Floret C, Pontannier R (1982) Aridity in pre-Saharan Tunisia. Works and documents of the ORSTOM. France, Paris
13. Fournet A (1960) Perimeter of Henchir Boua. Direction of Soils; Ministry of Agriculture, Tunis. Tunisia, Sebkhet Noual
14. Mori A (1969) Pedological study of the perimeter of Bled Sidi Mhadheb (complements) Direction des Soils. Ministry of Agriculture. Tunisia
15. Pontanier R (1969) Pedological study of Bou Hedma perimeter. Directorate of Soils, Ministry of Agriculture
16. Ballais J-L (1972) The Sebkhet Noual Depression: Geomorphological Study. Thesis of 3rd cycle of Univ. Paris I France
17. Ben Ouezdou H (1994) The southern part of the Tunisian Steppes; geomorphological study. Ph.D. thesis from Tunis University I
18. Slim H, Mahjoubi A, Belkhodja K, Ennabli A (2010) General History of Tunisia: Antiquity, vol I. South Editions. Tunisia, Tunis
19. Hajlaoui A, Fareh H, Riahi O (2016) Note on the ancient hydraulic installations of Oued Hadej (Bou Hedma/Tunisia centoméridionale). In. Proceedings of the Fifth International Symposium: Population, Territory and Matelural Cultures in the Mediterranean Sea. Kairouan 2004. Tunisia
20. Boukadi N (1985) Geometric and kinematic evolution of the north-south axis and Gafsa range interference zone (Maknassy Mazzouna and Jbal Bou Hidma). University of Louis Pasteur France
21. M'cherek A (2008) Oued Ouadran Valley (Ancient Vadara) a highly strategic tribal area from Corripe to Ibn Khaldoun. Proceedings of the 6th International Symposium on the History of Tunisian Steppes, in Sbeïtla. National Heritage Institute Tunis Tunisia
22. Zahar I (1998) Optimizing surface water management and upgrading runoff by traditional techniques in Tunisian arid climate. Water and soil fertility, two resources to manage together. Erosion Network Bulletin 18
23. Alaya K, Viertmann W, Waibel T (1993) The Tabias. GTZ Collection, Edition Directorate General of Forests (DGF) and National Forest Institute (NFI)
24. Akrimi N, Zaafouri M-S, Romdhane A, Jeder H (1996) Characterization of the natural resources and populations of the Hadej Park area. ROSELT/OSS document. la France
25. Floret C, Le Floch E, Pontanier R (1976) Map of sensitivity to desertification Central and Southern Tunisia (process degradation of soil and vegetation) Scale 1/100 00. In Soils of Tunisia. Bulletin of the Soil Division. Tunisia

26. Floret C, Pontanier R (1996) Desertification in Tunisia: a men's affair. Population and environment in southern countries. KARTHALA-CPED, Tunisia
27. Abdelkebir H (2004–2005) Elaboration of the CES development map and the fight against silting in the Hadej Bou Hedma Observatory using the GIS technique. Higher School of Engineers of Rural Equipment Medjez -El-Bab Tunisia

Chapter 16
A Historical Look at the Spatiotemporal Dynamics of Tunisian Wetlands by Earth Observation

Balkis Chaabane and Faiza Khebour Allouche

Abstract Wetlands are the transitional areas between terrestrial and aquatic systems who provide the world with natural storm barriers, environmental cleansers, and food and water resources for many forms of life. However, human interventions have disrupted these ecosystems and caused an important environment damage and 48% of wetlands in the Mediterranean basin have disappeared since 1970. This research aims to study the spatial and temporal evolution of a Tunisian wetland located in the Central-East using earth observation and GIS tools. The diachronic analysis of LULC temporal series of aerial photographs and satellite imagery between 1963 and 2018 showed that the wetlands of big Sfax city underwent important changes explained by coastal industrial activities and the urban sprawl. These factors have led, for example, to the reduction of 85% in the area of the seasonal brackish swamps in the north side of the study area and 55% in the south one, then the transformation of 'Ezzit' and 'El Haffera' streams to concrete canals.

Keywords Wetlands · Earth observation · GIS · Diachronic · LULC · Sfax city

16.1 Introduction

Wetlands are *"areas of marsh, fen, peatland or water, whether natural or artificial, permanent or temporary, with water that is static or flowing, fresh, brackish or salt, including areas of marine water the depth of which at low tide does not exceed six meters"* [1]. They are transitory areas between terrestrial and aquatic

B. Chaabane (✉)
High Institute of Agronomic Science-Chott Meriem, University of Sousse,
ISA CM BP 47, 4070 Sousse, Tunisia

F. Khebour Allouche
Laboratory of Phytopharmacy and Weed Science, Higher Institute of Agronomy-Chott
Meriem, University of Sousse, ISA CM BP 47, 4070 Sousse, Tunisia

GREEN-TEAM Laboratory (LR17AGR01), Higher Institute of Agronomic Sciences-Chott
Meriem, University of Sousse, Sousse, Tunisia

© Springer Nature Switzerland AG 2021
F. Khebour Allouche et al. (eds.), *Environmental Remote Sensing
and GIS in Tunisia*, Springer Water, https://doi.org/10.1007/978-3-030-63668-5_16

systems and. These complexed territory present a variable part of an ecological conditions ranging from permanently submerged to continuously exhausted and naturally drained environments [2, 3].

In the Mediterranean region, diverse types of wetlands exist with high ecological, economic and social values [4]. In this region, the wetlands are fewer than in moister regions, that why are special places for humans and wildlife [5]. The Tunisian wetlands are known by their faunal and floral diversity [6] and too generations are determined: intracontinental and coastal [7]. Due to the high variability of bioclimatic stages from humid in the north to Saharan in the south, several structural and functional types of wetlands exist in Tunisia [8].

The study of wetlands means the use of diverse and adapted methods [9]. Recently, the most widely used tool is remote sensing, who offers a range of data, aerial photographs, satellite and radar images and are increasingly used to study, characterize, and follow wetlands from local to regional scales [10]. It enables the monitoring of land use changes and the investigation of dynamic phenomena affecting these ecosystems [11].

The observation of landscape changes over time allows for the estimation of natural and anthropogenic processes involved and the evaluation of natural resource and territorial management issues [12]. Monitoring and studying the changes in wetlands present a great importance to assess and monitor their dynamics and to detect the factors of evolution. From example, climate factors contribute to changes in wetland soils and vegetation cover [13–16]. The changes in landscape structure allows the identification of different land use phases and induced landscaping changes [12], hence human activities are the main triggers for the disturbance of these ecosystems [17].

Several studies have demonstrated the relevance of wetland diachronic analysis in various fields of study. For example, in Algeria, Bouldjerdi et al. [18] have used aerial photographs from 1973 to 2004 to characterize the temporal and spatial dynamics of the vegetation in Beni-Blaid Ramsar site and the diachronic monitoring in this study revealed actions that threatened the environmental wetland state. In the same year, in Madagascar, Jacquin et al. [19] have analyzed time series of medium spatial resolution Terra-MODIS sensor imagery to characterize and track burned savannah surfaces and to ensure the best management and sustainable conservation of these ecosystems. In the west of Burkina Faso, Caillault et al. [20], have used aerial photographs and satellite imagery from 1952 to 2007 to identify landscape dynamics of and highlighthe importance of the relationship between spatial and temporal trends evolution of the LULC dynamics.

In Tunisia, different studies have been realised in this context: Bennasr [21] have studied the spatial organization of Sfax city and his urban movement. Dahech [22] have explored the evolution of the spatial distribution of air and surface temperatures in big Sfax between 1987 and 2010 through the use of satellite imagery and mobile measurement campaigns. The author concluded that the acceleration of the urbanization rate in 2010 was accompanied by a widening of the urban heat island crowns from 1987 to 2001. In Kerkennah archipelago, Etienne et al. [23] have detected the phenomenon of soil salinization and Sebkhas extension and considered that 18% of

Sbekhas extension on the site due to a climate change and anthropogenic human activities.

In this research, we propose to follow wetlands landscape dynamics through the visual interpretation of aerial photographs and satellite imagery in a semi arid coastal zone located in the center of Tunisia. The multi-date analysis of these wetlands allows us to detect the spatio-temporal dynamics of LULC and wetlands which are classified according to RAMSAR classification from 1963 to 2018 and to deduce the impact of human activities.

16.2   Tunisian Wetlands

16.2.1   Tunisian Wetlands Typology

On the scale of territory as Tunisia, with the strong variability of the bioclimatic stages, from the humid in the North to the Saharian one to the South, several structural and functional types of wetlands exist. The size and the morphology of these ecosystems are very dissimilar depending on the history origin of water supply (soft, brackish or salty) [24].

In this context, Aloui et al. [25] distinguish three types of wetlands in Tunisia:

1. Inland wetlands: these are, streams, oueds, lakes, peat bogs, marshes, garaats, sabkhas, chotts, and oases.
2. Artificial wetlands, dams, and hill reservoirs. At the end of 2003, they count 27 big dams, 182 hill dams, and 698 hill reservoirs. The hill reservoirs have a storage capacity of 66.6 millions of m^3 of water.
3. Marine and coastal wetlands:

    –   Coastal lagoons have of depth from 1 to 6 m (Bahiret), bound to their geographical and bioclimatic positions, going off the sub-humid to the arid, they contain a great variety of biotopes and occupy a total surface about 110,000 ha.
    –   Coastal sabkhas have very variables surfaces. They constitute an almost continuous band from the Cap Bon until the border Tunisian-Libyan.
    –   Maritime marshes and the oueds mouth. The Mejrda oued estuary.

16.2.2   Tunisian Wetlands Inventory

Further to the increased request in agriculture produces, the Tunisian ministry had mobilized surface water plans and favored the implementation of irrigation channels from wetlands drainage by agricultors [27]. However, many Tunisian wetlands were disappeared under the influence of the drainage and the construction of dams or affected by the industrial and the agricultural pollution [26]. That is why an inventory of the Tunisian wetlands was done to establish the balance sheet of the losses and the gains during the last hundred years and highlight their economic, hydrological, and ecological values [28].

In 1991, Tunisia has created the first wetlands inventory on the whole territory and updated in 1997. The information sources used in this inventory were topographic and hydrological maps and bibliographical documents (books, reports, published articles …).

The hydrological map on scale 1: 500,000 was used as a basic map to identify the wetlands which were published in 1986 by "Water Resources Direction" The dams and the reservoirs were localized on the maps of the "Hydraulic planning of the Northen and the Center of Tunisia" published in 1988 on the scale 1: 200,000 by Studies and Major Hydraulic Projects Direction.

The inventory of 1997 allowed to draw up the list of 261 natural and artificial wetlands which are localized in different coastal, continental, suburban, and rural sites. Karem [8] classified them in a different group of ecosystems: streams and dams; continental wetlands (sebkhas, chotts, garaâ, peat bogs); artificial wetlands (salt works,…) and coastal lagoons as follows:

- 27% of wetlands are constituted by oueds (streams) with permanently or wintry temporary flow whose the number exceeds 64 and distributed up on all the territory;
- 22% of wetlands are sabkhas with hydromorphone who can be permanent or temporary, an account of 54 sabkhas mainly located in the Sahel, the Center and the South of the country;
- 13% of wetlands are established by garaâts and freshwater lakes in number of 31;
- 12% of wetlands are constituted by the 28 big dams located essentially in the Center and in the North of the country;
- 7% of wetlands are occupied by 17 chotts characteristic of the Tunisian South;
- 5% of wetlands are represented by 13 coastal lagoons which strew the Tunisian coast.

In 2015, forty one Tunisian wetlands are recorded as Ramsar wetlands [29]. The national park of Ichkeul was the first Tunisian wetlands registered in the list Ramsar in 1981. These zones are divided into natural zones such as Sejoumi Sebkha (Fig. 16.1) and Ejerid Chott, artificial zones like the saltworks Sfax-Thyna (Fig. 16.2) and the oasian ecosystems such as the oasian wetlands of Kebili [27].

16.3 Study Area

The framework of this study is the coastal urban area of big Sfax's city. Its belongs to Sfax governorate which is located in the Center-East of Tunisia (Fig. 16.3).

The study area, 34° N 46' and 8° E 39', sprawls on approximately 556 km² with a population of 608.666 in 2014 [30]. It is dominated by a semi-arid climate with an average of rainfull did not exceed 237.8 mm per year, and 18.9 °C of temperature [31]. it is characterized by low monotonous relief, and not very rugged [31]. It corresponds

Fig. 16.1 Aerial view of Sejoumi Sebkha (@Google Earth modified)

Fig. 16.2 Aerial view of Saltworks Sfax-Thyna (@Google Earth modified)

to a vast coastal plain with a very low elevation, whose altitude does not exceed 10 m than from 3 to 4 km from the coast [32].

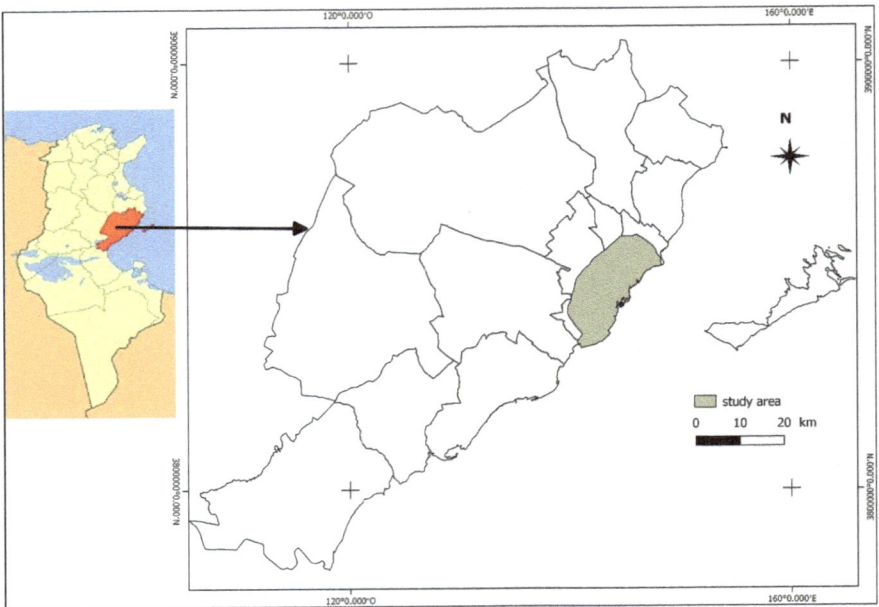

Fig. 16.3 Location map of study area

Table 16.1 Data used in the study

	Data	Scale/resolution	Date
Topographic map	SFAX	1/50.000	1939
	SFAX EL DJEM	1/200.000	1946
	Maritime map	1/50.000	
Panchromatic aerial photographs		1/25.000	1963
		1/10.000	1982
Satellite imagery	'IKONOS 2'	1 m*1 m	2003

16.4 Datasets and Methods

16.4.1 Datasets

Table 16.1 shows the différent data used in this study. Indeed, we used scanned topographic maps. The diversity of analysis scales and dates of these maps allowed to follow more or less finely the land use localized the continental wetlands, drawn the river systems. The maritime map serves to bound the marine wetlands by following bathymetry lines.

Aerial photographs are used in various research for mapping and studying wetlands. For information purposes, within the framework of Morocco wetlands inventory, Dakki and Hammada [3] are used aerial photographs, series of prospecting campaigns on the ground and the habitats typology 'MedWet' to map habitats at the Smir lagoon in the Northeast of Morocco. Also, a study made by Sawastschuk and Bioret [33] where they used multi-dates aerial photographs to map the vegetation on the Estuary of the Loire. These obtained cartographic data served as support to describe the consequences of the modifications stemming from human interventions. 10 panchromatic aerial photographs dating 1963 on the scale 1/25.000 and 31 panchromatic aerial photographs dating 1982 on the scale 1/10.000 were used (Fig. 16.4).

The 1963 mission is chosen as a historical landmark due to the lack of full coverage prior to this date. It also marks the beginning of the expansion and development of the city after independence. certainly, aerial photographs are constituted an accurate support to identify and delimited the wetlands for these dates.

For the year 2003, satellite imagery is taken from the satellite 'IKONOS 2' is used in this study. It is the first very high-resolution civilian sensor providing imagery comparable to aerial photographs allowing visualization of buildings, roads, and trees. These imagery are used in different fields of study, such as urban mapping, agriculture, natural space management. In this study, the 'IKONOS' imagery was used to digitilize wetlands.

The use of Google Earth satellite data in the absence or the lack of data seems to be an effective tool in the study and monitoring of certain natural (erosion) and human (urbanization) processes and in land use planning. Ozer [34] did a bibliographic study on the usefulness of using this imagery. He inferred that they were used as a support, for example, in monitoring coastal erosion, detecting, and mapping landslides, predicting areas at risk of flooding, and awareness of climate and environmental change issues among decision-makers. As a result, in our study resulting from a lack of recent data (high-resolution satellite images or aerial photographs), the

Fig. 16.4 Aerial photographs from 1963 (**a**) and 1982 (**b**) for the trading harbor

use of high-resolution archive imagery available in the free software Google Earth to map the wetlands of big Sfax in 2018 was used. Google Earth excerpts are an opportunity to update previous data (topographic maps, aerial photographs…).

16.4.2 *Methods*

The methodology adopted in this study is illustrated in Fig. 16.5. Indeed, it is based on the main steps that are: data collection, preprocessing and processing. These data were operated using the QGIS 2.18 software. It is based on the photo-interpretation of the different data and the analysis of wetlands spatial-temporal evolution.

The first, the areial photographs underwent a pre-processing action: scannarisation and georeferencing. The second, the processing action which consists of the recognition and the delimitation of landscape units presenting a certain homogeneity and resemblance. The raster analysis available is made according to the purpose of the research to extract the maximum information of the big Sfax wetlands.

Fig. 16.5 Workflow for GIS based wetland typology and LULC

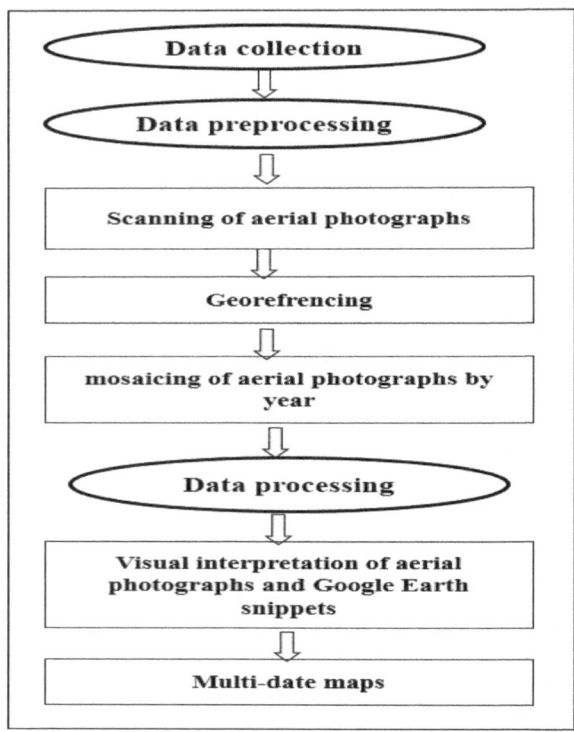

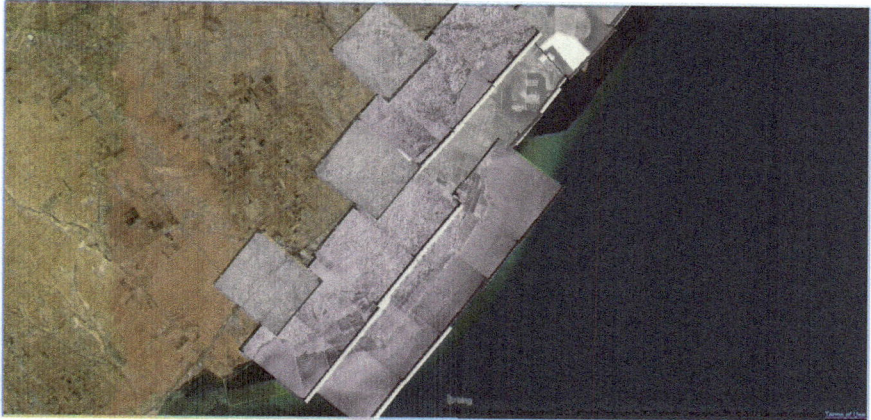

Fig. 16.6 Overview of the geo-referenced aerial photographs from 1982 (@QGIS2.18)

16.4.2.1   Data Preprocessing

- **Scanning of aerial photographs**

The aerial photographs collected for this study are in paper format. To obtain a digital image (Fig. 16.4) from a photograph (paper print), a scanner must be used. The quality of this device determines the final precision of the ortho-photography. In the case of our project, we have chosen a resolution of 600 dpi (dots per inch) which corresponds to the maximum optical capacity of the scanner.

- **Georefrencing**

The georeferencing of aerial photographs consist in recalculating to replace it in the projection geometry chosen and thus make it superimposable in all points on a map. The georeferenced photographs will then become connectable, thus allowing the realization of a mosaic [35]. The raster data are chock to the Bing image (layers extension under QGIS) using the same WGS 84/pseudo Mercator projection system so that the data from different dates can be perfectly superimposed (Fig. 16.6).

16.4.2.2   Data Processing

- **Landscape units identification**

The landscape units are the homogeneous spaces in terms of composition elements, landscape patterns, perceptions, and social representations [36]. Their limits are related to land use.

The identification of landscape units consists of delimiting the main components of land use (olives plantation area, orchards area, arabian medina and european district, harbor area, idustrial area, phosphogypsum, archeologic park) and the wetlands in the study area. The identification is based on photo-interpretation criteria. The criteria used are the hue, the shape, the size, the spatial pattern, the texture, the shadow and the association of the forms.

- **Wetlands digitization**

The task is to digitize the wetlands at different dates by type according to Ramsar calssification. Several new vector layers have been created namely 'seasonal brackish swamps', 'saltworks', 'permanent shallow marines'…, and other layers to present the surrounding environment of these areas as 'phosphogypsum', 'industrial area'.

Each vector layer contains its database created by entering information about each type in the attribute tables that contains specific indicators (soil, vegetation, length, area, …). Other options were used during digitization to extract the most information such as union, division, intersection, and grouping.

- **Multi-date maps**

After having examined the data from 1963, 1982, 2003 and 2018 and extract all the information required. these results will be put in the form of maps presenting the landscape units by date with the same color to compare them and study the evolution. It is done with QGIS software.

16.5   Results and Discussion

16.5.1   Big Sfax Wetlands Typology

The classification system adopted in this study is the Ramsar classification system. However, in this classification system the terms *'floodplain,' 'flood low zones,'* *'coastal depressions,' 'sebkha'* does not consider and identify as a specific type of wetlands, that is why we chose the type *'Ss' 'seasonal brackish swamps'* to identify them [28].

In the study area, three categories of wetlands are identified: marine wetlands, inland wetlands, and artificial wetlands [28].

The marine wetlands comprise the salt marshes and the permanent shallow marines waters. The salt marshes are constituted by salt meadows, kelp beds, and seagrass beds [37]. They spread out throughout the littoral strip of Sfax city before the Southeast extension of the 'saltworks of Thyna' and the installation of the Taparura zone. The width of the tidal area is variable and can attain 500 m.

The permanent shallow marines water is more important in the north coast than in the South. The isobath 5 m is in 3 km of the North coast while it is 750 m of the south coast.

The Inland wetlands are formed by irregular streams and seasonal brackish swamps. The streams have a temporary outflow, converge towards the city and unstop into the Mediterranean Sea. From north to south, we distinguish: 'streams Ezzit,' 'streams El Haffara,' 'streams Agareb' and 'streams El Maou.' The 'seasonal brackish swamps' lands (sebkha or chott) cover up the littoral fringe in 1963. From the 70 s, they are served as land reserves used for the southwest extension of saltworks and the installation of the industrial zones.

The saltworks of Thyna and the wastewater-treatment station of south Sfax presented the human-made wetlands. The basins of saltworks of Thyna have 1700 ha total surface. The basins of saltworks of Thyna have 1700 ha total surface. Since 2007, they are part of the Ramsar wetlands of Thyna [38]. The wastewater-treatment station created in 1983, covered 12 ha and localized in the South of phosphogypsum mountain of the Industrial Society of Phosphoric Acid and Fertilizers 'SIAPE'.

16.5.2 Spatio-Temporal Dynamics of Land Use/Land Cover from 1963 to 2018

The Interpretation of land use maps for the years 1963, 1982, 2003 and 2018 (Figs. 16.7, 16.8, 16.9, and 16.10) highlights changes in the spatial extension of big-Sfax wetlands during the period of 55-year. These variations are discontinuous over

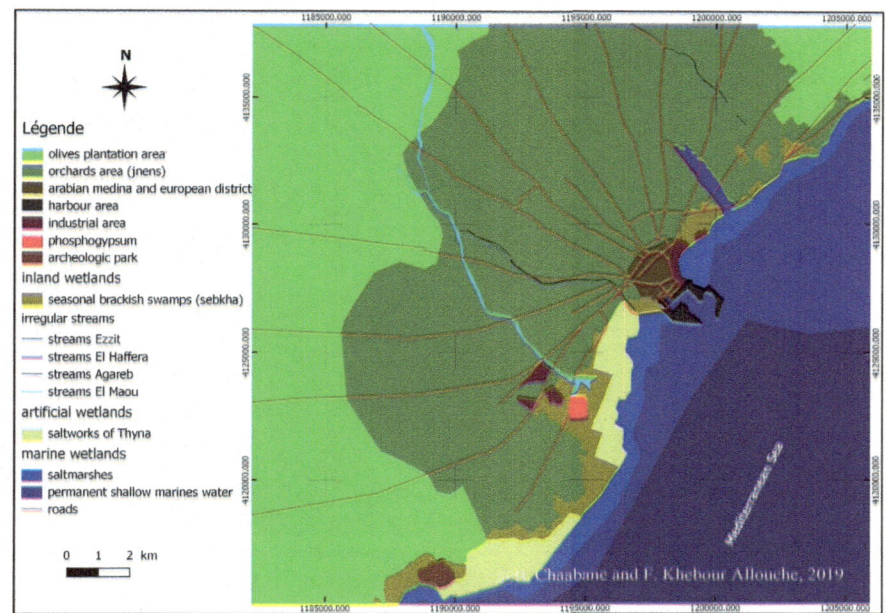

Fig. 16.7 LULC Big Sfax wetland map in 1963

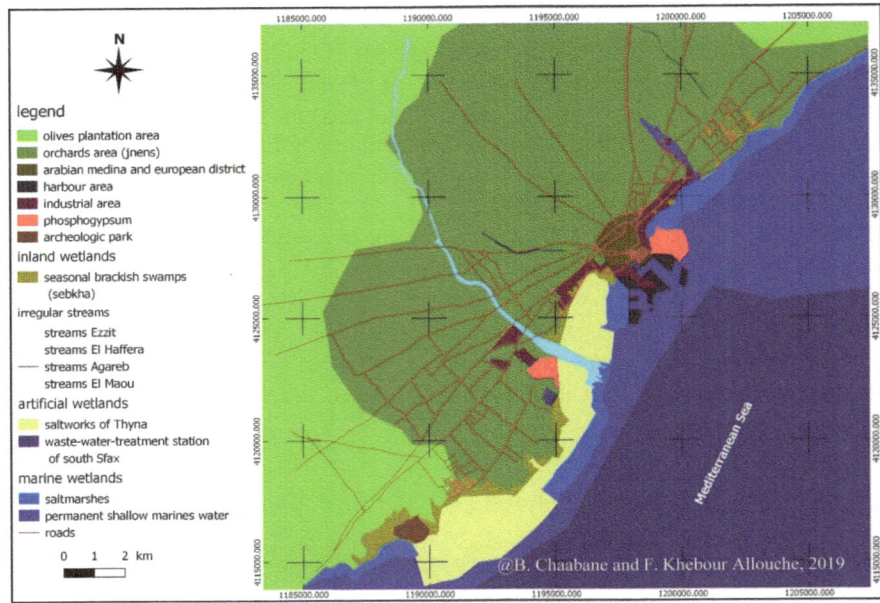

Fig. 16.8 LULC Big Sfax wetland map in 1982

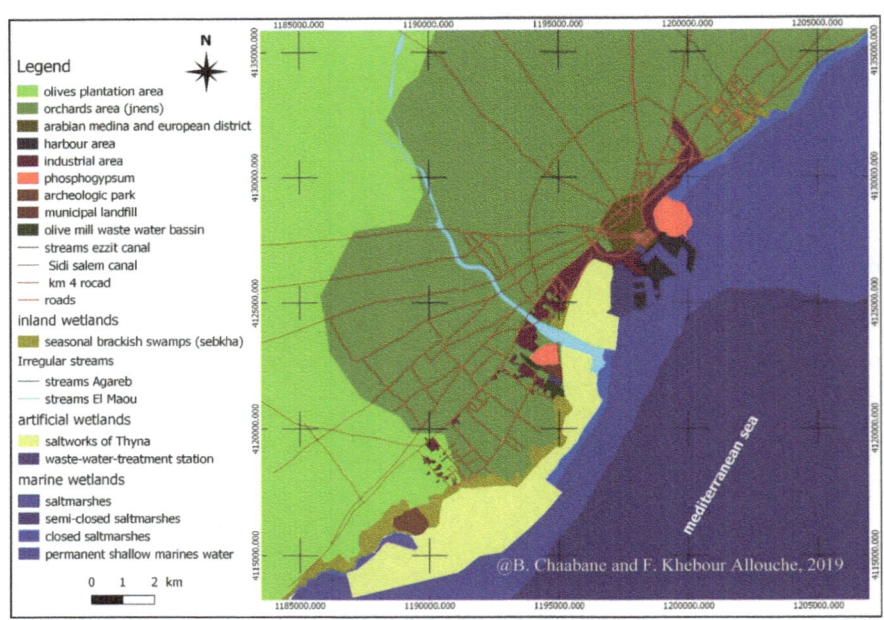

Fig. 16.9 LULC Big Sfax wetland map in 2003

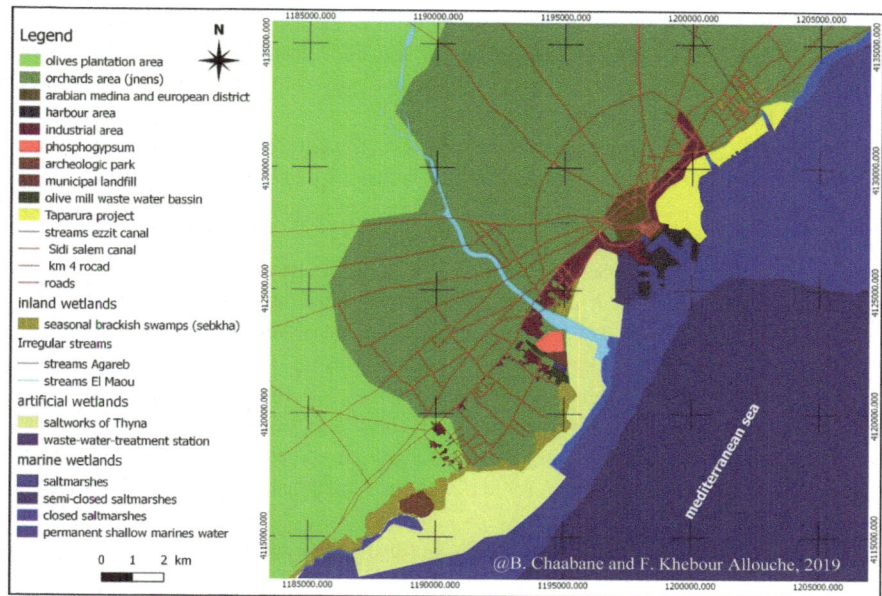

Fig. 16.10 LULC Big Sfax wetland map in 2018

time. There is a trend of regression and disappearance of certain types of wetlands, especially at the north coast of the study area around the 1970s. For example, the urban and industrial sprawls led to the regression of the most seasonal brackish marshes. However, the saltworks of Thyna are the only type of wetlands that have been extended at the expense of other wetlands types located in these surroundings.

The saltworks of Thyna, localized in the South of the study area, occupied grounds type of seasonal brackish swamps (Sebkhas). It is a coastal ground with no slop and the outlet of 'El Maou' and 'Agareb' streams and other small streams. In 1963, the saltworks of Thyna was installed on 1200 ha. Following the exceptional rains recorded in 1969, 'El Maou' stream caused an important ravages and materials damage. That is why the saltworks have had a spatial reorganization by adding new basins. This is an extension of 30% of the overall surface of the saltworks of Thyna from1963 to 2018 (of 1200 ha to 1700 ha), which affects not only the 'seasonal brackish swamps' entities but also the neighboring of saltmarshes. More-over, placing the city's economic interest at the top of the list, by creating industrial areas generally polluting but welcoming a large workforce, has greatly affected the city's spatial organization. This action led to the disappearance of 85% of seasonal brackish swamps area in the north of the city and 55% of the south one.

The coastline landscape dynamics are more advanced in the north side of the study area than in the south. For the past five decades, it has been subjected to a strong anthropogenic pressure following polluting industrialization (the Nitrogen, Phosphorus and Potassium manufacturing company 'NPK' and the industrial areas

Poudrier 1 and 2) and the creation of residential districts (the city of gardens, the city of gendarmes, the city of Bourguiba, the summer amphitheatre Sidi Mansour, the biotechnology centre of Sfax, etc.), which has caused the gradual disappearance of seasonal brackish swamps.

Contrariwise, the seasonal brackish swamps on the south side are bordered by the agricultural area 'Ain Falet' and occupied by saltworks of Thyna how blocking the access to the sea. This zone contains many sources of pollution including the Industrial Society of Phosphoric Acid and Fertilizers 'SIAPE', the phosphate processing plant and its enormous phosphogypsum deposit, industrial areas (industrial areas 'Sidi Salem' and 'Oued el Maou'), scattered industrial units not connected to the wastewater treatment, the wastewater treatment station of south Sfax and the municipal landfill. They have led to the degradation of environmental quality. This industrial activity has stimulated the demographic development. Some of these activities took place on seasonal brackish swamps, causing their decline. However, the registration of the saltworks of Thyna and the surrounding areas as important international wetlands in the Ramsar registry in 2007 has reduced urban sprawl movements on the south coast.

The installation of the Nitrogen, Phosphorus and Potassium manufacturing company 'NPK' on the north coast and these phosphogypsum discharges into the sea have led to the disappearance of 4 km of saltmarshes and the disturbance of maritime ecosystems. Restoration of this coast by depollution, backfilling and the creation of new beaches lead to the disappearance of the saltmarshes area from the north of the commercial harbor to the summer amphitheater 'Sidi Mansour' (the Taparura area). As for the south coast, the widening of the saltworks of Thyna led to a 60% reduction in the length of the saltmarshes.

The excessive urbanization in Sfax city has led to the disappearance and obstruction of several streams crossing the city, disruption of the flow regime and feeding of coastal seasonal brackish swamps (sebkhas), and soils waterproofing that contribute to increased runoff coefficient. Exceptional rainy events such as the 1963 floods and especially the 1982 floods caused important ravages, which led the public authorities to equip the city with flood protection infrastructures. The purpose of these protective structures was to intercept the runoff water to discharge it into the sea. Indeed, in 1984, the beds of 'Ezzit' and 'El Haffera' streams were recalculated, traced by lifting all the constructions that block it to reach the sea and transformed into concrete canals. 'El Maou' stream as it is the most important stream crossing the city, its southern part has been recalibrated, diked and reach the sea by segmenting the saltworks of Thyna in two parts. After these transformations, in 2003 and 2018, 'El Maou' stream which has preserved the same tracing of 1982 and 'Agareb' stream its southern part has been drawn.

16.5.3   Effect of Anthropogenic Factors on Wetland Landscape

The uniformity of topography and the highly privative nature of urban soil in the city of Sfax strongly influenced the architecture of the city [39]. Nevertheless, the urban development plans of Sfax city in 1961 and 1977 called for preferential development to the coast because of the disponibilité of land opportunities (maritime public domain, public land which are seasonal brackish swamps, etc.) and the protection of the orchards area 'Jnens', which should retain its semi-rural character. Almost all industrial areas (Poudrier 1 and 2 and Madagascar) are created by the industrial land agency since 1975 and have a coastal or sub-coastal location. Thus, progressively since the 1960s, all the coastline has been sacrificed to the industrial activity. Littoralization is not always justified and has turned into a capacious dump for the city: phosphogypsum deposits, municipal landfill, Margine deposit, etc. [39].

The growth of industrial activities was accompanied by the realization of many social housing projects: of the 25 estate projects carried out by public developers between 1972 and 1990, 23 of them were erected on the coastal and sub-coastal fringes [39]. This situation has eclipsed the Mediterranean Sea, a part of the city's history, from its urban landscape [39]. For these reasons, the population has turned its back on the coastline, which has become a socially devalued space. Everything was done as if the economic improvement was the antonym of a decent and clean living environment [21]. Bathing has been prohibited in all beaches of the city since 1978 [39]. However, the colonial management plan (1929) conscious of the importance of the coastal ecosystem has projected the creation of a cornice along the north and south coastline of the city and the development of artificial beaches [40]. Thus, it is the coastline that has suffered the negative consequences of industrialization. In addition, the sources of nuisances have increased, and industrial discharges have caused very high pollution of marine waters. The levels of polluting substances are indeed very high and often exceed tolerated standards and disrupt the biological life of the herbarium and the life of marine wildlife.

The two biggest phosphogypsum deposits that have marked the landscape are the most polluting solid discharges. The Metal levels in liquid SIAPE discharges are very important, such as fluorine is 3600 times greater than tolerated, phosphorus is 8670 times higher, cadmium 751 times more, and lead is twice as high [41]. These inputs also caused disturbances in pH, dissolved oxygen, increased turbidity, and significant amounts of metabolic compounds. The phosphogypsum caused a very large decrease in the areas occupied by the Posidonia meadows and their eradication in the bottoms of more than 10 m of depth.

Urbanization has caused a break in the balance between man and the natural environment. This environment destruction has unfortunate consequences even on catchment and runoff. In Sfax city, the fight against runoff was undertaken by the inhabitants themselves, building for centuries the vast maze of hedgerows surmounted by cactus 'Tabias' in the orchards area. But the modernization urban action undertaken by the municipality consisted of demolishing these fences; This translates into in the worsening of the flooding in the town as a result of increased runoff [42]. This urban extension and economic development have not been accompanied by adequate development of communal infrastructure. Similarly, industrial development has not been supported by appropriate pollution protection measures. One of the evolution of negative aspects is the increase in the volumes of gaseous, liquid, and solid releases. Finally, industrial and urban activities and even natural intake have put the city face to serious environmental problems that have a negative impact on the coastal environment essentially.

16.6 Conclusion

The importance of wetlands is not relative to their size but to their location in the landscape, their structure, and their functions [43]. These ecosystems are dynamic and threatened by human activities. The decline in their areas has been accompanied by ecological and environmental disturbances.

From 1963 to 2018, the wetlands of big Sfax has undergone enormous changes in these spatial extensions following the development of a polluting industrial activity and welcoming a large workforce. These variations are discontinuous over time. Urban and economic activities are concentrated on the coastline and in particular in the southern part. They have contributed, with the inadequate use of space to the degradation of life quality and environment. This inadequate space use has led to the embrittlement of wetlands and their disappearance in some cases.

Wetlands have undergone a regressive evolution and an anthropization tendency. The installation of industrial areas, since 1952, generally polluting and welcoming a large workforce, has affected the spatial organization of big Sfax and led to the disappearance of 85% of the seasonal brackish swamps area in the north and 55% in the south side during the period studied. The excessive urbanization in the study area has also played an important role in the spatial-temporal dynamics of these wetlands. It has led to the disappearance and obstruction of bed streams crossing the city, disruption of flow regime and feeding coastal seasonal brackish swamps (sebkhas), and soils waterproofing by increasing the runoff coefficient. In 1972, the southern part of streams 'El Maou' was recalibrated and diked to reach the sea after the flood of 1969 by dividing the saltworks of Thyna. This stream, in 2003 and 2018, preserve the same tracing of 1982 and 'Agareb' stream only its southern part has been drawn. Since 1984, the streams 'Ezzit' and 'El Haffera' were recalculated and transformed into concrete canals.

16.7 Recommendations

The use of high-resolution satellite images will enable the characterization and monitoring of wetland components with greater precision. The earth observation and the use of time series of aerial photography and satellite images helped us to draw wetlands history and assess their degrading or developing state. These kind of results can help the decision-makers to ensure sutaible management for wetlands sustainability.

Radiometric data will support the monitoring of wetlands through the use of radiometric indices such as soil moisture and also the seasonal monitoring of components such as vegetation status. A complemtary and supplementary data to optical and thermal sensors can be obtaiend by radar sonsors. Radar systems can provide information about soil, vegetation moisture content, surface roughness and sub canopy. Therefore, the use of this data can bring more specific information pertaining to wetland attributes as explain Henderson et al. [44]. The combination of different types of data with a DEM can improve the discrimination of wetland classes.

References

1. Ramsar convention secretariat (2013) The Ramsar convention manual: a guide to the convention on wetlands (Ramsar, Iran, 1971), 6th edn. Ramsar convention secretariat, Gland, Switzerland, p 112
2. Bouzillé J-B (2014) Issues related to wetlands. In Wetland ecology: concepts, methods, and demarches. Editions Lavoisier, Paris, French, pp 4–18
3. Dakki MF, Hammada S. (2005) Mapping of natural habitats of a Mediterranean coastal wetland: Smir marshes (Tetouan region, Morocco); sensitive coastal ecosystems of the Mediterranean: the case of Smir coastline, work of the scientific institute, Rabat, general series, n° 4, 9–15 (French)
4. Daoud-Bouattour A, Muller SD, Ferchichi-Ben Jamaa H, Ben Saad-Limam S, Rhazi L. Soulié-Marsche I, Rouissi M, Touati B, Ben Haj Jilani I, Gammar AM, Ghrabi-Gammar Z (2011) Conservation of Mediterranean wetlands: Interest of historical approach. Comptes rendues biologies 334(2011):742–756
5. Hollis GE (1990) Environmental impacts of development on wetlands in arid and semi-arid lands. Hydrol Sci J 35(4):411–428
6. Khemaissia H, Touihri M, Jelassi R, Souty-Grosset C, Nasri-Ammar K (2012) A preliminary study of terrestrial isopod diversity in coastal wetlands of Tunisia. Vie et milieu—Life Environ 62(4):203–211
7. Kamoun M, Khadraoui A, Ben Hamad A, Zaïbi CH, Langer MR, Bahrouni N, Ben Youssef M, Kamoun F, (2019) Impacts of relative sea level change and sedimentary dynamics on an historic site expansion along the coast between Sfax and Jebenina, Tunisia. Proceedings of the 1st springer conference of the Arabian journal of geosciences (CAJC-1) Tunisia 2018. Springer Nature Swizerland AG 2019, pp 141–143
8. Karem A (2014) A dot focal of Ramsar in Tunisia. Presentation on wetlands in Tunisia. Directorate of Forest Conservation. Ministry of Agriculture, Water Resources and Fisheries (French)
9. Michelot J-L (2005) Wetlands characterization, thematic notebook, PNRZH, 70 pages (French)
10. Rundquist DC, Narumalani S, Narayanan RM (2001) A review of wetlands remote sensing and defining new considerations. Remote Sens Rev 20(3):207–226
11. Mas JF (2000) A review of methods and techniques of remote sensing of change. Can J Remote Sens 26(4):349–362 (French)
12. Skupinski G, Tran DB, Weber C (2009) Multi-date spot images and spatial metrics in the study of urban and sub-urban change, the case of the Lower Bruche Valley (Bas-Rhin, France). Eur Geogr Rev (French)

13. Malclon Conly F, Van Der Garth K (2001) Monitoring the hydrology of canadian prairie wetlands to detect the effects of climate change and land use changes. Environ Monit Assess 67:195–215

14. Papastergiadou ES, Retalis A (2008) Environmental monitoring of spatio-temporal changes using remote sensing and GIS in a mediterranean wetland of northern Greece. Water Resour Manage 22:579–594

15. Klemas V (2015) Remote sensing of wetlands: case studies comparing pratical techniques. J Coastal Res 27(3):418–427

16. Gallant Alisa L (2015) The challenges of remote monitoring of wetlands. Remote Sens 7:10938–10950

17. Djibel M, Labar S, Medjani F, Boufia I (2013) Study of the ecological changes of wetlands in desert environments using Landsat imagery and GIS. Int J Environ Water 2(5):81–87. ISSN:2052-3408 (French)

18. Bouldjerdi M, Bélai G, Mayache B, Muller SD (2011) Threats and conservation of North Africa wetlands: the case of the Beni-Belaid Ramsar site (NE algérien). Biological reports n°334, pp 757–772. Sciences Academy. Published by Elsiver Masson SAS (French)

19. Jacquin A, Cheret V, Sheeren D, Balent G (2012) Determination of fire regime in the middle of Savane in Madagascar from time series of MODIS images. Int J Remote Sens 32(24):9219–9242 (French)

20. Caillault S, Ballouche A, Delahaye D (2012) Towards the disappearance of the bushes? Multiscalar analysis of landscape dynamics in western Burkina Faso since 1952. Cybergeo: Eur J Geogr (French)

21. Bennasr A (2005) Sustainable urban planning and governance: the case of Sfax. In Colloque SYFACTE/ GREGUM 'Cities facing the challenge of sustainable development'. 15 pages (French)

22. Dahech S (2012) Evolution of the spatial distribution of air and surface temperatures in the Sfax agglomeration between 1987 and 2010. Impact on energy consumption in summer. Climatology, Special Issue, 'Climate and Climate Change in Cities, pp 11–33 (French)

23. Etienne L, Dahech S, Beltrando G, Daoud A (2012) Recent dynamics of the coastal sebkhas in the Kerkennah archipelago (South-Central Tunisia): contribution of remote sensing. Remote Sens J, Contemporary Archives Editions, 11(1):273–281 (French)

24. Barnaud G, Fustec E (2007). Wetalnds conservation: why? How? educargi editions and Quae edidtions, 291 p (French)

25. Aloui A, El Hamrouni A, Souissi A, Neff N, GroBmann A (2005) (ecosystems/wetlands). Climate change: effects on the Tunisian economy and adaptation strategy for the agricultural sector and natural resources (French)

26. Aini R, Bedhief S (2010) Worldwide forest resources assessment. National Report, Tunisia, 57 p (French)

27. Ramsar sites information service. Annotated list of wetlands of International Importance. Tunsia. http://ramsar.org/fr/zone-humide/tunisie

28. Chaabane B, Boujelben A (2017) Identification, classification, and characterization of big Sfax wetlands. LARHYSS J 30:45–65. ISSN:1112-3680 (French)

29. OTEDD (Observatoire Tunisien de l'Environnement et du Développement Durable) (2009) Indicators of sustainable forests. 32p (French)

30. INS, Tunisian National Institute of Statistics (2019) http://www.ins.nat.tn/

31. Missaoui M, Zerai K, Fridhia A, Baroudi M, Baroudi J (2013) Atlas of Sfax governorate. Ministry of Equipment and Environment. Directorate-General for Regional Planning, 105 p (French)

32. Bouzid J (2006). SMAP III project—Tunisia (2006–2008). Integrated management strategy for the southern coastal area of big Sfax. data Collection. Preliminary report, 116 p (French)

33. Sawastschuk J, Bioret F (2012) Diachronic analysis of the spatial dynamics of vegetation in the Loire Estuary. Photo-Interprétation Eur J Appl Remote Sens 3:15–28. Editions ESKA (French)

34. Ozer P (2014) Natural disasters and land use planning: from the interest of Google Earth images in developing countries. Int J Trop Geol; Geog Ecol 38:209–220 (French)

35. Muraz J, Durrieu Labbe S, Andreassian V, Tangara M (1999) How to add value to aerial photos in GIS? engineering—E A T, pp 39–58 (French)
36. Landscape Atlases. Method for identification, characterization and qualification of landscapes (2015) (French)
37. Chaabane B, Khebour Allouche F, Bouzid I, Boujelben A (2017) Mapping wetlands of the big Sfax (Centre-East of Tunisia): Tool for decision making. LARHYSS J 31:41–49. ISSN:1112-3680 (French)
38. Karem A (2007) Ramsar wetlands descriptive sheet. Saltworks of Thyna. Tunisian Ministry of Agriculture and Hydraulic Resources. Directorate General of Forests, 11 p (French)
39. Megdiche T (2005) The evolution of the social division of space in Sfax (Tunisia). Symposium cities in the challenge of sustainable development: what control of urban space and associated segregation? Nov 24–27, 2005 in Sfax (French)
40. Bennasr A (2011) Sfax: from the regional city to the metropolis project. Globalization and Urban Change, University Publishing Center, pp 79–95 (French)
41. Zairi M, Rouis MJ (1999) Environmental impacts of phosphogypsum storage in Sfax (Tunisia). Bulletin of the Bridges and Roads laboratories—129, pp 29–40 (French)
42. Fakhfakh M (1979) Development and environment along the Tunisian coasts, 3rd series, volume 35; man and his natural environment in the Maghreb, pp 75–79 (French)
43. Fustec E (2000) Wetlands functions: acquired and lacuna, chapter in wetland functions and values, technical industries, environment series, pp 17–38 (French)
44. Henderson M, Lewis J (2008) Radar detection of wrtlands ecosystems: a review. Int. J Remote Sens 29(20):5809–5835. 20 Oct 2008

Part VII
Conclusions

Chapter 17
Conclusions and Recommendations for "Environmental Remote Sensing and GIS in Tunisia"

Faiza Khebour Allouche, Abdallah Gad, and Abdelazim M. Negm

Abstract This concluding chapter contains the significant conclusions and recommendations from all chapters to provide perspective and guidance for student, teachers, researchers, engineers, decision makers, etc. The book topics extend to cover analyses of land degradation, dryland, drought, as well as to climate change, risks, groundwater vulnerability, and wetlands. Conclusions of different Tunisian case studies related to these phenomena using remote sensing and GIS tools are presented and various methodological approaches have been exposed. The presented recommendations strengthen the link between remote sensing and GIS by opening the doors to future applicable and universal research work. Especially recommendations which consist in sensitizing the readers to a technology applied to environmental studies which does not stop evolving at a high speed. The essence of all conclusions and recommendations can be distilled into threads that run throughout this chapter.

Keywords Monitoring · Spatial distribution · Decision support · Mapping · Vulnerability · Risks · Anthropogenic · Degradation · Drought · Geo-information · Remote sensing · Earth observation · GIS · Management · Environmental · Wetland · Arid

F. Khebour Allouche (✉)
Laboratory of Phytopharmacy and Weed Science, Higher Institute of Agronomy-Chott Meriem, University of Sousse, ISA CM BP 47, 4070 Sousse, Tunisia

GREEN-TEAM Laboratory (LR17AGR01), Higher Institute of Agronomic Sciences-Chott Meriem, University of Sousse, Sousse, Tunisia

A. Gad
National Authority for Remote Sensing and Space Sciences (NARSS), 23, Josef Tito St., 1564, Cairo, Egypt
e-mail: agad@narss.sci.eg

A. M. Negm
Faculty of Engineering, Water and Water Structures Engineering Department, Zagazig University, Zagazig 44519, Egypt
e-mail: amnegm@zu.edu.eg

© Springer Nature Switzerland AG 2021
F. Khebour Allouche et al. (eds.), *Environmental Remote Sensing and GIS in Tunisia*, Springer Water, https://doi.org/10.1007/978-3-030-63668-5_17

17.1   Update

This chapter begins with an update of the themes applied to environmental studies in Tunisia.

The evaluation of the dynamic parameters of the soil and vegetation helps in the evaluation of erosion but the field measurements make it possible to determine the presence of a clear relationship between the observed changes and the accelerated erosion of the soil [1]. Tunisia is the smallest country of Maghrebian states. However, a few spaces of this country are unequally populated. This chapter emphasizes that dasymetric method, based on the use of ancillary data, mainly on Remote Sensing and GIS should enhance the representation of population density in Tunisia at a small scale. It's an interesting alternative to the classic choropleth map. It can help various categories of users as planners and geographers to implement spatial decisions when and where the accurate location of the population is needed. Therefore, the intention was to apply an equation formula summarizing the overall calculation of the density for each new space unit corresponding to one land use class and implying the possibility of drawing dasymetric maps. The remotely sensed dataset MRSID (Multi-resolution Seamless Image Database) 30 m and a Basemap THRS 2017 were employed for that purpose and [2] land use classes inspired from Corine Land Cover taxonomy but adapted to the Tunisian case were chosen. Procedure applied to the governorate of Tozeur gives more realistic dasymetric map and gives an idea of the real distribution. Comparing to the satellite image, it fits the land use classes. It also refines the original enumeration units often using the administrative apportionment.

During the twentieth century, urban planning allowed permanently to integrate the planted spaces in the policies and the urban practices, in particular, in the developed countries [3]. A study on employing the unsupervised classification technique have been used to map four states of the urban sprawl of Grand Sfax in 1987, 2001, 2006 and 2014, using Landsat images as well as the Google Earth image. Results show that in 27 years, the built-up areas have increased from 7.3 to 35.2% of the total territory of Grand Sfax. This artificialization of soils in Sfax city has aggravated the physical vulnerability of the city to the risk of flooding [4]. This industrial city is subject to intense urban changes that reveals more and more clearly its negative externalities especially as' it generates environmental issues such as the scarcity of green spaces. Then, the attention is attracted to human well being through the assessment of green spaces using GIS tools. Results show that the Sfaxian population concerned by the lack of green spaces, the mesh method has been used in this chapter and it shows that more than 50% of the total population suffers from the absence of any nearby public green space.

In Tunisia, the Meskat system, an anti-erosion technique used by farmers to collect runoff water that has been used since Roman times. Water is directed through distributors to irrigate the olive trees planted in downstream plots, called Mankaas. At this level, the runoff is intercepted by mounds known as "Tabias" and directed from one plot to another through small openings "majref" [5]. However, in the last decades, this technique has faced a dropout phenomenon. So, it is important to assess the saptio-temporal dynamic of meskat system, specially in the Sahel of Tunisia. The supervised classification of a Landsat series was performed by the Geomatica software PCI was done. Results show that the Meskat system is mainly divided among three water-sheds: Sabkha Halek El Menjel, Wadi Laya El Hammam and Wadi Hamdoun. The diachronic analysis revealed an expansion of urban areas at the expense of the Meskat system. Indeed, urban areas recorded an increase of 1, 8, and 4%, while areas occu-pied by Meskat system were decreased by 6, 8, and 13% for the Sabkha Halek El Menjel, Laya El Hammam Wadi and Hamdoun Wadi watersheds, respectively.

In the contexte of the use of GIS tools in rural applications, we found in this book a noval application in the field of invasive weeds. In Tunisia, many plant species from other parts of the world have been introduced for a range of purposes as garden ornamentals, as crop species, for stabilizing sand dunes and as barrier and hedge plants. From example, *Solanum elaeagnifolium* Cav. (#SOLEL) is native to North-east Mexico and Southwest USA [2, 6]. In Tunisia, since 1985, SOLEL started to become a noxious weed [7]. The invaded area is increasing and this weed is becoming a potential threat to thousands of hectares of irrigated fields in arid and semiarid regions [8]. Giving the SOLEL negative impacts in the agro-ecosystems, Tunisian researchers and authorities have increased their interest in its management. Several research programs were conducted to support the implementation of an appropriate control approach against SOLEL [9]. In this chapter, authors attend to map and characterize the spatial distribution of this weed using the GIS technique. Results confirm the wide expansion of this species; more than 60% of the Al Alem irri-gated perimeter was infested by *Solanum elaeagnifolium*. Although, SOLEL distri-bution pattern characterization reveals that the patchy pattern was the most common distribution type of the weed. It's ongoing invasion process in agriculture lands and considering its high invasiveness potential, this alien weed can also invade adjacent natural or semi-natural areas.

The water sector is one of the most sensitive domains in Tunisia because of its scarcity and its economic and social importance. Furthermore, demographic growth, urbanization and the important sanitation development in Tunisia contribute to a continuous increase of this nonconventional source of water. Thus, the aquifer recharge with this water constitutes an interesting solution of its storage and use for irrigation during the peak periods. A methodology was proposed in this chapter to help selecting the best suitable sites for aquifer recharge with reclaimed water. The tools used were Multicriteria Analysis and the Geographic information system. Grombalia shallow aquifer was chosen as pilot site and the effluent of Bou Argoub wastewater treatment plant as reclaimed water. This application prove that the weighting of the thirteen criteria (Slope, soil salinity, soil texture, land use, geological formations, groundwater quality, etc.) to rank the suitable sites using AHP method [10] states that the distance to the residential areas and to the WWTP are the two most influential criteria for site selection decision. The combination of GIS and the multicriteria PROMETHEE II method offer an efficient tool to rank the suitable sites for Grombalia aquifer recharge with reclaimed water.

The peninsula of Cap Bon in Tunisia is an area containing a much contrasted land use. It is divided into large natural subareas next to the mountains and "artificialized" spaces scattered now along the coast. Thus, the main goal of this research is to develop a methodology that can identify, locate and map, at different scales and by using various modes of presentation, the land use as an overall information system. It contains different phases bringing together a useful synergy between GIS and remote sensing. The first step is based on the design of a Conceptual, Logical and Physical Data Model using the HBDS (Hypergraph Based Data Structure) method, respecting the rules of the graphics semiology (especially when dealing with colors and color-values). The second one, represents the conception of a Geodatabase fitting the described CDM and using Arc Catalog software. Then, the creation of datasets from remotely sensed imagery and visual image interpretation approach. After, the creation of a geographic database of the land uses area (BADOS), multi-source, multi-scalar and updatable using the software ArcGIS is done. At the end, the conduction of a multiscalar mapping procedure figuring the land cover and land use (urban, agricultural land, natural areas, wetlands, water bodies...) for the Cap Bon is applied. This conceptual work favors downstream a thematic cartography very rich and very variable. Therefore the LUIS is not just a stack of layers, but rather hierarchical levels of land use land cover that can be displayed and manipulated If the user needs to perform a large number of actions.

The North African agricultural production has increased the risk of water scarcity and groundwater quality deterioration [11]. This is the case of the Oued Laya phreatic aquifer located in the Centre East of Tunisia and characterized by an important urban and agricultural pollution. This chapter emphasized the use of three parametric approaches standard DRASTIC [12] as intrinsic vulnerability, Pesticide DRASTIC [13] and Susceptibility Index SI [14], has specific vulnerabilities to assess Oued Laya aquifer vulnerability to pollution and map the sensitive zones for contamination. However, this assessment was carried out by using the Geographic Information System as a tool for environmental studies and geospatial modelling of natural phenomena. Results show that for the three methods, the most vulnerable parts of the aquifer corresponds to areas subjected to the agricultural pollution from the irrigated lands and to urban contamination, which comes from the wastewater treatment plant (WWTP) of Kalaa Sghira and the uncontrolled landfill located in the Oued Laya river bank close to Akouda. The groundwater protection of Oued Laya aquifer against these pollutions requires removing the uncontrolled landfill and improving the wastewater treatment performance of Kalaa Sghira WWTP.

In the modern world, impacts of disaster events on human lives and the economy are increasing every year [15]. The Abdeladim watershed located in the west of tunisia, is distinguished by a set of bio-physical and anthropogenic characteristics that make it a fragile natural equilibrium. Indeed, climate factors (lower semi-arid), and anthropogenic (generally inadequate cultivation techniques and anti-erosion management) make it a vulnerable environment to degradation. In this study, remotely sensed images such as aerial photographies, thematic maps and Landsat satellite image are used. The study of the evolution of the river system between 1963 and 2016 showed a tendency to increase the number of wadis in length and width. This development is mainly achieved by the processes and mechanisms of water erosion carried out by concentrated flow (linear, regressive and lateral erosion). Therefore, a growing sensitivity to the degradation of its agricultural potential watershed both by soil loss and by the reduction of quality of the latter is deduced. Climatic conditions and anthropogenic inappropriate to the potential of soils are the two main factor of degradation.

Urban risk is expected to rise in the future as urban areas continue to expand into hazard-prone areas, while climate-related hazards such as floods, storms, heat-waves or sea level rise increase in frequency and intensity in many parts of the world [16, 17]. Tunisia faces major risks associated with natural hazards and climate change. Rising temperatures and changing precipitation patterns combined with the increasing frequency and intensity of floods and droughts are threatening the agricultural sector, human health and the local economy [18].

The Tunisian Nationally Determined Contributions (NDCs) which were submitted to the United Nations Framework Convention on Climate Change (UNFCCC) in 2015 mentioned mitigation, vulnerability, and adaptation challenges. Moreover, the NDCs declared disaster risk management (DRM) as a part of the national objectives defined in the future Master Plan. The chapter presents a methodology for urban risk assessments using a multi-temporal remote sensing data from aerial photographs, WorldView3 and Landsat applied to a coastal region, Monastir. Results show that the flash flood model showed a particular concentration of water depth in specific areas of Monsatir city and residential areas located in the eastern part are affected for the 100-year return period event. However, the coastal regions of the outer parts of the municipality are often more severely damaged and more exposed to erosion processes. Population exposure showed significant clusters in the old medina and the eastern part of the city center. Both areas are densely populated and affected by the flood extent, with one census sub-unit registering up to 57% of its population as exposed. All these data analyses are made available in the web GIS platform useful for decision makers and stakeholders.

Climate change is a very serious phenomenon and has become a major global issue in recent years, especially in developing countries strongly affected by its impacts [19]. This natural variation is accentuated by anthropogenic actions [20] and degradation (LD) is among the major environmental issues driven by a change in land use-land cover (LULC) and climate change worldwide especially in semi-arid regions [21]. The problem of the degradation of natural resources is highly felt in Southern Mediterranean countries with semi-arid and arid climates, where water and soil resources are scarce and vulnerable [22].

The integration of remote sensing and GIS technologies is an effective approach for analyzing land use and cover changes [23]. Land change models can help scientists and users to understand change processes and design policies to reduce the negative impact of human activities on the earth system at scales ranging from global to local [24]. In this chapter two Landsat satellite images (2007 and 2014) and ISOCLUST for classification method Land Change Modeler (LCM) have been used and applied to a southern region, Zarzis to monitor its LULC changes. Gain and loss analysis show in this chapter that a net profit in terms of olive trees (good vigor) with a large amplitude of positive change (+47.54% of change) between the year 2007 and 2014 while for olive groves on silted soil decreased, the highest percentage of negative change (−36.7% of change). Net change by category results show that the area for olive trees (Good vigor) increased by 11386 hectares with a percentage of positive change (+32.86%) while the olive groves on disturbed soil decreased by 7535 hectares with a percentage of negative change (−27.38%). Although, transition from olive groves on non-disturbed soil to olive groves on disturbed soil shows that transitional regions are limited to highly sandy regions riparian to the sebkha and areas. However, the transition of olive groves, from disturbed soil to non-disturbed soil, is valuable in the studied region since the olive trees have been cultivated for centuries for its socioeconomic benefits for farmers and consumers.

Drought types are classified into meteorological drought due to a lack of precipitations over a period, hydrological drought with inadequate water resources required for established water uses, agriculture drought due to vegetation water stress, and socioeconomic drought with the failure of water resources systems to meet water demands [25]. However, the availability of free time series satellite images such as MODIS sensors since the year 2000 had allowed the exploration of various models for drought assessment where multispectral images are combined to derive drought indicators of the vegetation in a Tunisian Mediterranean ecosystem. In this context, the present chapter, applied to a subhumid region in Tunisia, the Khroumirie ecoregion, exposed LAI-MODIS time-series images processing and integrated into a water balance model for the simulation of water stress spatio-temporal dynamic. Results show that with an LAI that is hypothetically invariable in space (i.e. no adjustment between vegetation and soil), stress is strongly conditioned by the intrinsic characteristics of the soil. With LAI-FIX, 37% of the vegetation with 1-year of water stress during the period 2003–2008 is found on shallow soil, whereas it increases to 41% if the simulations were done with LAI varying from year to year. The potential of integrating time-series products such as LAI in water balance models for a better understanding of the ecohydrological ecosystem functioning.

Around 10–20% of drylands and 24% of the world's productive lands are degraded [26]. Unfortunately, soil degradation in dryland is irreversible [27, 28] and represent a severe and long-lasting disturbance that will prevent ecosystem recovery in the absence of comprehensive artificial restoration measures. Southern Tunisia as a part of the North African steppe is knew by its significant degradation since the eighties. The combination of severe droughts and an exponential increase in livestock had a catastrophic impact on pastoral resources [29, 30]. Nevertheless, satellite remote sensing is commonly used to monitor land degradation because it provides spatially continuous, replicable and homogeneous information in a cost-effective manner and over large areas. Moreover, long-term changes can be derived from time series data and explained by radiometric indexes. Thus in this chapter an overview of methods used for the assessment of dryland degradation vulnerability is exposed and an estimation of soil and water conservation techniques efficiency in the watershed of Wadi Bouhamed in south Tunisia using remotely geo-information data was applied. Results reveals that the 1988 year was a dry one during which the natural vegetation degradation has been enhanced by increased pasturing needs. This drought was followed by a relatively wet period explaining the regression rate of 25% of over pastured zone in 2000. The over pastured zones changed drastically between the three periods, from 1988 to 2000 they decreased (34.6%) but in of 2011 the area increased (20%). Tree crop zone areas remain of the same range, respectively 25, 32 and 27% for 1988, 2000 and 2011. This fact could be explained by the state soil management strategy that was based on olive trees and eucalyptus trees plants all over the Saharan border. The regression of wheat crop areas is explained by both local agriculture practices changes and by farmer abandonment of parcels. Although, it's remarkable that sandy zone areas have extended twice between 1988 and 2011, this is due to wind effect strengthened by the vegetation degradation that lets large opened areas.

Tunisian oasis forests cover about 40000 ha, with diverse and intensive production systems. Thus, these ecosystems are altered due to adjacent human activity and excessive land use. Hence, there is an urgent need to develop methodologies for assessing and monitoring oases from multiple perspectives and especially by using remote approaches. Monitoring, evaluating, and predicting the spatiotemporal changing patterns of oases are essential for managing oases for sustainability [31]. This study combined remote sensing and landscape metrics to monitor the dynamics of Tunisian oasis landscape located in the Nefzaoua and Djerid regions. Spatio temporal NDVI changes extracted from Landsat imagery series show that in 1979 oases were much more concentrated in the North of Nefzaoua. In 1984, there was a major change in the Djerid region. In 1995, we observed an increase of oasis density with newly irrigated perimeters in many areas. In 2009, there was a continuous increase of oases, particularly in Rjim Maatoug and Hazoua. And, in 2015, irrigated areas were still increasing in surface and number in the study area but more remarkably in Nefzaoua. Results reveal an unequally distributed oasis progression in space, with a slightly different rate over the period considered. The increased oasis area was mainly due to uncultivated land cover. However, the analysis of spatial metrics suggests a transition from an accelerated fragmentation phase of the oasis landscape, to an expansion phase by the continuous spreading of existing oasis surfaces. A noval relationship explained by the presence of a correlation between metrics and human's oasis dynamic dimension was presented, and GIS using Inverse Distance Weighting (IDW) tools to predicted map for future oasis extension was applied.

Moreover, Bou Hedma National Park belongs to a region that is a typical example of a landlocked region surrounded by mountainous terrain almost everywhere. It is part of the Long-Term Ecological Monitoring Network Observatories (ROSELT network), implemented by the Observatory of the Sahara and the Sahel [32].

Through the use of a multi-source processing and exploitation data (descriptive, map, field surveys, statistics, history, etc.), the chapter explain the implementation of a GIS database for characterizing and monitoring ancient and recent environmental sensitivity in BouHedma national park. This Park is home to the last *acacia raddiana* forest north of the Great Sahara, which has given it significant ecological value and allowed it to be classified as a biosphere reserve by UNESCO. The analysis of the biophysical and anthropogenic characteristics shows that it is threatened by two types of potential degradation. The first is the potential degradation of soil quality. The second is the potential degradation through water erosion. This sensitivity calls for a number of measures to be taken to safeguard and enhance the natural environment. It is essentially a question of changing the way in which land is used and.

From the Pleistocene to the beginning of the historical period, environmental dynamics were guided by the interaction of climatic fluctuations, lithology and morphometry of watersheds. However, during Roman times, the current Bou Hedma National Park cone thongs have experienced a phase of deposition of fine and fertile materials. Then, in late Antiquity-High Middle Ages to the colonial period, the terraces accumulations of crops become the seat of attack processes that gradually led to the loss of fine and loose materials. More recently, current dynamics are explained by the erosion areolar, regressive and lateral and the sensitivity to the potential degradation of soil quality in Bou Hedma National Park is the result of inadequate anthropogenic use of soil potential in areas of temporary human exploitation. Nevertheless, the sensitivity to water erosion degradation has shown that more than 1/4 of the area of Bou Hedma National Park is sensitive, highly sensitive and extremely sensitive. They concern most of the piedmont of the Bou Hedma.

The Tunisia wetlands are known by their faunal and floral diversity [33] and situated in diverse sites coastal and continental. Due to the high variability of bioclimatic stages from humid in the north to Saharan in the south, several structural and functional types of wetlands can exist [34]. This research chapter aims to study the spatial and temporal evolution of the wetlands of the big Sfax using earth observation and the GIS tools. The analysis of LULC temporal series of aerial photographs, satellite imagery between 1963 and 2018 showed that these wetlands underwent important changes explained by the installation of a polluting coastal industrial activity and the urban extension. These factors have led, for example, to the reduction of 85% in the area of the seasonal brackish swamps in the north and 55% in the south side of the seasonal brackish swamps and to the transformation of 'Ezzit' and 'El Haffera' streams to concrete canals. These variations are discontinuous over time. There is a trend of regression and disappearance of certain types of wetlands especially at the north coast of the study area around the 1970s. Big Sfax wetlands have undergone a regressive evolution and an anthropization tendency.

17.2   Conclusions

The use of remotely-sensed data in natural resources mapping and as source of input data for environmental processes monitoring has been popular in recent years. In Tunisia, different fields of research have been conducted. The book examines monitoring systems designed based on the use of scientific principles to exposure both at the individual and population levels, and also focuses on the development of monitoring systems at the local regional and global scales related to the assessment of various risks, for instance, drought, pollution, degradation and the consequences of climate change and natural resource management. The use of remote sensing and GIS for environmental and natural resources mapping and data acquisition is presented.

Application examples in urban and rural studies are discussed and different methods combined to the use of GIS and remotely data are proposed.

Two case studies have applied to urban landscapes, the first one has concluded that the dasymetric method in mapping population densities reveals very useful when compared to the classical methods and describes accurately the Human environment. The second one indicates that the use of remote sensing and GIS tools intends to comprehend the spatial effect of urban mutations on green spaces. However, applications in rural landscapes are based to the use of GIS tools. This technology is considered the foundation for the development of a long-term strategic management plan to protect ecosystems biodiversity and prevent *Solanum elaeagnifolium* and other alien plant species invasion in Tunisian irrigated perimeter. The combination of these tools to the multicriteria PROMETHEE II method has carried out a useful decision support tool to help the decision maker to better choose the suitable sites for phreatic aquifer recharge with reclaimed water in the different region of Tunisia. The application of GIS with remote sensing techniques has contributed effectively in the identification and separation of land use units such us meskat systems conducted in semi-arid zones and combined with the use of Standard Drastic, Pesticide Drastic and Susceptibility Index, it can greatly assess vulnerability and risks to pollution in groundwater. In the same context, research works cited in this book have concluded that remote sensing and GIS have the potential to generate relevant information to support urban disaster risk and to assess environmental risk degradation under climate stress and anthropogenic pressure.

The use of time-series satellite combined to the EWTCAN approach provides more reliable results for drought assessment. Likewise, these temporal series using radiometrix inedexes and GIS tools contributed effectively in the identification and separation of components such us soil units, land use and helps researchers to assess and monitor dryland vulnerability, LULC and wetlands dynamics. Results are useful to modelize and predict future changements, to highlight prototypes and to help decision makers to apply sustainable strategies. Endly a noval relationship have been highlighted by founding a correlation between metrics and human's oasis dynamic dimension. The obtained predicted map for future oasis extension will be useful for decision makers and planner to propose scenarios for sustainable management of these oases in the long term.

17.3 Recommandations

Recommendations are presented for Environmental Remote Sensing and GIS in Tunisia. These recommendations emerged from 16 chapters and six areas of research were distinguished at this book.

Much progress has been made in the use of remote sensing and GIS tools in Tunisia recent decades, especially in environmental applications. The primary goal of this book is to highlight the diversity of applications and the secondary one, but nonetheless important, outcome is to reaffirm the utility of these tools for addressing key science goals.

1. In the field of urban application, authors recommended that the dissymmetric approach should be adopted for different kind of landscapes and then produced maps may be achieved to portray the great disparities characterizing urban spaces' occupation. However, green spaces must be a key consideration in urban planning strategies using GIS tools.

2. In rural landscapes, sample applications presented in this book prove the usefulness of these tools to assess and monitor meskat system distribution, to rank the suitable sites for aquifer recharge with reclaimed water, to pay attention toward strategic of the Tunisian national plan to fight against SOLEL spread and to strengthen governance and institutions to manage disaster risk.

3. Mapping and modeling applications carried out presents useful decision support tool that could help decision makers to better choose the suitable sites for phreatic aquifer recharge with reclaimed water in different regions and to predict the vulnerability and groundwater pollution risk more accurately.

4. The upscale process using remote data and GIS is recommended to assess urban and natural risks in a national level and oo popularize the information with the population, rationalize land use and implement adequate erosion control in risk areas.

5. The application of remote sensing data based on Land Change Modeler has proven useful technology to monitor and assess LULC and soil degradation in arid regions. It can be helpful for local agricultural development agencies for decision makings on combating desertification in the area.

6. Ecohydrological indicators could be used in water and carbon budgets as well as in climate change models in limited water resources environments and for operational applications; findings could also bring useful information in forest management and prevention of fires.

7. Spectral Mixture Analysis, object-based oriented classification and Change Vector Analysis are recommended as a most suitable method for monitoring and mapping dry land vulnerability in arid and semi-arid environment.

8. Multi-scale time-series investigations of satellite images on patch-mosaic gradient and graph network models can go beyond simply describing dynamics. Recommends are oriented for application other landscape metrics to exploit the presence of other metric-human dimension relationships.

9. The use of a high resolution optical imaging series combined with a DEM and polarimetric radar image variables to improve discrimination of wetland classes and to characterize flood risks is recommended in future work.

10. A common thread among different recommendations is the need of web application for territory planning and better governance. Significant progress is anticipated, especially through the improved availability of airborne and ground-based remote sensors and Radar Data.

References

1. Vrieling A (2007) Mapping erosion from space. Doctoral Thesis, Wageningen University— with ref.—with summaries in English and Dutch. 167 p
2. Brunel S (2011) Pest risk analysis for Solanum elaeagnifolium and international management measures proposed. EPPO Bull 41:232–242
3. Mehdi L, Weber C, Pietro F, Selmi W (2014) Evolution of the place of plants in the city, from the green space to the green network. VertigO. https://doi.org/10.4000/vertigo.12670
4. Daoud A (2013) Feedback from the floods in the agglomeration of Sfax (Southern Tunisia) from 1982 to 2009: from prevention to the territorialisation of risk. Rev Geogr Rev East 53:1–2
5. Majdoub R, Khlifi S, Ben Salem A, M'Sadak Y (2013) Impacts of the Meskat water harvesting system on soil horizon thickness, organic matter, and canopy volume of olive tree in Tunisia. Desalinization Water Treat 52:2157–2164
6. Mekki M (2007) Biology, distribution and impacts of Silverleaf Nightshade (Solanum eleaeagnifolium Cav.). EPPO Bull 37(1):114–118
7. Chalghaf E, Aissa M, Mellassi H, Mekki M (2007) Control of the spread of Solanum elaeagnifolium Cav. in the governorate of Kairouan (Tunisia). EPPO Bull 37:132–136
8. Mekki M (2006) Potential threat of Solanum elaeagnifolium Cav. to the Tunisian fields. Invasive plants in Mediterranean type regions of the world. In: Proceedings of the first international symposium on invasive plants in Mediterranean type regions of the world. Environmental Encounters Series no. 59:165–170, Council of Europe Publishing
9. Mekki M (2011) Distinction between weed control and invasive alien plant management approaches: case study of Solanum elaeagnifolium management in North african countries. In: Proceedings of the international symposium on system intensification towards food and environmental security, organized by the Crop and Weed Science Society and Bidhan Chandra Krishi Viswavidyalaya, 24–27 February, 2011, Turkey
10. Saaty TL (1980) The analytic hierarchy process. McGraw-Hill, New York, NY
11. Ben Moussa A, Zouari K, Valles V, Jlassi F (2012) Hydrogeochemical analysis of groundwater pollution in an irrigated land in Cap Bon Peninsula North-Eastern Tunisia. Arid Land Res Manage 26(1):1–14
12. Aller L, Lehr JH, Petty R, Bennett T (1987) DRASTIC: a standardized system to evaluate groundwater pollution using hydrogeologic settings. J Geol Soc India 29(1):23–37
13. Rodney CS (2006) Groundwater vulnerability to agrochemicals: a GIS-based DRASTIC model analysis of Caroll, Chariton, and Saline counties. University of Missouri, Columbia, MO, USA, p 147
14. Ribeiro L (2000) SI: a new index of aquifer susceptibility to agricultural pollution, Internal report, ER-SHA/CVRM Lisbon, Portugal
15. Petiteville I, Ward S, Dyke G, Steventon M, Harry J (2015) Satellite earth observation intérieure support of disaster risk reduction. Special 2015 WCDRR. Edition: CEOS and ESA. 84 p. http://www.eohandbook.com/eohb2015/files/CEOS_EOHB_2015_WCDRR.pdf
16. Garschagen M, Romero-Lankao P (2015) Exploring the relation-ships between urbanization trends and climate change vulnerability. Clim Change 133(1):37–52. https://doi.org/10.1007/s10584-013-0812-6
17. IPCC (2012) Managing the risks of extreme events and disasters to advance climate change adaptation. A special report of working groups I and II of the intergovernmental panel on climate change
18. USAID (2018) Climate risk profile Tunisia. Available from: https://www.climatelinks.org/sites/default/files/asset/document/Tunisia_CRP.pdf
19. Adham A, Wesseling JG, Riksen M, Ouessar M, Ritsema CJ (2019) Assessing the impact of climate change on rainwater harvesting in the Oum Zessar watershed in Southeastern Tunisia. https://doi.org/10.1016/j.agwat.2019.05.006
20. Khaldi A (2005) Impacts of drought on the regime of underground flows in the limestone massifs of western Algeria "Monts de Tlemcen - Saida". Oran University (French)

21. Mashame G, Akinyem F (2016) Towards a remote sensing based assessment of land susceptibility to degradation: examining seasonal variation in land use-land cover for modeling land degradation in a semi-arid context. ISPRS Annals of the Photogrammetry, Remote Sensing and Spatial Information Sciences, Volume III–8. XXIII ISPRS Congress, 12–19 July 2016, Prague, Czech Republic

22. Trabelsi K (2011) Mapping of land use of the Oued Chiba watershed in 1987 and 2010 from Landsat 5 TM images. End of studies project, Mograne Higher School of Agriculture Tunisia

23. Chaudhry A, Sharma S (2015) Remote sensing and GIS based approaches for LULC change detection-a review. Int J Curr Eng Technol 5(5):3126–3137

24. Camacho Olmedo MT, Paegelow M, Mas JF, Escobar F (2017) Geomatic approaches for modeling land change scenarios. An introduction. Lecture notes in geoinformation and cartography, 1–8. https://doi.org/10.1007/978-3-319-60801-3_1

25. Wilhite DA, Glantz MH (1985) Understanding the drought phenomenon: the role of definitions. Water Int. 10:111–120

26. EDL (2014) A global initiative for sustainable land management, The economics of land degradation. http://www.eld-initiative.org

27. Van de Koppel J, Rietkerk M, Weissing FJ (1997) Catastrophic vegetation shifts and soil degradation in terrestrial grazing systems. Trends Ecol Evol 12:352–356

28. Rietkerk M, Van de Koppel J (1997) Alternate stable states and threshold effects in semi-arid grazing systems. Oikos 79:69–76

29. Mtimet A (2001) Soils of Tunisia. In: Zdruli P, Steduto P, Lacirignola C, Montanarella L (eds) Soil resources of Southern and Eastern Mediterranean countries. Bari: CIHEAM, Options Méditerranéennes: Series B. Studies and Research 34:243–262

30. Hirche A, Salamani M, Boughani A, Belala F, Essafi B, Gashut EH, Hourizi R, Grandi M, Ain Hamouda T (2017) Land degradation and restoration: The North African experiences. Geophysical Research Abstracts 19, EGU2017-11898

31. Yuchu X, Gong J, Sun P, Gou X (2014) Oasis dynamics change and its influence on landscape pattern on Jinta oasis in arid China from 1963a to 2010a: integration of multi-source satellite images. Int J Appl Earth Obs Geoinf 33:181–191

32. ROSELT-OSS (2004) Biodiversity study at the Haddej pilot observatory in Bou Hedma, Tunisia. ROSELT Collection—OSS,CT n°7 Montpellier

33. Khemaissia H, Touihri M, Jelassi R, Souty-Grosset C, Nasri-Ammar K (2012) A preliminary study of terrestrial isopod diversity in coastal wetlands of Tunisia. Vie milieu—Life Environ 62(4):203–211

34. Karem A (2014) A dot focal of Ramsar in Tunisia. Presentation on wetlands in Tunisia. Directorate of Forest Conservation. Ministry of Agriculture, Water Resources and Fisheries

Printed by Printforce, the Netherlands